Kenichi Miyamoto • Kenji Morinaga
Hiroyuki Mori
Editors

Asbestos Disaster

Lessons from Japan's Experience

Springer

Editors
Kenichi Miyamoto, Ph.D.
Professor Emeritus
Osaka City University
3-3-138 Sugimoto, Sumiyoshi-ku
Osaka 558-8585
Japan
and
Shiga University
1-1-1 Banba, Hikone
Shiga 522-8522
Japan

Kenji Morinaga, M.D., Ph.D.
Advisory Medical Doctor
Department of Asbestos Health
Damage Relief
Environmental Restoration and
Conservation Agency of Japan
Myuza Kawasaki Central Tower 9F
1310 Omiya-cho, Saiwai-ku
Kawasaki, Kanagawa 212-8554
Japan

Hiroyuki Mori, Ph.D.
Professor
College of Policy Science
Ritsumeikan University
56-1 Tojiin Kitamachi, Kita-ku
Kyoto 603-8577
Japan

ISBN 978-4-431-53914-8 e-ISBN 978-4-431-53915-5
DOI 10.1007/978-4-431-53915-5
Springer Tokyo Dordrecht Heidelberg London New York

Library of Congress Control Number: 2011923334

Printed on acid-free paper

Springer is part of Springer Science+Business Media (www.springer.com)

Preface

Although the asbestos disaster is not a recent phenomenon, comprehensive studies have not been developed with an interdisciplinary approach to elucidate the problem. Most asbestos studies have been conducted in medical science, but social science has not produced satisfactory investigations. Because asbestos issues as well as those in other kinds of pollution have been decisively influenced by social, economic, and political structures, the social sciences should contribute to studying them by integrating natural sciences. Even in social science, segmented specialties have hindered comprehensive understanding of asbestos issues. For example, although the asbestos disaster comprises both labor and environmental issues, they have been divided into economic and legal aspects.

This book originated from the asbestos research project at Ritsumeikan University, based in Kyoto, beginning in 2005. The project has been a unique interdisciplinary study as it has incorporated medical science, economics, political science, law, architecture, and environmental engineering.

The work focuses on asbestos issues in Japan as it relates each chapter to diverse sciences. This approach reflects an interdisciplinary perspective as well as a comprehensive understanding of Japan's asbestos problems, but it should be applicable to studies in other countries.

Chapter 1 introduces a political economy approach to explore asbestos problems as a subject of social science and to investigate the methodology of interdisciplinary study of asbestos issues. Chapter 2 characterizes the asbestos disaster as a "complex stock disaster" that demands new scientific approaches, and it provides a comprehensive review of asbestos issues in Japan. Chapter 3, through a historical review of asbestos issues in Japan, discusses the factors in the spreading of asbestos, such as social application of medical knowledge, activities of governments and industries, economic advantages of asbestos products, and the knowledge gap among stakeholders. Chapter 4 describes asbestos-related diseases with historical data on the use of asbestos in Japan. Chapter 5 focuses on industrial asbestos pollution caused by the Kubota Corporation in Amagasaki, Hyogo Prefecture. Chapter 6 explores the asbestos disaster in the Sennan District of Osaka on the basis of a historical analysis of political economy, and discusses the government's responsibility in the disaster. Chapter 7 describes the realities facing asbestos victims in these areas, with interviews of victims, and the dynamic processes involving those victims and the supporting organizations, companies, national and local governments, lawyers, and others. Chapter 8 deals with the international aspects of the asbestos industry and disaster involving Japan and South Korea in a historical perspective. Chapter 9 focuses on delayed public policies in Japan with attention to the divisions of responsibility of, and actions taken by, various ministries and other governmental organizations. Chapter 10 traces the political process in legislation of the Act on Asbestos Health Damage Relief, primarily through an analysis of information provided by newspapers. Chapter 11 centers around asbestos used in construction and the problems in maintenance and demolition management, with accident data on construction workers suffering asbestos-related diseases. Chapter 12 explores countermeasures enacted by local governments against asbestos problems in construction and proposes an obligatory policy of cooperation between national and local governments. Chapter 13 comprehensively summarizes asbestos-related lawsuits and points out decisive legal aspects of the asbestos disaster in Japan from a broad perspective. Finally, Chapter 14 introduces the French asbestos compensation system FIVA as a complement to the relief measures and compensation systems existing in Japan.

It is hoped that this volume will serve as a textbook on asbestos issues for all countries, especially where there is widespread use of asbestos. The lessons of asbestos problems and policies in Japan are particularly important for developing countries in Asia, Central and South America, and southern Africa. Knowledge about asbestos damage, its causes and the responsibility for it, compensation and relief systems, relief activities in emergencies, and future measures to be taken is imperative to prevent the proliferation of asbestos disasters in those regions.

The studies reported in this book were subsidized with Grants-in-Aid for Scientific Research by the Japan Society for the Promotion of Science (title number: 18310036, 21241014).

Kenichi Miyamoto
Kenji Morinaga
Hiroyuki Mori
Editors

Contents

Contributors

Hiroyuki Mori, Ph.D. *(Chapters 1 and 6)*
Professor
College of Policy Science, Ritsumeikan University, 56-1 Tojiin Kitamachi, Kita-ku, Kyoto 603-8577, Japan

Kenichi Miyamoto, Ph.D. *(Chapter 2)*
Professor Emeritus
Osaka City University, 3-3-138 Sugimoto, Sumiyoshi-ku, Osaka 558-8585, Japan; Shiga University, 1-1-1 Banba, Hikone, Shiga 522-8522, Japan

Takehiko Murayama, Ph.D. *(Chapter 3)*
Professor
Faculty of Science and Engineering, Waseda University, 51-05-07A, 3-4-1 Okubo, Shinjuku-ku, Tokyo 169-8555, Japan

Yuji Natori, M.D. *(Chapter 3)*
Chief
Asbestos Center, Z-Building 5F, 7-10-1 Kameido, Koto-ku, Tokyo 136-0071, Japan

Kenji Morinaga, M.D., Ph.D. *(Chapter 4)*
Advisory Medical Doctor
Department of Asbestos Health Damage Relief, Environmental Restoration and Conservation Agency of Japan, Myuza Kawasaki Central Tower 9F, 1310 Omiya-cho, Saiwai-ku, Kawasaki, Kanagawa 212-8554, Japan

Yasushi Shinohara, Ph.D. *(Chapter 4)*
National Institute of Occupational Safety and Health, 6-21-1 Nagao, Tama-ku, Kawasaki, Kanagawa 214-8585, Japan

Norio Kurumatani, Ph.D. *(Chapter 5)*
Professor
Department of Community Health and Epidemiology, School of Medicine, Nara Medical University, 840 Shijyo-cho, Kashihara, Nara 634-8521, Japan

Shinji Kumagai, Ph.D. ***(Chapter 5)***
Associate Professor
Department of Environmental Management, School of Health Science, University of Occupational and Environmental Health, 1-1 Iseigaoka, Yahatanishi-ku, Kitakyushu, Fukuoka 807-8555, Japan

Masafumi Kato ***(Chapter 7)***
Editorial Writer and Senior Staff Writer
The Kobe Shimbun, 1-5-7 Higashikawasaki-cho, Chuo-ku, Kobe, Hyogo 650-8571, Japan

Shinjiro Minami, Ph.D. ***(Chapter 8)***
Postdoctoral Fellow
Ritsumeikan Global Innovation Research Organization, Ritsumeikan University, 56-1 Tojiin Kitamachi, Kita-ku, Kyoto 603-8577, Japan

Norio Obata, Ph.D. ***(Chapter 9)***
Professor
College of Policy Science, Ritsumeikan University, 56-1 Tojiin Kitamachi, Kita-ku, Kyoto 603-8577, Japan

Michiya Mori, Ph.D. ***(Chapter 10)***
Associate Professor
Graduate School of Public Policy, Ritsumeikan University, 1 Nishinokyo-Suzaku-cho, Nakagyo-ku, Kyoto 604-8520, Japan

Kazuhiko Ishihara ***(Chapter 11)***
Professor
College of Policy Science, Ritsumeikan University, 56-1 Tojiin Kitamachi, Kita-ku, Kyoto 603-8577, Japan

Kazuhisa Hiraoka ***(Chapter 12)***
Professor
College of Policy Science, Ritsumeikan University, 56-1 Tojiin Kitamachi, Kita-ku, Kyoto 603-8577, Japan

Katsumi Matsumoto, Ph.D. ***(Chapter 13)***
Professor
School of Law, Ritsumeikan University, 1 Nishinokyo-Suzaku-cho, Nakagyo-ku, Kyoto 604-8520, Japan

Gakuto Takamura, Ph.D. ***(Chapter 14)***
Associate Professor
College of Policy Science, Ritsumeikan University, 56-1 Tojiin Kitamachi, Kita-ku, Kyoto 603-8577, Japan

Abbreviations

AFA	Asbestos Compensation Fund, Belgium
ANDEVA	National Association for the Defense of Asbestos Victims, France (Association Nationale de Défense des Victimes de l'Amiante)
ANOVA	One-way analysis of variance
BANJAN	Ban Asbestos Network Japan
Busan MBC	Busan Munhwa Broadcasting Corporation
CI	Confidence interval
CNAMTS	National Health Insurance Fund for Salaried Workers, France (Caisse Nationale de l'Assurance Maladie des Travailleurs Salariés)
CT	Computed tomography
DPJ	Democratic Party of Japan
EPA	Environmental Protection Agency
ERCA	Environmental Restoration and Conservation Agency of Japan
EU	European Union
EWG	Environmental Working Group
FCAATA	Asbestos Workers Early Retirement Fund, France (Fonds de l'Allocation de Cessation Anticipée d'Activité des Travailleurs de l'Amiante)
FDMA	Fire and Disaster Management Agency
FIVA	Special Funds of Compensation for Asbestos Victims, France (Fonds d'Indemnisation des Victimes de l'Amiante)
GDP	Gross domestic product
GHQ (SCAP)	General Headquarters (Supreme Commander for the Allied Powers)
IARC	International Agency for Research on Cancer
ICD	International Classification of Diseases and Related Health Problems
ILO	International Labour Organization

INSERM	National Institute of Health and Medical Research, France (Institut National de la Santé et de la Recherche Médicale)
JKCC	Japan–Korea Cooperation Committee
JSOH	Japanese Society for Occupational Health
LDP	Liberal Democratic Party of Japan
METI	Ministry of Economy, Trade and Industry
MEXT	Ministry of Education, Culture, Sports, Science and Technology
MHLW	Ministry of Health, Labour and Welfare
MIC	Ministry of Internal Affairs and Communications
MITI	Ministry of International Trade and Industry
MLIT	Ministry of Land, Infrastructure, Transport and Tourism
MOE	Ministry of the Environment
NCI	National Cancer Institute
NGO	Non-Governmental Organization
OIT	Office of International Trade
SD	Standard deviation
SMR	Ratio of observed deaths to expected deaths
TEM	Transmission electron microscope
WHO	World Health Organization
WTO	World Trade Organization

Chapter 1
A Political Economy of Asbestos Disasters

Hiroyuki Mori

1.1 The Impact of Asbestos Disasters

Near the end of the twentieth century, asbestos disasters evolved into a social problem in many advanced industrialized nations. Now that the twenty-first century has arrived, it has begun to appear that asbestos may develop into one of the largest global industrial calamities of all time.

Asbestos is a naturally occurring fibrous mineral with physical properties ranging from heat resistance, heat insulation, sound insulation, and friction resistance to chemical resistance, electrical insulation, and corrosion resistance. Further, because it is well-suited for bonding with textiles as well as other materials, and is also inexpensive, asbestos was treated as an ideal material resource for the purposes of modern industrialization.

On the other hand, asbestos dust in the workplace setting and in the wider environment was also known to be a cause of serious asbestos-related diseases in both workers and local residents. Those asbestos-related diseases include pulmonary asbestosis, lung cancer, and mesothelioma, all of which present with severe symptoms.

Table 1.1 summarizes, in chronological order, the growing body of knowledge that developed with regard to the major asbestos-related diseases. As the table suggests, asbestos was a known cause of pulmonary asbestosis even prior to World War II, and was observed in the 1940s and 1950s to be a cause of lung cancer and in the 1960s to be the principal cause of mesothelioma.

Even in Japan, large-scale epidemiological studies of asbestos textile workers were carried out by the Ministry of Home Affairs as early as the 1930s, followed

H. Mori (✉)
College of Policy Science, Ritsumeikan University, 56-1 Tojiin Kitamachi, Kita-ku, Kyoto 603-8577, Japan
e-mail: hmt23243@sps.ritsumei.ac.jp

K. Miyamoto et al. (eds.), *Asbestos Disaster: Lessons from Japan's Experience*,
DOI 10.1007/978-4-431-53915-5_1, © Springer 2011

Table 1.1 Asbestos-related chronology, 1898–1964

1898	Factory inspectors in the UK identify the "evil effects" of asbestos and the "easily demonstrated" danger to the workers' health
1899	Asbestos worker admitted to a London hospital suffering from pulmonary fibrosis. He dies the following year
1924	In Britain, the first inquest on an asbestos worker leads to the first medical description of asbestosis
1931	British government introduces dust control regulations in the asbestos industry
1935	Asbestosis and lung cancer cases appear in the medical literature in the US and Britain. A pathologist suggests to British government medical officers that the diseases might be linked
1938	German pathologist declares for the first time that lung cancer is an occupational disease of asbestos workers. In Britain, the government notes a significant rise in lung cancers in asbestosis cases
1943	German government recognizes asbestos-induced lung cancer as a compensable occupational disease. One German pathologist links asbestos with rare pleural cancers.
1947	British government statistics note a high percentage of asbestosis cases with lung (including pleural) cancers
1955	First epidemiological study of a group of British asbestos workers confirms a lung cancer risk
1960	Landmark study published by South African researchers shows a linkage between mesothelioma and both occupational and nonoccupational exposure to asbestos
1964	Catastrophic cancer mortality demonstrated among American insulation workers and publicized at a conference in New York

Source: McCulloch and Tweedale (2008:8)

by similar studies after World War II. Investigations into asbestos-induced lung cancer began to appear in the 1950s, and more knowledge and information on mesothelioma had begun appearing in the literature by the 1960s. One may conclude from these details that a knowledge base on asbestos-related diseases was developing at about the same pace in most of the advanced industrial nations.

Nonetheless, in the meantime, Japan's government continued to promote the use of asbestos. From the early postwar reconstruction phase through the years of the major economic boom, asbestos yarn and fabrics were utilized as an essential material to sustain the operations of the leading industries that functioned as the engines of economic growth in fields ranging from fertilizers, shipbuilding, and rail to electric power, steel, petrochemicals, and automobiles. To this end, the government assisted, both directly and indirectly, in ensuring that the industrial sector had access to a supply of inexpensive asbestos products. While urbanization during the boom years created a demand for fire-resistant building designs and a shift to high-rise structures, construction materials containing asbestos were promoted under the provisions of the Building Standards Law and related ordinances as inexpensive materials suitable for these purposes. The former Kubota Corp. Kanzaki Factory, which triggered the Kubota Shock in 2005, was a facility engaged in the manufacture of asbestos cement pipe products. The principal clients for these products were contractors engaged in publicly funded water pipe-laying projects implemented by local municipalities nationwide.

Eventually, the Japanese government began to adopt legal measures aimed at dealing with the hazards of asbestos. In 1960, it established the Pneumoconiosis Law, thus demonstrating for the first time that it considered asbestos to be a material which was hazardous to human health. This law stipulated that factories engaged in processes involving the use of asbestos should take steps to curb worker exposure to asbestos dust and begin providing their workers with health examinations. In 1971, it enacted the Ordinance on the Prevention of Hazards Due to Specified Chemical Substances, thus mandating that business operations handling asbestos should install on-site exhaust ventilation systems and monitor environmental concentrations of asbestos dust in their indoor workplaces. The Chemical Substances Ordinance was revised in 1975, when steps were taken to reinforce regulations that treated asbestos as a carcinogen. In addition to these labor policies, in the early 1980s the government began environmental monitoring of health effects relating to asbestos and decided that the risk from asbestos was small. The Air Pollution Control Law was revised in 1989 to regulate all facilities producing asbestos dusts, and to set criteria for the dust density at the borders of such facilities. However, the government overlooked some serious industrial pollution, such as that at the Kubota factory. This is a failure of the "PPM" principle that measures the levels of polluting substances rather than investigating human health.

However, most of these legal measures suffered from serious flaws. The obligations mandated by the Pneumoconiosis Law applied to large corporations, but lacked feasibility when applied to the many small-scale businesses running asbestos textile operations. Furthermore, business operators lacked incentives to actually improve occupational health and safety because no specific, in-depth regulations on dust concentrations or related matters had been implemented. In addition, worksites (construction sites, etc.) that did not readily fit into the "indoor workplace" category were effectively outside the scope of regulation under the Ordinance on the Prevention of Hazards Due to Specified Chemical Substances. Also, under certain conditions, applications of sprayed asbestos were allowed for use on building pillars and other steel-frame components. Products containing less than 5% asbestos by weight were not banned. Among these various exclusions and exemptions from regulation, companies were also spared an obligation to install on-site exhaust ventilation equipment if compliance was deemed to be difficult or the project operations were only temporary. What is more, the limited scope of the Chemical Substances Ordinance as a set of regulations applying only to workplace settings actually encouraged acceptance of the manufacture and sale of asbestos-containing products.

Later, in the 1980s, asbestos gained status as a social problem mainly in response to several isolated events, including the illegal dumping of asbestos scrap from operations to dismantle a US aircraft carrier, and an asbestos scare stemming from revelations over the use of asbestos in school buildings. In 1991, the Socialist Party of Japan and the Socialist Democratic Federation submitted a joint bill to the Diet for the establishment of an "Asbestos-Containing Product Control Act," a legislative proposal prepared by Ban Asbestos Network Japan (BANJAN) that would ban asbestos in principle and promote a set of comprehensive asbestos countermeasures. However, the Liberal Democratic Party was against the bill and had it withdrawn prior to any deliberations.

It was the Kubota Shock of 2005 that transformed asbestos into a political issue. This event involved the disclosure that many residents around the former Kanzaki factory in downtown Amagasaki had contracted mesothelioma. At that time, nearly 200 residents were already ill with the disease, and the total scale of the disaster was closer to 300 citizens if one included the many Kubota workers that had also become victims. As a case involving harm from environmental exposure to asbestos, this would eventually be known as the single largest asbestos disaster of its kind in the world.

In 2006, prompted into action by the Kubota Shock, the Japanese government brought in the Act on Asbestos Health Damage Relief, thus creating a new safety net to provide coverage for victims of environmental asbestos exposure, and also for other asbestos victims not covered by the national workers accident compensation insurance framework. However, the new safety net was flawed because it provided significantly less compensation than the amounts paid through workers accident compensation insurance, and did not provide any coverage for patients with pulmonary asbestosis, although serious asbestosis has come to be covered by the Act since July 2010. On top of that, from the perspective of social justice, it was flawed because it sought its funding through the collection of tax revenues from all businesses nationwide instead of adhering faithfully to the polluter-pays principle.

Asbestos-related accidents and disasters in Japan eventually developed into an issue for the courts. In one high-profile case, the government was sued for damages over charges of failure to exercise its regulatory authority. In May 2010, a Sennan lawsuit seeking damages from the government culminated in victory at the Osaka District Court for a group of 30 plaintiffs, including the victims and families of victims who had contracted asbestosis after exposure to asbestos in asbestos textile factories, and the government appealed to a higher court. In addition, a large group of construction workers has filed a class-action lawsuit against the government and construction material manufacturers for joint inaction, and another lawsuit against the government and Kubota has been filed to assign liability for health problems caused by environmental exposure to asbestos. To date, asbestos-related companies (manufacturers, distributors, property management companies, etc.) nationwide have been taken to court in numerous cases involving litigation and conciliation.

1.2 A Social Scientific Perspective on Asbestos Disasters

1.2.1 Complex-Stock Disasters

As first designated by Miyamoto (2006), asbestos disasters are complex-stock disasters. Their social scientific context has been summarized by Miyamoto (2007:256–259). Two implications are derived from the social scientific perspective applied here.

The first is "complexity." Asbestos disasters combine workplace accidents with air pollution, assorted forms of pollution stemming from material consumption (including the use of residential structures and office buildings, the utilization of automobiles, maritime transport, and rail services, and the utilization of other goods and services that incorporate the use of asbestos), and pollution resulting from the disposal or dumping of waste materials. Asbestos has been utilized in as many as 3 000 different categories of commercial merchandise, and on top of that, it is a substance that has found its way into the most common everyday settings where people reside and work. If a building that contains asbestos construction materials is remodeled or demolished, the construction or demolition workers involved in those activities, as well as citizens residing nearby, will all be exposed to asbestos dust. Asbestos can be rendered harmless only through ultra-high-temperature dissolution. However, in Japan there are currently only six facilities nationwide which are equipped to perform that process. Consequently, most asbestos scrap materials end up buried in landfill dump sites. However, this method of disposal exposes the surrounding areas to a new risk: that of contamination from dumped materials.

These are features that describe a complex-stock disaster, whereby hazards to human health are generated at every economic stage of the product life cycle, from raw material resource mining or extraction to product manufacture, distribution, consumption, scrapping, and final disposal. In Japan and many other countries around the globe, the hazards and health problems associated with asbestos have developed into enormous complex-stock disasters of this kind. The asbestos-linked health problems caused by crocidolite mines in South Africa and the Libby mine in Montana – which caused the largest asbestos disaster in the US – are classic examples of complex-stock disasters that were caused at the mineral extraction stage.[1] The asbestos disaster in the Sennan community was triggered by industrial textile operations that were engaged in the processes of pulverizing raw asbestos and producing asbestos yarn and fabric. The Kubota factory manufactured asbestos cement pipes using crocidolite, causing immense harm in its workplace and to the surrounding residential environment. Workers at factories that manufactured brake linings also became victims of asbestos exposure. Teachers and workers who contracted mesothelioma following exposure to asbestos in construction materials used in school facilities and hotel buildings have already been approved to receive workers accident compensation insurance benefits. Some of the construction workers that were engaged in the demolition of buildings that had collapsed during the Great Hanshin-Awaji Earthquake of 1995 later contracted mesothelioma. They, too, were approved for workers accident compensation insurance benefits on the grounds that exposure to asbestos dust during demolition work caused their medical conditions. In Japan, most scrap asbestos materials have been disposed of in landfill sites, partly to sidestep the high cost of the treatment process required to render asbestos harmless. However, not all landfill sites containing buried asbestos materials have been properly managed, and in some cases they already pose an imminent risk of causing new health problems.

The second social scientific implication is the "stock pollutant" nature of asbestos. Illnesses or medical conditions resulting from relatively short-term exposure to air or water contamination are typically treated as the outcome of "flow pollution." Asbestos disasters, on the other hand, are an example of stock pollution. The symptoms of health problems stemming from exposure to asbestos take from 10 to 50 years to become manifest. Furthermore, even if society were to completely ban the manufacture and utilization of all asbestos products, it is likely that asbestos will continue to cause health problems as long as accumulations of this material remain in our environment. In other words, we describe the harmful effects that arise after long-term exposure to toxic materials that have accumulated in human tissues, products, or the environment as a phenomenon of "stock pollution." Stock pollution is a form of pollution that can be expected to increase in the years ahead as an outcome not only of asbestos disasters, but also of contamination from waste scrap materials and the effects of global warming.

1.2.2 *The Issue of Responsibility for Asbestos Disasters*

Solutions to the problems of asbestos as a complex-stock disaster will demand a different societal focus on the issue of liability than would apply to cases of flow pollution. Granted the complexity spanning all economic stages, the causal relationships underlying the health problems that arise at each stage and the responsibility or blame for those problems will presumably be matters for debate. As another potential issue, foreign governments might seek to assign blame for health problems traceable to the international trade in asbestos and products containing asbestos. As the sources of pollution, manufacturers and corporations share responsibility for workplace disasters and environmental pollution. However, the responsibility for product contamination will presumably be determined in keeping with the principle of extended manufacturer liability.

Compensation for damage from asbestos disasters must be based on the polluter-pays principle. Kubota has paid relief money which is comparable in scale to accident compensation insurance benefits to more than 200 mesothelioma patients who were victims living near its asbestos factory. However, it has provided those payments nominally as relief funding rather than as compensation on the grounds that the causal links between the airborne asbestos dust from its factory and the health problems of the victims remain unclear. Hence, unless and until it accepts legal responsibility and pays damage compensation, Kubota will not have provided a solution to the problem even though it may have paid victims relief funding which is comparable in scale to the workers accident compensation insurance benefits currently received by eligible patients.

The Japanese government acted quickly to establish the Act on Asbestos Health Damage Relief and accordingly succeeded in sidestepping social controversy over the issue of asbestos-induced health problems, while also avoiding certain responsibilities in the process. Unlike the Law Concerning Pollution-Related Health Damage Compensation and Other Measures, the new asbestos law only provides for relief funding and does not replace compensation payments for civil damage.

Under the provisions of the Act on Asbestos Health Damage Relief, it was estimated that approximately 9.05 billion yen in relief payments will be needed during the period from 2007 through 2010. Of that total in operating expenses, the national government was expected to furnish 750 million yen, local governments 920 million yen, and businesses in the private sector 7.38 billion yen. The private-sector component was further broken down into a share to be paid by special business entities and another share to be paid by ordinary business entities. The group of special businesses consisted of four companies selected on the basis of the following criteria: they had a history of asbestos use totaling 10 000 metric tons or more; in the vicinity of their operations, the incidence of mesothelioma deaths was higher than the national average; and at least ten of their employees had asbestos-linked lung cancer and were eligible for workers accident compensation. This group of four firms was obligated to pay 338 million yen, with the remainder of approximately 7 billion yen to be collected from maritime business owners and businesses enrolled in the workers accident compensation insurance program. However, this was not a collection approach based on the polluter-pays principle. Its logic was based on the idea of assigning a value to the profit made by the asbestos industry as a whole, and collecting payments from all companies with no regard for their individual levels of liability in cases of asbestos pollution. Furthermore, funding contributions by the national and local governments were deducted from general tax revenues without any explanation of the relationships linking asbestos-induced health problems and fiscal resources. In terms of assigning responsibility, this approach is even weaker than that applied by the Law Concerning Pollution-Related Health Damage Compensation and Other Measures, which treats general collections as funding from companies with SOx emissions, and bases the national government burden on the automobile weight tax.

Epidemiological studies will be of the utmost importance in future efforts aimed at developing and refining the asbestos damage compensation framework. The percentage of asbestos victims approved for workers accident compensation is lower than the levels seen in general mortality tables. That finding points to a need for employees and owners of asbestos-related businesses to undergo health examinations. Further, epidemiological studies of past and current residents living near the Kubota factory and other asbestos-related facilities must also be performed. National and local government agencies have refrained from implementing full-scale epidemiological studies to date, citing budgetary constraints from the fiscal crisis or a lack of personnel. However, considering that the asbestos problem has already surfaced and come to the attention of the courts, and that news organizations and the public at large have developed a strong interest in this issue, the opportunity to perform such studies may be lost unless it is acted upon soon. Epidemiological studies will be essential to the prevention of further harm from asbestos.

1.3 An Economic Analysis of Occupational Safety and Health

The study of asbestos disasters will demand a synthesis of knowledge from the social and natural sciences. The exact structure of that process will conceivably comprise a consolidation of social science research upon a base of medical understanding. In this chapter, we will follow a comprehensive approach that starts with the perspectives of political economy. As a common platform for the following chapters, our attention will be focused on the political economy of occupational health and safety.

1.3.1 The Labor Economics Approach

To date, labor economics has been virtually the only field of economics to devote serious study to the issue of asbestos disasters. Mainstream labor economists have treated occupational health and safety as a good that is interchangeable with wages, and have pursued analytical studies by applying the same logic as the indifference curve used in conventional microeconomics. In other words, the choice of whether to engage in relatively low-wage labor at a workplace equipped with facilities ensuring high standards of occupational health and safety, or in relatively high-wage labor while tolerating poor conditions of occupational health and safety, is considered to be a matter of preference for the worker. Should one choose to engage in a high-risk occupation, one will be rewarded with high wages; in the reverse situation, i.e., with a low-risk occupation, one will have to accept low wages. In the field of economics, this is referred to as the theory of compensating wage differentials.

Assuming that companies also make these types of choice, companies and workers together will enjoy optimal labor market equilibrium provided that companies hire workers that have their own preferences. Should the government step in and regulate companies in an effort to prevent asbestos-related health problems, the utility level of workers that prefer high wages in high-risk occupations will be diminished.

Figure 1.1 graphically illustrates the logic behind the theory of compensating wage differentials. Here, the vertical axis represents the wage level and the horizontal axis represents the probability (scale of risk) of an occupational disaster. On the same indifference curve, all points indicate the same level of utility. In short, measures of utility level are constant along the same indifference curve. The utility level of a worker rises as we move from indifference curve U1 to curve U2.

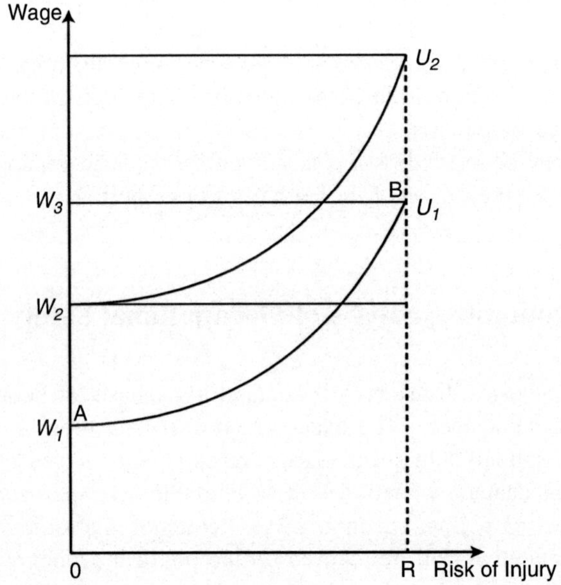

Fig. 1.1 Indifference curves for wage and risk

Let us look more closely at indifference curve U1. A worker that seeks this utility level will only enjoy wage W1 if they opt to engage in risk-free work (point A). To earn a higher wage, W3, at the same utility level, the worker must engage in work that poses higher risk, i.e., at the level of risk of injury R (point B). The assumption that individual workers freely choose these different combinations of risk and reward is one of the basic tenets underlying the theory of compensating wage differentials.

One of the underlying preconditions for this theory is that rational workers in the labor market make free choices based on the information they need, and that as a result, the risk of occupational disasters does not become a huge issue. In addition, if workers have not been provided with the information they need in order to make their choices, conveying that information then becomes the role of public policy. Furthermore, occupational health and safety is treated purely as a singular good, the level of which is determined on the basis of the "price" of the worker's health or life and the probability of harm.

The economist Peter Dorman posits that two entirely different viewpoints have been applied to occupational health and safety. One is the orthodox and empirically based view that unnecessarily risky labor is a form of abuse that must be regulated in accordance with principles of justice. The other is the economics-derived view that is based on the ideas of a perfect market, free choice, and compensating wage differentials. Dorman then summarizes points for the revision of the model presented by mainstream economists themselves, as described below (Dorman 1996:42–50).

First is the issue of labor market friction. In economic theory, it is hypothesized that workers are able to change their occupation with perfect freedom. However, in practice, this hypothesis is unrealistic because changing one's occupation also entails various costs, including the cost of finding alternative work, the cost of acquiring specific skills and experience, and the cost of changing to a new career.

Second, the information supplied to workers on matters of occupational health and safety is not always adequate, for several reasons. One is that working conditions at risk of causing difficult-to-diagnose occupational diseases are outside the scope of consideration. In addition, in many industrial disasters, information is only available to the workplace involved, thus requiring that an enormous amount of time and labor has to be expended to identify and understand their causes. This represents an exceptionally high cost for many workers, and especially for workers that earn low wages in high-risk occupations. What is more, the information provided to workers may be completely useless in situations where corporate employers have a strong incentive to supply workers with erroneous information with the ulterior goal of avoiding compensation costs.[2] The strategic manipulation of information in this way is also a phenomenon that applies to national governments.

Third, occupational health and safety usually have an impact on many workers at the same time and should be treated as a public good.

Other issues have to do with the way mainstream economic models take into account the role of labor unions and *ex-post* compensation following an industrial disaster. However, not all of these issues demand a fundamental shift to the use of the labor market model. In fact, Dorman notes that the indifference curve-based correlations between wage levels and levels of occupational health and safety predicted by this model have not been completely substantiated by statistical data, and points out that labor performed under the more dangerous working conditions is in reality rewarded only with lower wages, and thus questions the very premise on which the theory of compensating wage differentials is based. Dorman (2000) writes as follows on these points: "As a generality, one can say that the most dangerous jobs are the ones lowest in the economic hierarchy: precarious employment, informal employment, work in small and medium enterprises (SMEs), and work performed by groups subject to discrimination and marginalization" (Dorman 2000:5).

He reinforces his criticism with the point that people normally do not engage in behaviors that are preconditioned on a monetary trade-off for life, health, or similar values (Dorman 2000:97–101).

Dorman notes that given considerations of the fear of unemployment, labor market segmentation, and imbalances in the rights of labor and management, workers cannot be assured of adequate wage compensation through reliance on the labor market alone, and thus highlights a need for the establishment of workers' rights (including the right to have access to information and to reject dangerous labor), the implementation of public regulatory frameworks that fulfill the dual roles of defining minimum tolerances and ensuring information disclosure, and the role of public health experts in supporting these measures (Dorman 2000:208–228).

In his thesis, Dorman applies a broad-based perspective to elaborate on his criticism of inherent flaws in the approach of mainstream economics. However, that is not to imply that he has abandoned the theory of compensating wage differentials, which is one of the concepts that he criticizes. On the contrary, Dorman views risk-compensating wages as a factor that encourages fairness in that they reward workers that engage in dangerous labor, and also as something that provides corporate employers with an incentive to develop a safer work environment (Dorman 2000:219). In other words, it is his assertion that market-defined levels of compensating wages are not enough, and that complementary structures need to be engineered.

The aforementioned doctrine of compensating wage differentials is flawed in terms of the constant emphasis it places on consumer choice. The question is whether humans lead such a rational existence that they are able to accept as an outcome of personal choice the injury of a worker in an occupational disaster even if that worker performed dangerous work with full knowledge of the risk involved. It seems more likely that people would have serious regrets if they considered the future realities for the injured worker, including economic hardship and the impact on the livelihood of their family, and would then agree with the need for preventive measures that promote occupational health and safety. Given this scenario, the theory of compensating wage differentials cannot be described as something that realistically accounts for the limits of worker rationality.

1.3.2 The Post-Keynesian Economics Approach

The approach taken by post-Keynesian economics provides a different analytical framework from that applied by mainstream economics to the subject of occupational health and safety.

Post-Keynesian economics treats preferences as lexicographic preferences. Because hierarchical relationships exist within each group to which a variety of categories of goods belong, the ability to substitute one group of goods for another is thought to be extremely limited. In microeconomics textbooks, it follows that theories premised on the ability to substitute apples for oranges are considered to be extremely limited in their realistic scope. This is termed a lexicographic preference. As such, the concept raises serious doubts about compensating wage differentials, which treat wages interchangeably with occupational health and safety.

Mark Lavoie criticizes the mainstream economics practice of converting all goods into a utility as a framework that allows for the indiscriminate comparison of individual wants. Then, drawing on the hierarchical or pyramid structures of needs proposed by the classical economists Karl Menger and Alfred Marshall and the psychologist Abraham Maslow, he asserts the need for an economic theory that is based not on undifferentiated wants, but on differentiated needs. Lavoie demonstrates the impossibility of substituting needs that belong to different groups, and the limited scope of indifference curve-based selection theory as used in price theory (Lavoie 1992:Chap. 2).

In connection with the subject of lexicographic preferences, Arild Vatn has built on a theory that applies to environmental goods. He states that the feasibility of substituting one group of goods for another is extremely limited (the principle of irreducibility), and that a hierarchical structure delineates different needs (the principle of subordination). In the light of these principles, Vatn emphasizes that price has meaning only for goods within a given group, and criticizes the use of orthodox economic principles in the analysis of environmental goods (Vatn 2009). Although this constitutes a discussion within the context of environmental goods, it naturally also applies to occupational health and safety. In effect, the theory of lexicographic preference as stressed by post-Keynesian economics may be interpreted as a scathing criticism of mainstream economics, which strives to measure wage levels and occupational health and safety in terms of the logic of worker choice as based on an indifference curve.

Assuming a lexicographic preference has been established for worker choices relating to wages and occupational health and safety, it follows that these cannot be treated as interchangeable goods. In blunt terms, this amounts to an expression of the impossibility of putting a price on human life. As such, it may be applauded as a re-articulation of the previously discussed common-sense approach to mainstream economics within the context of a new economic doctrine.

"Uncertainty" may be cited as yet another vital conceptual pillar of post-Keynesian economics. Suffice it to say that an awareness that reality is full of uncertainties is a common thread of post-Keynesian economic thought. Lavoie states that theories rooted in the idea of uncertainty differ decisively from theories that are based on the concepts of certainty or the quantification of risk (Lavoie 1992:92). Regarding the relationship between the concept of uncertainty and environmental economics, Andrew Mearman considers it necessary that optimal pollution models based on public and private costs and benefits should be able to identify future costs and benefits accurately. Nonetheless, he acknowledges that this is unrealistic, and states that a corresponding probability distribution curve may not even exist. Mearman then asserts that in the light of this notion of uncertainty, cost–benefit analysis – one of the most widely applied methodologies in the field of environmental economics – should be abandoned (Mearman 2009:38).

Acceptance of the post-Keynesian economics-tendered concepts of lexicographic preference and uncertainty has confronted mainstream labor economics with an acute problem. In particular, with regard to scenarios of the asbestos problem as discussed here, it provides a useful framework for analysis.

The theories and discourse of post-Keynesian economics have generated a valuable body of insights and knowledge that now surpass the scope even of mainstream economics. Nonetheless, it may be noted that they still face the following problems.

First, one cannot scientifically demonstrate how lexicographic preferences take form. Lexicographic preferences can be extremely powerful as a criticism of the indifference curve and other analogies enlisted by mainstream economics. Nonetheless, they do not provide a satisfactory explanation of why individuals may have strong preferences for goods that cannot be readily replaced by conventional private goods.

Second, they also fail in terms of explaining how the asbestos problem may be consistently viewed as a series of industrial disasters and cases of environmental pollution. The reason why it is difficult to understand the asbestos problem using the tools of the social sciences is because they tend to break down the problem into separate components, i.e., problems relating to workplace environments on the one hand and environmental contamination on the other. Post-Keynesian economics itself has not overcome this limitation.

1.3.3 The Institutional Economics Approach

Analytical approaches based on the institutional and Marxian economic schools of thought appear to have overcome the aforementioned limitations of the mainstream and post-Keynesian schools. Institutional and Marxian economists share the view that occupational disasters and pollution are problems embedded into the very fabric of market economic systems powered by the profit-motive-oriented behavior of capitalist corporations. Further, they believe that the profit motive is not a passive mechanism that responds purely to trends in prices, but rather a force that encourages companies to seek endless systemic reforms for their own self-interest. We will first discuss the research work of K. William Kapp, a leading institutional economist.

In his book *The Social Costs of Private Enterprise*, Kapp discusses a variety of social costs. In addition to environmental pollution, which has become a representative theme of his research, Kapp further focuses on an array of problems including unemployment, idle resources, and technological change. He also discusses the social cost-related dimension of occupational disasters.

Kapp describes social costs as “all those harmful consequences and damages which third persons or the community sustain as a result of the productive process, and for which private entrepreneurs are not easily held accountable” (Kapp 1950:14).

Utilizing this concept of social costs, Kapp portrays, as an inseparable component of the capitalist economic system, the embedded mechanism by which private companies socialize their private costs. These private costs are shifted onto the shoulders of workers, consumers, local residents, and the national government. In other words, the economics of Kapp portrays the phenomena of environmental pollution, occupational disasters, and other social costs not as independent events, but as differences in the forms of expression characterizing common elements in the modes of behavior exhibited by capitalist corporations. Furthermore, Kapp notes that although social costs lead to an awareness of a growing gap between economic and social value, this underscores the point that corporate behavior and the market economy are not isolated systems, and that they must be identified in terms of a never-ending complementary relationship with the social system and the natural environment.

The approach taken by Kapp theoretically and empirically sheds light on the sheer gravity of the diversity of social costs generated by corporate behavior and a market economy. However, it also provides an important perspective for social scientific examinations of asbestos disasters. In other words, it furnishes a framework for an understanding that the concrete manifestations of corporate behavior in the labor and reproductive processes, and the problems in the work environment and the general environment that those processes influence, are connected phenomena attributable to a single source, namely, social costs.

This perspective has fueled sharp criticism of mainstream economics, which seeks to understand the phenomena of social costs in terms of an imperfect market. The reason is that because social costs are an intrinsic element of the market economic system, even under the conditions of a perfect market, the act of passing those costs onto society's shoulders will in no way come to an end.

Many have commented that in one sense Kapp's theory of social costs seems to comprise an assortment of arguments. Such criticism has been especially common among environmental economists. To be sure, in one context, that criticism was essential to the establishment of environmental economics as an academic field – a field that Kapp helped to pioneer. However, the drawback of establishing a discretely defined academic field of this kind was that it ignored the essence of the social cost theory that Kapp had articulated. Pietro Frigato comments as follows on this point: "Kapp's lesson on this fundamental type of social costs has been totally ignored in socio-economic epidemiology and in occupational risk economics…. Occupational safety and health economics is more compatible with Kapp's analysis of the impairment of the human factor of production, but its approach tends to be rigidly circumscribed and separated from other analyses of social losses. It may therefore fail to elaborate coherent proposals for public policy" (Frigato 2006:181).

1.3.4 The Marxian Economics Approach

As a Marxian economist, Kenichi Miyamoto began applying Marx's law concerning the tendency of constant capital saving from an early stage, and proposed a theoretical framework as a shared foundation for occupational health and safety and the environment. The "tendency of constant capital saving" articulates the principle that investments of capital in facilities and equipment that are not directly involved in production – including, for example, facilities for occupational health and safety and environmental protection – are detrimental to an enterprise's profit margin, and are therefore to be avoided or curtailed as much as possible. In his book *Social Overhead Capital*, Miyamoto states that "factories and mining facilities under the influence of capitalist accumulation have a tendency to constantly refrain from making investments in measures designed to ensure worker safety and prevent workplace disasters. Additionally, as noted earlier, due to this tendency to avoid spending on socially shared means of consumption, general conditions for urban workers are destroyed and the following conditions of economic hardship and social turmoil arise." Miyamoto then cites the intensification of conditions of pollution (air and water pollution, noise and vibration pollution, land subsidence, water shortages, etc.) (Miyamoto 1967:161–162).

As a causal factor, Miyamoto points to the tendency of capital saving as articulated by Marx and comments as follows. "A situation characterized by efforts to conserve spending on socially shared means of consumption is the same as conserving the constant capital component of spending for worker health and safety in the factory (or more broadly speaking, in the workplace). In their thirst for profit, capitalist corporations endeavor endlessly to expand constant capital. Incidentally, constant capital includes fixed equipment (for employee welfare and the prevention of pollution, etc.) that is not directly involved in production processes. Accordingly, conserving constant capital is one condition for increased profit. In particular, there is a general tendency to conserve constant capital for improvements in worker health and safety. . . . Because this is the way companies have conserved constant capital, avoiding capital investments that are in the interest of workplace safety, it is no surprise that they have shown no consideration whatsoever for constant capital investments in safety outside the workplace. In the roughly 300 years that have elapsed since the development of capitalism, most companies have avoided capital spending on safety equipment outside the factory setting. As a consequence, air pollution and other forms of pollution were largely ignored until after World War II" (Miyamoto 1967:163–164).

In *Environmental Economics*, Miyamoto further pursues this line of reasoning. "This tendency (to conserve constant capital) was clearly evident in the structure of industrial capital investments that were made during Japan's economic boom years.... Every effort was made to conserve investments in constant capital meant to ensure worker safety in the workplace (investments in pollution and disaster prevention). As will be discussed later, except for only a few companies, the corporate sector made almost no investments in pollution prevention (expenditures for pollution prevention) up to around 1970." Miyamoto then states, "This phenomenon was all the more apparent in countries engaged in the rapid capitalist accumulation known as an 'economic boom' and in countries that ignored human rights and valued economic growth above all else" (Miyamoto 2007:52–53). As also elaborated by Kapp with his theory of social costs, this tendency may be described as the true nature of capital in a capitalist economy.

Noting their dependency on capital, Miyamoto explains that whenever occupational disasters or episodes of pollution have occurred, capitalist nations do not automatically intervene from a position of neutrality, but instead show a strong interest in contributing to the expansion of corporate profits and economic growth, and that institutional reforms aimed at boosting occupational safety or environmental protection would never show progress in the absence of labor movements, citizen-led campaigns, and other social movements. Although this represents an understanding of capitalist nations that is in line with the Marxian tradition, Miyamoto's argument gains its persuasive power from the fact that it was derived from a historical analysis of environmental problems in Japan. Furthermore, this interpretation of the nature of the nation-state diverges significantly from the views espoused by Kapp and other institutional economists.

Drawing from Japan's experience with environmental pollution, Miyamoto has also proposed features of, and necessary concepts for, the modalities by which pollution-induced health problems become manifest.

First, health problems tend to be more concentrated among patients that are physiologically weak. Such individuals include children, the elderly, and women. The reverse implication is that workers tend to be comparatively less affected by pollution, and Miyamoto focuses on this point as one reason why measures against pollution show little progress.

Second, health problems tend to be more concentrated among the socially disadvantaged. This group includes low-income families and citizens living in poverty. Episodes of environmental pollution tend to occur in or near communities with a relatively large number of residents living in poverty. One implication is that self-reliant steps in prevention or rehabilitation are significantly more difficult for victims in such communities, thus underscoring the importance of environmental policies.

Third, episodes of pollution entail irreversible and absolute losses and thus call for a preventive approach. Irreversible and absolute losses are losses that cannot be reversed once they have taken place or that are extremely difficult to reverse, and include the loss of human health or life, the loss of nonrenewable natural environments, and the loss of irrecoverable cultural assets. This is a perspective shared in common with the lexicographic preference-based approach of post-Keynesian economics, but that differs from the notion of lexicographic preference because it is based on clearly defined criteria.

Miyamoto's perspectives on environmental economics apply well to the asbestos-induced environmental pollution caused by the Kubota factory. This is underlined by the characteristically high risk of harm to women from environmental asbestos exposure. Furthermore, the impact of an asbestos-induced occupational disaster appeared in its most classic form in a relatively impoverished district and in low-income occupations: namely, the Sennan community and among construction workers (workers engaged at construction sites in particular). These factors clearly explain the reasons why the government's policy response to asbestos victims was so slow.

As indicated by the foregoing discussion, the institutional and Marxian schools of environmental economics place fundamental importance on certain perspectives that receive little attention from orthodox schools of economics: namely, that the factors triggering occupational disasters and environmental devastation are built into the very mechanisms of capitalist economic systems, and that economic mechanisms must be understood in terms of their complementary relationship with social systems and the environment. Institutional economics has demonstrated an optimistic stance regarding the analysis of public policies imposed through state intervention. As a school of thought that emphasizes its insights into class systems and social classes, Marxian economics has presented reasons for the slow policy response in certain episodes of environmental pollution and occupational disasters. Furthermore, these schools of economics are also to be credited for the valuable knowledge they have accumulated in proposals for a preventive approach in principle, based on the concept of irreversible and absolute loss.

1.4 Toward a Political Economy of Asbestos Disasters

World-wide, virtually no studies of asbestos disasters have been performed in the social science field. One key reason is that analytical studies of asbestos problems demand an interdisciplinary approach.

In this chapter, we have sought to define the features of asbestos disasters in keeping with several established economic schools of thought, with attention to the fact that patterns of asbestos use have traditionally been shaped by complementary relationships between industry and public policy. Asbestos disasters cannot be explained solely in terms of economic theory in the strict sense. Indeed, it is only possible to understand asbestos disasters by combining multiple economic approaches. Specifically, occupational disasters and environmental pollution need to be analyzed independently from one another and then identified as a series of connected events. To that end, there seems no choice but to establish an entirely new field of political economy that can aid our understanding, and eventual solution, of the unprecedented social problem known as a complex-stock disaster. Further, by consolidating knowledge and insights from peripheral social sciences and the fields of medicine and architecture, this new political economy of asbestos disasters would be expected to prepare the social sciences for asbestos-related research that is genuinely valuable.

Notes

1. The outbreak of asbestos-related health problems among Libby mine workers and their family members was attributed to the asbestos found in the vermiculite ore that was being mined but in general it may be considered an example of an asbestos disaster (Mori 2008).
2. As a specific example, Dorman cites the corporate withholding of information on causal relationships between asbestos and lung cancer that medical researchers had discovered in the 1930s. That information became widely known by the 1970s, but in the interim enormous numbers of workers were exposed to the risk of contracting asbestos-related illnesses due to corporate decisions not to disclose such information (Dorman 1996:43).

References

Dorman P (1996) Markets & mortality. Cambridge University Press, Cambridge

Dorman P (2000) The economics of safety, health, and well-being at work: an overview. In: Focus program on safework, International Labour Organization, The Evergreen State College. http://www.ilo.org/wcmsp5/groups/public/---ed_protect/---protrav/---safework/documents/publication/wcms_110382.pdf. Accessed 5 May 2005

Frigato P (2006) Social costs and human health: Kapp's approach and its growing relevance today. In: Elsner W, Frigato P, Ramazzotti P (eds) Social costs and public action in modern capitalism. Routledge, New York

Kapp K (1950) The social costs of private enterprise. Schocken Books, New York, Paperback edition

Lavoie M (1992) Foundations of post-Keynesian economic analysis. Edward lgar, Cheltenham

McCulloch J, Tweedale G (2008) Defending the indefensible: the global asbestos industry and its fight for survival. Oxford University Press, Oxford

Mearman A (2009) Recent developments in post Keynesian methodology and their relevance for understanding environmental issues. In: Holt R, Pressman S, Spash C (eds) Post Keynesian and ecological economics. Edward Elgar, Cheltenham

Miyamoto K (1967) Social overhead capital (in Japanese). Yuhikaku, Tokyo

Miyamoto K (2006) Toward the Sustainable Society (in Japanese). Iwanami Shoten, Tokyo

Miyamoto K (2007) Environmental economics. New edition (in Japanese). Iwanami Shoten, Tokyo

Mori H (2008) The asbestos disaster in Libby Montana. In: Policy Science Association Ritsumeikan University (ed) Policy science supplement: special edition on asbestos problems (in Japanese). Policy Science Association Ritsumeikan University, Kyoto, pp 185–201

Vatn A (2009) Combining post Keynesian, ecological and institutional economics perspectives. In: Holt R, Pressman S, Spash C (eds) Post Keynesian and ecological economics. Edward Elgar, Cheltenham

Chapter 2
An Exploration of Measures Against Industrial Asbestos Accidents*

Kenichi Miyamoto

2.1 Industrial Asbestos Accidents and the Status of Current Countermeasures

2.1.1 The Status of Asbestos Exposure and Relief Measures

In June 2005, three mesothelioma-afflicted residents of the community surrounding Kubota Corporation's Kanzaki Factory, together with their support groups, filed charges demanding that Kubota assume responsibility for their exposure to asbestos. Although Kubota had offered relief money for work-related illnesses stemming from asbestos exposure, it was shocked by the level of asbestos exposure and harm that local residents had experienced, and was therefore prompted to go public with details surrounding the extent of asbestos-related contamination from its manufacturing facilities to date, and the status of related relief efforts. Popularly dubbed "the Kubota Shock," this event set the stage for a sweeping public policy response. From the 1970s to the present day, there had been repeated incidents of harm from exposure to asbestos in Japan, a health issue that called for all-out countermeasures. Despite that reality, within less than half a year in each case, coverage of these events in newspapers and other media faded away, and fundamental remedial measures were never taken. However, the Kubota Shock itself compelled not only Kubota, but also many other companies involved in the asbestos industry, to publicly disclose details about incidents of industrial contamination, study the impact on local citizens, and implement relief measures accordingly. As of March 2010, 391 people had

*Translated and revised from the Japanese original, Miyamoto, K (2008). An Exploration of Measures against Industrial Asbestos Accidents. *Policy Science* (in Japanese), Supplementary Volume, 167–184, with permission of the Policy Science Association of Ritsumeikan University.

K. Miyamoto (✉)
Osaka City University, 3-3-138 Sugimoto, Sumiyoshi-ku, Osaka 558-8585, Japan
and
Shiga University, 1-1-1 Banba, Hikone, Shiga 522-8522, Japan
e-mail: miyaken@kyoto.zaq.ne.jp

K. Miyamoto et al. (eds.), *Asbestos Disaster: Lessons from Japan's Experience*, DOI 10.1007/978-4-431-53915-5_2, © Springer 2011 become victims of asbestos contamination by Kubota, and 344 had died. (Of the total number of victims, 221 were local residents near Kubota facilities, including surviving family members, and 190 had already died.) Kubota negotiated with a group of victims and provided individual relief money packages which were equivalent to those usually paid out to the victims of industrial accidents, and which ranged in scale from 25 to 46 million yen each.

The industrial accident reports released by Kubota left local citizens in a state of shock. In brief, from 1957 to 1975, Kubota had utilized around 90 000 metric tons of crocidolite (blue asbestos) in its operations, and of the 251 workers who had been engaged in its asbestos cement pipe manufacturing process for a period of 10 years or more, 120 had become ill and 61 had died. In an announcement released at the time of its interim financial report in March 2010, the company disclosed that 144 former employees had died from asbestos-related causes and that another 26 were undergoing treatment. These numbers underscored a state of devastation virtually analogous to total annihilation in a war zone. Further, studies by Norio Kurumatani and Shinji Kumagai demonstrated a clear link between cases of mesothelioma among local residents and the asbestos contamination caused by Kubota. However, Kubota has denied responsibility for the contamination, and has offered relief funding instead of damage compensation.

In a related development, in February 2006 the government, responding to the outcome of public opinion polls, enacted the "Act on Asbestos Health Damage Relief" (hereinafter referred to as the "Asbestos Relief Act" or "New Act"). This constituted a legislative step to provide blanket protection for cases of asbestos contamination and victims thereof who were not deemed to be eligible for workers accident compensation. From FY 2006 through FY 2009, 6,205 patients were approved for coverage under the New Act, and of those, 2,969 died. These examples of relatively quick action by Kubota and the national government demonstrated that they had learned lessons from the Minamata disease and other renowned cases of industrial pollution from years past, and had effectively deflected the societal condemnation that may have resulted from a drawn-out legal dispute. Nevertheless, Kubota and the government have not acknowledged asbestos contamination to be a case of industrial pollution or accepted legal responsibility, and have only offered money as a form of relief rather than as payment for damages. These details illustrate that the lessons of Minamata disease and other environmental disasters have yet to be fully applied, and highlight many questions surrounding government policy on future asbestos disasters which are expected to surface in the years ahead. This chapter seeks to shed light on recent industrial asbestos accidents and the status of relief measures following enactment of the New Act.

Table 2.1, which was compiled by Ban Asbestos Network Japan (BANJAN), lists recent trends in compensation and relief for patients afflicted with mesothelioma or asbestos-induced lung cancer. The mesothelioma data in the table are estimates because no statistical records exist prior to 1994. Further, in keeping with the Helsinki Criteria, asbestos-induced cancer rates are estimated to be double those for mesothelioma. Although the resulting relief compensation rate under the New Act has increased, the approval and relief compensation rate for patients who have died of mesothelioma has leveled off at 36.5%. For lung cancer, the approval and relief

Table 2.1 Trends in compensation and relief for patients afflicted with mesothelioma or asbestos-induced lung cancer (until March 2008)

Mesothelioma

Death	*–1994*	*–2004*	*2005*	*2006*	*2007*	*Unknown*	*Total*	*Percentage*	*Death before 2007*	*Relief %*	*Death (unknown year)*	*Relief %*	*Death in 2008*	*Survivor*
	3685	*7013*	*911*	*1050*	*1068*	–	*13727*			*100%*		*100%*		
Workers' compensation	83	419	502	1001	500	–	2505	40.2 52.3	1906	13.9%	95	14.6%	24	480
Ship workers' compensation	–	4	19		8	–	31	0.5	25	0.2%	3	0.2%	0	3
Past National Railway worker' compensation	–	2	11	20	18	45	96	1.5	79	0.6%	0	0.6%	6	11
Other compensation	–	1	–	4	2	2	9	0.1	7	0.1%	2	0.1%	0	0
"New Act" litigation relief	–			570	46	–	616	9.9	616	4.5%	0	4.5%	0	0
"New Act" death cases				1538	279	–	1817	29.2 47.7	1817	13.2%	0	13.2%	0	0
"New Act" survival cases				627	525	–	1152	18.5	–	–	464	3.4%	–	688
Total of compensation or relief	**83**	**426**	**513**	**3779**	**1378**	**47**	**6226**	**100 100**	**4450**	**32.4%**	**564**	**36.5%**	**30**	**1182**

Asbestos-induced lung cancer

Death	*–1994*	*–2004*	*2005*	*2006*	*2007*	*Unknown*	*Total*	*Percentage*	*Death before 2007*	*Relief %*	*Death (unknown year)*	*Relief %*	*Death in 2008*	*Survivor*
	7370	*14026*	*1822*	*2100*	*2136*	–	*27454*			*100%*		*100%*		
Workers' compensation	120	234	213	783	502	–	1852	70.1 85.5	1069	3.9%	159	4.5%	15	609
Ship workers' compensation	–	–	14		10	–	24	0.9	17	0.1%	0	0.1%	0	7
Past National Railway worker' compensation	–	–	–	10	17	33	60	2.3	45	0.2%	0	0.2%	5	10
Other compensation	–	–	–	1		1	2	0.1	1	0.0%	1	0.0%	0	0
"New Act" litigation relief	–			272	49	–	321	12.2	321	1.2%	0	1.2%	0	0
"New Act" death cases				52	41	–	93	3.5 14.5	93	0.3%	0	0.3%	0	0
"New Act" survival cases				172	117	–	289	10.9	–	–	116	0.4%	–	173
Total of compensation or relief	**120**	**234**	**213**	**1304**	**736**	**34**	**2641**	**100 100**	**1546**	**5.6%**	**276**	**6.6%**	**20**	**799**

(continued)

Table 2.1 (continued)

Total (mesothelioma and asbestos-induced lung cancer)														
	–1994	*–2004*	*2005*	*2006*	*2007*	*Unknown*	*Total*	*Percentage*	*Death before 2007*	*Relief %*	*Death (unknown year)*	*Relief %*	*Death in 2008*	*Survivor*
Death	11 055	21 039	2 733	3 150	3 204	–	41 181			100%		100%		
Workers' compensation	203	653	715	1 784	1 002	–	4 357	49.1 62.2	2 975	7.2%	254	7.8%	39	1 089
Ship workers' compensation	–	4	33		18	–	55	0.6	42	0.1%	3	0.1%	0	10
Past National Railway worker' compensation	–	2	11	30	35	78	156	1.8	124	0.3%	0	0.3%	11	21
Other compensation	–	1	–	5	2	3	11	0.1	8	0.0%	3	0.0%	0	0
"New Act" litigation relief	–			842	95	–	937	10.6	937	2.3%	0	2.3%	0	0
"New Act" death cases				1 590	320	–	1 910	21.5 37.8	1 910	4.6%	0	4.6%	0	0
"New Act" survival cases				799	642	–	1 441	16.3	–	–	580	1.4%	–	861
Total of compensation or relief	203	660	726	5 083	2 114	81	8 867	100 100	5 996	14.6%	840	16.6%	50	1 981

Source: Data from Ban Asbestos Network Japan (BANJAN)
Note: Data of "Past National Railway Worker' Compensation" is until 7 November 2008

compensation rate for asbestos-induced lung cancer patient deaths has not risen beyond 6.6% relative to Helsinki Criteria estimates, which is partly due to the difficulties involved in screening. The comparable estimate for both illnesses combined is 16.6%. As this table illustrates, details about a previously hidden fraction of the population victimized by asbestos contamination have come to light in the aftermath of the Kubota Shock. Whereas the number of patients approved for benefits under accident compensation insurance totaled 42 (25 mesothelioma patients and 17 lung-cancer patients) in 1999, the corresponding total in 2004 came to 186 (128 mesothelioma patients and 58 lung-cancer patients). Furthermore, in 2006 the combined number of patients approved for benefits under workers accident compensation insurance rose to 1 783 (1 000 mesothelioma patients and 783 asbestos lung-cancer patients), and under the provisions of the New Act, the combined total has reached 3 230. Accordingly, the total for mesothelioma and asbestos-induced lung cancer patients together with deaths caused by asbestos rose to 5 013. This represents a 120-fold increase on the corresponding figures from 7 years earlier, and a 27-fold increase on the figures from just 2 years earlier. Under the New Act, relief money has also been provided to 841 patients who would otherwise have been ineligible because they had received relief money under the provisions of the earlier statute of limitation. Although there was a surge in applications for approval when the act initially came into effect, it is now known that both illnesses claim the lives of at least 2 000–3 000 patients per year. One explanation is that the effects of asbestos exposure are not as readily apparent as the effects of other forms of industrial pollution owing to the relatively lengthy period from initial exposure to the development of symptoms in those affected. On a more fundamental level, though, the statistical trend seems to highlight how seriously negligent the corporate sector and national government have been with regard to implementing effective countermeasures. What is more, there is much that still remains beneath the surface. Needless to say, in the current situation, few surrounding residents are certified as asbestos pollution victims, and perhaps only 10% of all mesothelioma patients are approved for workers accident insurance benefits. It is believed that many patients with asbestos-induced lung cancer have been treated as though their use of tobacco was the prime cause of their illness (see BANJAN 2007a, b and Table 2.1).

The manifestations of asbestos exposure and consequent harm were this extensive after the Kubota Shock and the implementation of follow-up countermeasures. It may be one of the iron-clad empirical rules of conduct in cases of industrial pollution that unless and until victims speak up, seek recognition of their status as victims of industrial pollution, and bring charges against the perpetrators, the industrial pollution (or more broadly, industrial accidents) in question will not be brought to light. This state of affairs illustrates the societal discrimination that victims experience and the courage they need in order to assert their human and civil rights. In addition, within the context of asbestos contamination and disasters, insights into the actual numbers of victims are hindered by the facts that symptoms typically begin to appear long after the asbestos exposure, and that patients often die soon after their symptoms become manifest. The three courageous victims who initially spoke out in the Kubota case have already passed away, but their courage has aided many other victims to obtain care and assistance.

Given that backdrop, we now consider the factors that created so many victims. Table 2.2 lists factory facilities with emissions of specified particulates, and industrial sectors that experienced industrial accidents prior to 2004. Cases of certified workers accidents (in FY 2005 and FY 2006) spanned numerous industries, with 1 414 cases in the construction sector, 1 680 cases in the manufacturing sector, 97 cases in the transport sector, 34 cases in the power, gas, water, and heating sectors, and 149 cases in other fields. The manufacturing sector had many certified cases, with shipbuilding operations (211 businesses) accounting for 459 cases, ceramics operations (128 businesses) having 303 cases, the transportation equipment sector (88 businesses) having 205 cases, machinery and equipment manufacturing operations (81 businesses) having 140 cases, and chemical manufacturing (69 businesses) having 117 cases. A total of 2 514 businesses reported certified workers accidents involving asbestos exposure in FY 2006. Table 2.3 lists certified cases reported under the New Act, prefecture by prefecture, from April 2006 through July 2010.

Table 2.2 The number of compensation or relief cases about asbestos exposure work by industry (April 2005 – March 2007)

	Business establishments	Workers' compensation	New Act relief	Total of compensation and relief
Construction	1 356	1 130	284	1 414
Mining	8	3	5	8
Manufacture	918	1 159	521	1 680
Transport	83	74	23	97
Electricity, water, or energy supply	26	22	12	34
Other	123	112	37	149
Total	2 514	2 500	882	3 382

Source: Ministry of Health, Labour, and Welfare in March 2008

Table 2.3 The state of relief (New Act) by prefecture in Japan (April 2006 – July 2010)

Prefecture	Patient application	Death before act enforcement	Death or no application	Total
Hokkaido	109	114	9	232
Saitama	180	175	5	360
Chiba	104	109	8	221
Tokyo	257	254	16	527
Kanagawa	194	211	12	417
Shizuoka	73	82	7	162
Aichi	138	112	8	258
Osaka	325	299	18	642
Hyogo	336	306	5	647
Hiroshima	76	89	4	169
Fukuoka	130	111	7	248
Total in all prefectures (including the other 36 prefectures)	2 912	3 061	173	6 146

Source: Environmental restoration and conservation agency of Japan in August 2010

Accidents with 100 or more victims occurred chiefly in metropolitan areas, with Hyogo Prefecture accounting for the largest number, 647 cases, followed by Osaka with 642 cases, Tokyo with 527 cases, Kanagawa with 417 cases, Saitama with 360 cases, and several other prefectures with 100 cases or more. Victims of such workers accidents were reported in all 47 prefectures nation-wide. A breakdown of the number of victims reported to the Ministry of Health, Labour, and Welfare (MHLW) encompassed an exceptionally broad cross-section of industry. Accordingly, occupational exposure was not limited to workers or employees in manufacturing companies directly involved in the utilization of asbestos, such as Kubota or the NICHIAS Corporation. Victims of exposure also included dock workers and truck drivers involved in the handling or transport of asbestos cargoes, workers who handled merchandise containing asbestos, workers engaged in the repair of buildings, joinery, and electrical, gas, and plumbing fixtures, bakers and other workers engaged in food preparation, structural demolition workers and other specialized workers engaged in the disposal and processing of spent asbestos products, and even stage technicians. As mentioned earlier, the official statistics indicate that only a small fraction of workers harmed by asbestos exposure have been approved to receive benefits under workers accident compensation regulations. Other than a small number of individuals living in the vicinity of the Kubota or NICHIAS facilities, little is known about family members of workers who have not been approved for benefits, or of other indirect victims or victims of industrial pollution. In addition, only a small percentage of actual cases have been brought to light owing to the tendency among members of the corporate community to cover up incidents of industrial accidents or pollution, as exemplified by the incidents involving the Chugoku Electric Power Co. Inc., or residents living in the vicinity of the NICHIAS Corporation's Hashima factory in Gifu Prefecture. Furthermore, if the public should lose interest in such events, there is a risk that studies of the levels of contamination and harm to local residents may lose momentum or come to a standstill (Tables 2.2 and 2.3).

The New Act constitutes an emergency first-aid measure and is significantly flawed in certain areas because it was designed as a measure to win public support. Table 2.4, which was compiled by BANJAN, compares workers accident compensation benefits with assistance provided under the New Act. The single largest issue was that the New Act limited assistance to victims of mesothelioma and asbestos-induced lung cancer, and excluded victims of pulmonary asbestosis who were eligible for benefits under workers accident compensation insurance. However, asbestos exposure together with a considerable disorder of pulmonary function was added to the subjects covered by the New Act in July 1, 2010. Pulmonary asbestosis patients in the Sennan district of Osaka Prefecture have filed a lawsuit seeking damage compensation from the national government. Their relief amounts are extremely small compared to the relief money and accident compensation amounts received by victims of Kubota's asbestos contamination. Immediately following enactment of the New Act, many victims filed applications for relief. After 1 year, though, the flow of applications slowed. One reason is that applicants with asbestos-induced lung cancer are required to submit documentation delineating how their illness differs from other forms of lung cancer. Gathering such documentation is

Table 2.4 A comparison of workers accident compensation and relief money under the New Act

		Relief money under the New Act	
	Workers accident compensation	Survival cases	Death cases prior to enactment of the law
Funding source	Workers accident compensation insurance	Asbestos health damage relief foundation (general + special contributions)	
	National government	Environmental Restoration and Conservation Agency of Japan (ERCA)	
Implementing agency	Labor standards supervision office	Organization office, local environmental management offices, community health centers	
Applicable period	In principle, applied retroactively from the initial date of diagnosis; no provisions for fixed duration	Applied from date of application for an effective duration of 5 years after approval. (Renewable for patients with no prognosis of cure)[b]	
Medical conditions covered.	(1) Mesothelioma, (2) lung cancer, (3) pulmonary asbestosis, (4) benign asbestos pleural effusion, (5) diffuse pleural thickening, (6) other medical conditions that are clearly linked to work involving exposure to asbestos	Designated diseases: (1) mesothelioma, (2) lung cancer, and (3) any other disease caused by inhaling asbestos and designated by Cabinet Order: (a) pulmonary asbestosis with considerable disorder of pulmonary function, and (b) diffuse pleural thickening with considerable disorder of pulmonary function[b]	
Medical expenses	Full amount covered	Patient responsible[b]	None[a]
Outpatient expenses	In principle, actual expenses covered in full	None[a]	None[a]
Compensation for lost work	Approx. 330 000 yen per month (80% of average monthly wage)	Uniform 103 870 yen per month as allowance for medical treatment[b]	None[a]
Funeral service fee	Approx. 820 000 yen (average wage for 30 days plus 315 000 yen, or average wage for 60 days)	Uniform 199 000 yen (time limit: 2 years, same as under workers accident compensation insurance)[b]	Uniform 199 000 yen[b]

Lump payment to family	Uniform 3 million yen (plus lump payment of 13.7 million yen [average wage for 1 000 days] to families ineligible for pension benefits)	If a patient with a medical condition prior to enactment of the law dies within 2 years after enactment of law, the sum listed on the left shall be paid as an adjustment if the total in medical expenses and allowances for medical treatment is less than the 2.8 million yen sum listed on the right[a,b]	Uniform 2.8 million yen as special survivor condolence money[b]
Survivor pension	Approx. 2.75 million yen (average wage for 153 days for surviving family consisting of one dependent, 201 days for two dependents, 223 days for three dependents, 245 days for four dependents or more). If period for relief money has expired, many beneficiaries receive a small accident compensation pension benefit of less than 2.4 million yen[b]	None[a]	None[a]
Education assistance	12 000 yen monthly for day-care center for children or elementary school student, up to 38 000 yen monthly for university student	None[a]	None[a]

Source: Ban Asbestos Network Japan (BANJAN)

Note: Despite eligibility for approval under the criteria of the New Act, there have not been any cases of relief money being provided to patients who in the past were not approved under the criteria of the Workers Accident Compensation Insurance framework

[a] "Cracks" in relief money

[b] Lack of "fairness"

troublesome, and this probably explains the limited number of applications actually submitted or approved. Although flaws in the administrative approval process had been cited earlier during the handling of applicants with Minamata disease, one must wonder whether similar flaws are impeding the handling of asbestos-related applications. To identify applicants with Minamata disease, the screening committee relied on the symptoms seen in severe cases, and took Hunter–Russell syndrome as one of their approval criteria. This approach impeded a full understanding of Minamata disease and had the effect of excluding certain classes of patients from assistance. Similarly, it would appear that applying the severe symptoms of mesothelioma as the core criteria for approval of patients with asbestos-related diseases would excessively restrict eligibility for relief money. Furthermore, is not the scope of damage compensation limited by the composition of the screening committee itself, which is a body consisting solely of physicians with no legal professionals? Does not the decision to exclude residents with pulmonary asbestosis point to an over-reliance on the limited perspectives of the professional medical community (Table 2.4)?

2.1.2 Epidemiological Studies (and Projections)

Epidemiological studies of asbestos-related industrial accidents face numerous difficulties because they must look back in time to the initial point of exposure. For workers accident compensation purposes, corporations still in business could examine their payroll records to create registers of employees who were engaged in their operations during the period of time they utilized asbestos, and could recommend that those individuals receive medical examinations. The 2514 businesses cited earlier should be keeping records of the names of their employees and the duties in which each employee is or was actually engaged. In cases of industrial pollution, it will presumably also be necessary to register the names of residents living in the neighborhoods surrounding plant facilities that were engaged in the manufacture of asbestos products. In FY 2006, assessments of the health risks associated with asbestos were performed by Osaka Prefectural Government, Saga Prefectural Government, and the City of Amagasaki. In that study, surviving family members of patients listed as having died of mesothelioma in selected demographic mortality tables over the 3-year period from 2003 through 2005 were interviewed about each patient's occupational record and history of residence in Osaka Prefecture (Osaka City, Takatsuki City, and Higashiosaka City) and Saga Prefecture. In addition, the study also confirmed the residential addresses listed in each patient's record of registered domicile through the period from 1955 to 1974. Medical records of these patients kept by health-care institutions were then updated with their respective occupational histories. Industrial facilities that handled asbestos products from 1955 to 1974 – the period of heaviest asbestos use in Japan – were also identified. While no professional medical appraisal of this study can be offered, it does raise several noteworthy points about Osaka Prefecture that should be taken into account. These points are discussed below.

Of the 263 deceased patients considered by the study, interviews were held with the consenting family members of 130. Of that number, it was learned that 32 (24.6%) had been approved for workers accident compensation; a relatively high percentage. In addition, 108, or 83%, of the total were subject to at least one path of exposure associated with their workplace, whether it involved an industrial accident, their occupational field, exposure within their home, or entry into facilities that handled asbestos. The pathway of exposure for 22 others could not be identified. Of 338 industrial facilities known to have handled asbestos, 109 are concentrated in the City of Osaka, 47 in the City of Sennan, and 46 in the City of Hannan. Although they tend to be spread throughout the Cities of Osaka, Sennan and Hannan, they are mostly concentrated in the lowland or flatland areas. The epidemiological study concluded that it was not possible to uniformly identify the types or amounts of asbestos these facilities handled, the periods during which they used asbestos, the asbestos form-factors or package formats they handled or processed, or whether they released asbestos into the general environment during the two decades from 1955 to 1974. The study chiefly plotted the physical locations of the asbestos-handling facilities and the residential addresses of the targeted patients, and investigated the correlations between the two. It focused in particular on the correlations with patients who were thought to have been exposed to, and contaminated by, asbestos that had been released into the general environment, but concluded that no correlations could be drawn for those who resided in the Amagasaki area (as will be considered in more detail later), and that further study of that topic was warranted. In addition, albeit perhaps because it was limited to a focus on mesothelioma, the study also concluded that no correlations could be found in the Sennan and Hannan districts between asbestos-handling facilities and local deaths from mesothelioma attributable to asbestos exposure through general environmental pathways (Osaka Prefecture, Osaka City, Sakai City, Takatsuki City, and Higashiosaka City 2007).

The risk assessment conducted by the Osaka Prefectural Government in the Sennan district was aimed at individuals aged 40 and over who had lived in the area prior to 1990 and were considered to be at risk of being exposed to asbestos through general environmental pathways. A total of 309 participants in the assessment underwent diagnostic interviews and examinations. Of 168 subjects (54.4%) who had medical indications, 30 (9.7%) were found to be ill. Of these 168 subjects, 142 (84.5%) were also found to have occupational or family occupational histories related in some way to asbestos, whereas 26 (15.5%) did not have a directly identifiable history of asbestos exposure. Of the 168 who had medical indications, 130 (77.4%) had pleural plaque; of the 30 determined to have a disease condition, 5 (16.7%) had pulmonary asbestosis, 3 (10.0%) had lung cancer, and one subject in this latter group had both pulmonary asbestosis and lung cancer. Of the 26 subjects who did not have a directly identifiable link to asbestos exposure, 16 (61.5%) had pleural plaque, 7 (26.9%) had enlarged lymph nodes, and one had lung cancer. A follow-up of the lung-cancer patient has been scheduled in order to determine whether his illness is due to asbestos exposure. The risk assessment report does not offer any conclusions regarding links with the industrial facilities that have handled asbestos. Nonetheless, the risk that local residents have been exposed to, and contaminated by, asbestos pollution cannot be denied given that the assessment did identify typical disease symptoms even in subjects who had no directly identifiable links to asbestos exposure (Osaka Prefectural Government 2007).

The Ministry of the Environment (MOE) and the City of Amagasaki have issued an interim report on the findings of a joint epidemiological analysis and survey of asbestos exposure. In that study, approximately 180 000 Amagasaki citizens with a history of local residence between 1955 and 1974, and a continuous record of domicile up to the end of 2001, were selected and divided into several observation groups. Of 50 Amagasaki citizens who had died of mesothelioma between 2002 and 2004, 42 who had a record of domicile in the specified time-frame were selected for study. Of this group, 16 were determined to have been exposed to asbestos through an industrial accident or their occupation, and 5 through exposure in their home; the exposure pathway for another 10 subjects could not be identified, and the remaining 11 were not actually studied. Based on the findings of this study, the standardized mortality ratio (SMR) was high for Amagasaki residents up to 1965, and highest of all for those who became Amagasaki citizens between 1955 and 1959. Although many subjects in the sample did not readily fit into any of the administrative classifications, the SMR for men ranged from 10.6 to 21.1, and was the highest (21.1) for those who resided in the Oda district (Amagasaki) until December 31, 1969. The SMR for women living in the same Oda district was also at its highest (68.6) in the same time-frame, compared with a range of 29.6–68.6 for all time-frames. The Oda district is the neighborhood surrounding the Kubota facility (The Ministry of the Environment and Amagasaki City 2007).

Given the relatively small scale of the subject sample in this risk assessment and the lack of data on past sources of exposure, further study will be needed. Nevertheless, even with the limited data, the findings of this study still support the previously cited conclusions of Kurumatani and Kumagai. Despite the inadequacies of risk studies of this kind, it seems undeniable that the Sennan district in Osaka Prefecture and the Oda district in the City of Amagasaki have been hit by an asbestos disaster.

2.1.3 Regulations and Follow-Up Countermeasures for Asbestos Businesses and Related Facilities

Immediately following the Kubota Shock, the Ministry of Health, Labour, and Welfare disclosed details of its administrative guidance for asbestos-related business establishments. However, out of 124 establishments that received guidance, as many as 57 (46%) were cited for violations of regulations designed to prevent asbestos-related diseases. In particular, 30 (62.5%) out of 48 business establishments involved in processing goods that contained asbestos were cited for such violations. The highest violation rate was for a failure to implement asbestos-related health examinations (24.2%), followed by violations of rules on the installation and use of protective respirator equipment (21.0%), and failure to perform required measurements and prepare records thereof (20.2%). In addition, over 10% of the establishments were cited for failure to install washroom or disposal facilities or implement restricted-area practices. These violations point to a strong likelihood of future asbestos-related industrial accidents. The current level of noncompliance with established laws and regulations illustrates how difficult it may be to enforce future restrictions on structural demolition projects and other operations involving the risk of asbestos exposure.

In on-site inspections of 389 facilities subject to inspection in 15 prefectures, the Ministry of Internal Affairs and Communications (MIC) Administrative Evaluation Bureau found that 36 had degraded asbestos stocks that were at risk of airborne dispersal and creating an exposure hazard. Of those 36, it was determined that 15 had not sealed off locations that used asbestos or taken other steps to prevent asbestos exposure. High facility clean-up costs have been cited as the prime reason for their failure to act. One project to decontaminate a privately operated parking tower in Kagawa Prefecture was canceled owing to the estimated 60 million yen clean-up cost. There have also been reports of facilities resorting to the dangerous practice of routine sweeping of contaminated areas rather than implementing exhaustive clean-up measures. Furthermore, according to the Ministry of Land, Infrastructure, Transport, and Tourism (MLIT), approximately 32 000 out of some 42 000 large-scale facilities in Tokyo with floor-space exceeding 1 000 m^2 have not been inspected. Even inspections of a sample of smaller facilities outside the scope of the regulation detected asbestos exposure at 16.7% of the structures inspected (Asahi Newspaper, December 12, 2007, and Nihon Keizai Shimbun Newspaper, December 12, 2007).

As these findings indicate, not enough has been done thus far to prevent asbestos-related accidents.

In 2006, the national government announced an all-out ban on the use of asbestos. Until that time, asbestos had been described as a wonder material and treated almost as if the national economy would not survive without it. However, did Japan actually face any turmoil or pandemonium after the ban took effect? The truth is that in practically every industrial sector, economic productivity has not been impeded to any significant extent by the use of substitute materials. This suggests that the benefits of asbestos derive not from its properties as a material, but rather from the fact that it is less expensive than its substitutes. That point is further demonstrated by the fact that dangerous asbestos textile industries, such as the one based in Osaka's Sennan district, have traditionally run their operations using low-wage labor. One question worth asking at this point is how much asbestos remains in industry stockpiles. Investigations of the roughly 3 000 products that contain asbestos have not even been initiated. In December 2003, the Japan Asbestos Association released quantitative data on the construction materials in which asbestos is used most extensively. From 1971 to 2001, the industry produced a cumulative total of over 4 015 million m^3, or 43.42 million metric tons, of construction materials containing asbestos, with an estimated scrap or disposal rate of 5%, or 2.17 million tons, at the time of use. By the association's estimates, existing structures incorporate over 3 814 million m^3, or 41.25 million tons, of construction materials containing asbestos, with an estimated 5.41 million tons of asbestos being utilized (Table 2.5).

Table 2.5 Estimated (projected) quantities of asbestos-containing construction materials in existing structures

Year	Shipped quantities: Floor space (1 000 m^2)	Shipped quantities: Weight (metric tons)	Scrap rate (%) at time of use	Amount scrapped at time of use (metric tons)	Estimated quantities of asbestos-containing construction materials in existing structures: Floor space (1 000 m^2)	Estimated quantities of asbestos-containing construction materials in existing structures: Weight (metric tons)	Estimated quantities of asbestos (metric tons)
1971	123678	1229826	5	61491	117494	1168335	190933
1972	137528	1321029	5	66051	130652	1254978	208618
1973	176208	1724671	5	86234	167398	1638437	270475
1974	137293	1350705	5	67535	130428	1283170	212273
1975	119399	1172095	5	58605	113429	1113490	185194
1976	132239	1319281	5	65964	125627	1253317	208650
1977	134801	1351686	5	67584	128061	1284102	212088
1978	132574	1328947	5	66447	127845	1262500	208312
1979	152962	1517406	5	75870	145314	1441536	236550
1980	147552	1444330	5	72216	140714	1372114	214395
1981	120729	1251092	5	62555	114693	1188537	167810
1982	124206	1276821	5	63841	117996	1212980	172344
1983	121990	1254728	5	62736	115891	1191992	163471
1984	131060	1342755	5	67138	124507	1275617	173824
1985	150255	1494169	5	74708	142742	1419461	185980
1986	150982	1518656	5	75933	143433	1442723	173904
1987	167524	1709219	5	85461	159148	1623758	187892
1988	182727	1849651	5	92483	173591	1757168	209028
1989	177081	1803727	5	90186	168227	1713541	209070
1990	184212	1862501	5	93125	175001	1769376	207869
1991	178699	1857209	5	92860	169764	1764349	195376
1992	141376	1636397	5	81820	134307	1554577	173026
1993	116571	1462937	5	73147	104498	1359790	153792
1994	109998	1431246	5	71562	104498	1359684	151581
1995	108629	1463480	5	73174	103198	1390306	141929
1996	107316	1465438	5	73272	101950	1392166	142599
1997	97802	1343287	5	67164	92912	1276123	131082
1998	75272	1042259	5	52113	71508	990146	91821
1999	71462	1005021	5	50251	67889	954770	87991
2000	59971	898780	5	44939	56972	853841	80589
2001	41593	689931	5	34497	39513	655434	64279
Total	4015689	43419282	–	2170962	3814904	41248320	5412652

Source: Data from the Japan Asbestos Association

Figure 2.1 illustrates the projected trend in the volume of scrap construction materials containing asbestos if the average service life is estimated at 30 years. Based on that formula, the projected volume will peak at close to an estimated 1.8 million tons per annum in the year 2020. By that time, the estimated cumulative floor space of demolished structures will surpass 170 million m^2, and the estimated

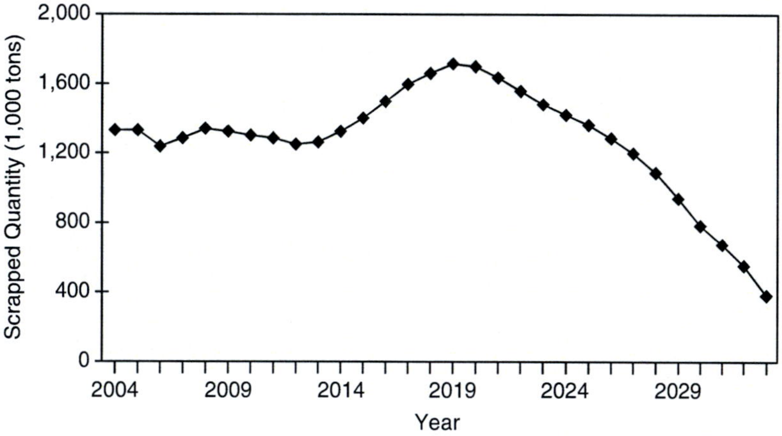

Fig. 2.1 Projected trend in the volume of scrapped asbestos-containing construction materials
Source: Japan Asbestos Association

volume of scrap asbestos alone will exceed 200 000 metric tons. One point that must be borne in mind here is the sheer quantity of asbestos in use by this sector: around 10 million tons. Even if other industrial sectors use only 1% or 0.1% of that amount, in quantitative terms that still equates to a substantial 100 000 tons or 10 000 tons, respectively. This is why it will be necessary to establish an accurate record of asbestos use in the 3 000 products containing asbestos cited earlier (Fig. 2.1).

Regarding the different types of scrap asbestos, the government has established strict handling criteria, and mandated that industrial users should assign special industrial waste product managers to supervise the handling of types of particulate asbestos. Nonparticulate scrap asbestos products, such as those in formed board shapes, are considered safe provided they are not pulverized or cut apart. The Ministry of the Environment delineates these separately from other types of scrap product, requires that they be covered with sheeting or placed into bags for storage or transport purposes, and stipulates that as a rule they must not be pulverized or cut apart. Although these materials may be disposed of in safe disposal sites traditionally approved for the disposal of building rubble, if they are disposed of in landfill sites, it is required that this must be done in designated locations, and that the landfill materials must be covered with a surface layer of soil at the end of each working day.

As these examples suggest, strict regulations have been established for the disposal of scrap asbestos products. However, when an enormous volume of waste material has to be disposed of, the question is whether disposal site operators will be prepared to devote the care and attention required for proper disposal in line with regulations, or to provide the oversight framework needed in order to blow the whistle on violators of the previously cited preventive regulations against asbestos contamination. Above all, the question remains whether asbestos-containing scrap materials can be safely disposed of in an age where candidate landfill sites for the disposal of industrial waste in general are becoming increasingly scarce.

Intermediate processing is one way of ensuring the safer disposal of particulate asbestos materials, but not enough facilities for dissolution processing have yet been built. Although interviews with the Ministry of the Environment heard reports that dissolution processing facilities were available in 16 locations, interviews with the Japan Asbestos Association determined that ultra-high-temperature dissolution facilities were available in only six locations operated by the Nippon Steel Corp. and other companies, and that in any event, not many facilities of this kind exist. Processes for chemical dissolution have not yet been perfected for commercial use. Although it would be worthwhile to clearly specify the amount of asbestos in construction materials, procedures for special handling would be complicated. One fear is that homes and other small-scale structures will be bulldozed at the time of demolition, thus aggravating the extent to which asbestos-containing materials are pulverized or crushed.

Interviews with the Japan Asbestos Association reportedly found that the transition to asbestos substitutes is almost complete. As demonstrated by cases of silica litigation currently under way in the US, the safety of material alternatives to asbestos has been brought into question, and will be a matter for future study. In addition, as echoed by revelations that the NICHIAS Corporation cheated on inspections by dousing its fire retardant construction materials in water prior to tests, questions about the actual effectiveness of asbestos substitutes still linger.

Although public concern about the asbestos issue clearly subsided after enactment of the New Act, the harm caused by asbestos can be expected to continue to mount over the coming half-century. Counter-measures to deal with the asbestos issue, including reforms and innovations under the New Act, are only now beginning to take shape. Given that backdrop, we now explore world-wide trends and some of the theoretical issues in order to gain a clearer picture of the asbestos crisis.

2.2 The Asbestos Crisis and Modern Political and Economic Systems

2.2.1 International Trends in Asbestos Use and Contamination

Utilizing statistics on world asbestos production and consumption published by the US Geological Survey, I selected 15 countries for comparison, and explored quantitative trends in asbestos use. Table 2.6 lists the quantities used in 10-year increments, starting in 1920. The latest year for data was 2003. Asbestos use is estimated to have reached an actual cumulative total of around 180 million metric tons from 1920 to the present day, the period for which statistical records are available. As far as rough estimates permit, correlations can be drawn between the quantity of asbestos used and the scale of GDP in the advanced industrial countries up to the time that asbestos use was limited or banned. Further, the quantities used peaked during periods of accelerated economic growth. As these trends in consumption show, the quantities of asbestos used declined first in Sweden, then in the UK, and then in the US after early restrictions on asbestos use came into effect in the 1980s. Comparable declines began to appear in Germany, France, and Italy in the 1990s, with Japan effectively the last industrial power to follow suit in the first decade of the twenty-first century. Conversely, in the developing countries, the quantities of asbestos used actually began climbing sharply in the 1990s and thereafter, particularly in the Asian countries of China (to 500 000 tons/year in the 2000s), India (300 000 tons/year in the same period), and Thailand (100 000 tons/year in the same period), and in the South American country of Brazil (200 000 tons/year in the same period). Among countries transitioning from socialist to capitalist market economies, Russia exhibited an unusual increase in asbestos use from 1.47 million tons in 1980 to 2.15 million tons in 1990. Even now, it is still a heavy consumer, using over 400 000 tons/year. These countries have not yet implemented any asbestos countermeasures; that will be a future challenge. As these trends illustrate, asbestos was a material which was essential in the drive for rapid industrialization and urbanization. Banning the use of asbestos is possible once the pace of industrialization and urbanization has begun to slow, but to date very few countries have banned its use during an accelerated growth phase aimed at building an economic structure based on energy-intensive consumption (Table 2.6).

In keeping with the definition of complex stock pollution, citizens face the risk of exposure and harm at all stages of the asbestos product cycle, from manufacture and distribution to consumption and disposal (for more details, see Miyamoto 2006, 2007).

In other words, the future victims of asbestos contamination can be expected to be found in many workplaces and households, both nation-wide and world-wide. Moreover, their symptoms will begin to manifest themselves anywhere from 10 to 50 years after being exposed. Given this scenario, it is possible that the business establishments that were responsible for their exposure to asbestos will no longer be readily known. It is believed that exposure to asbestos is the underlying cause in 80–90% of all cases of mesothelioma. However, the effects of tobacco use, dietary habits, and environmental conditions should also be factored in as possible causes of lung cancer and other cancers of the internal organs. For this reason, government agencies tend to be reluctant to declare asbestos contamination a clear-cut cause. Owing to a mixture of influences, including the uncertainties of, and potentiality for, multiple causal agents and sources of responsibility, flaws in medical diagnostic procedures, and the slow pace of progress in the field of epidemiological science, the true extent of the harm from the massive, long-term use of asbestos world-wide is not yet known.

In the United States, approximately 10 000 citizens currently die each year from illnesses attributed to asbestos exposure. This annual total is expected to reach its peak in the year 2015. Various corporations have already been ordered by the courts to pay around $65 billion (nearly 7 trillion yen) in damages. In France, yearly deaths from asbestos-related causes average about 3 000, and the cumulative loss of human lives is expected to reach up to 100 000. In 2000, the French government established

Table 2.6 Asbestos consumption by country (units: metric tons)

Country/year	1920	1930	1940	1950	1960	1970	1980	1990	2000	2003
China	–	315	140	102	81 288	172 737	241000	185 748	382 315	491 954
India	1 847	34	5 554	11 160	23 652	49 792	96 892	118 964	145 030	192 033
Japan	4 965	11 193	26 673	12 245	92 483	319 473	398 877	292 701	85 440	23 437
South Korea	–	–	5 589	610	631	36 664	46 641	76 083	30 124	23 799
Thailand	–	–	–	–	6 433	21 272	58 756	116 652	109 600	132 983
Canada	–19824	41 433	15 162	41 157	45 327	95 374	106 369	76 014	–4965	19 781
USA	151 766	192 454	237 071	660 129	643 462	668 129	358 708	32 456	1 134	4 634
Sweden	336	1 061	2 878	10 002	17 079	18 646	1 181	595	–12	–
UK	21 199	23 217	95 008	107 606	163 019	149 895	93 526	15 731	268	22
France	445	–	19 130	38 921	83 385	152 357	125 549	63 571	–26	–5
Germany	6 647	13 709	11 181	93 842	167 408	226 703	440 045	15 084	212	102
Italy	3 838	6 942	13 471	24 813	73 322	132 358	180 529	62 407	40	–
Australia	839	83	14 821	18 182	43 834	68 902	66 485	1 706	1 424	19
Russia (USSR)	1 629	38 332	71 127	136 458	453 384	680 589	1 470 000	2 151 800	449 239	429 020
Brazil	205	136	559	9 279	26 906	37 710	195 202	163 238	172 560	78 403
World total	183 868	388 541	522 282	1 266 929	2 178 681	3 543 889	4 728 619	39 963 873	3963873	2 108 943

Source: US Geological Survey, *Worldwide Asbestos Supply and Consumption Trends from 1990–2003*, Supplement Policy Science Special Issue on Asbestos Problems, 2008 a special indemnification fund for asbestos victims (*FIVA, Fonds d'Indemnisation des Victimes de l'Amiante*), and began managing the fund in 2002. It is estimated that funding in the range of 26.8–37.2 billion euros (roughly 4.5–6.2 trillion yen) will be needed over the coming 20-year span. Between 2002 and 2006 in the UK, under the compensation framework for industrial accidents, approximately 12 000 citizens were recognized as patients with legally defined occupational illnesses stemming from asbestos exposure. Of that total, 6 420 were mesothelioma patients.[1]

Japan has utilized an estimated 10 million metric tons of asbestos to date. Takehiko Murayama projects cumulative deaths from mesothelioma at around 100 000 over the coming 40 years. Extrapolating from that estimate, mesothelioma deaths world-wide may reach as many as 1.8 million, considering that the global use of asbestos has totaled approximately 180 million tons to date. Under the Helsinki Criteria, nearly twice as many deaths are believed to stem from asbestos-induced lung cancer. That is equivalent to roughly 3.6 million lives world-wide. Together, these estimated totals may surpass five million lives. If we also take into consideration the many patients who have suffered and died from pulmonary asbestosis, it would seem no exaggeration to describe the harm from asbestos as a catastrophe on an unprecedented scale. Moreover, large quantities of asbestos are still being utilized by the developing world. Even if the future brings advances in preventive methods and medical treatment, asbestos-induced illnesses are likely to rank alongside automobiles as one of the top causes of social loss attributable to normal economic activity throughout the rest of this century. What is known now is only the tip of the iceberg. Further efforts must be made through epidemiological research and the identification of asbestos victims in order to gain clear insights into the full scale of the asbestos catastrophe. How could this catastrophe have been allowed to occur in the first place, and then be so largely neglected?

2.2.2 The Asbestos Crisis as a Product of Flaws in Modern Socioeconomic Systems

Reflecting its popularized image as a miracle or wonder material, asbestos does excel in various properties, including heat and fire resistance, suitability to a variety of applications, and physical stability. Furthermore, it has been in wide use for many years. However, the heavy levels of asbestos use commonly observed now did not appear until after the Industrial Revolution was underway. As Table 2.6, trends in asbestos consumption, illustrates, it was the American way of life, with its mass consumption and urbanization, that set the trend toward heavy asbestos use into motion. Suffice it to say that the arms race (and the mechanization of warfare with the development and manufacture of battleships, fighter aircraft, tanks, and other vehicular military machinery) further fanned the flames behind this trend. This was highlighted by a sharp jump in asbestos use within the US to almost 200 000 tons a year starting around 1930, a trend which was sparked in the 1920s and thereafter by the invention of the Model-T automobile, and fueled by the heavy consumption of energy to supply the mass market with automobiles, electrical machinery, and other durable consumer goods, the trend toward urbanization, the widespread construction of dwellings as another class of consumer durable, and the steady build-up of new steel and steel-reinforced concrete high-rise structures alongside conventional stone architecture buildings. Many other countries have also demonstrated a sharp increase in consumption in parallel with modernization efforts influenced by the US economic model. Following the shift into its own phase of rapid economic growth in the 1960s, Japan also experienced a steep surge in asbestos use coinciding with its efforts to urbanize and adopt American-style mass consumerism, particularly with the manufacture of automobiles, the construction of concrete high-rise buildings, the development and spread of a modern water supply and sewerage infrastructure, a supporting base in a heavy chemical industry and large power-generating facilities, and other elements of energy infrastructure.

From a different perspective, energy-intensive manufacturing facilities and consumer goods require asbestos for its properties as a heat-resistant material. Urbanization has been accompanied by the concentrated build-up of factories and other business establishments together with residential zones, which in turn raise the risk of fire hazards. These trends presumably encouraged the increased use of asbestos to take advantage of its qualities as an effective fire-resistant, soundproofing material. From that perspective, one may conclude that modern economic systems driven by energy-intensive forms of mass production and consumption, and spatial urbanization strategies that strive to harness the benefits of infrastructure build-up, provided the key source of demand for intensive asbestos use. Suffice it to say that demand for asbestos climbed when such systems were being built. Once the structure is in place, however, it should be possible to make the transition to asbestos alternatives.

Although many developing countries are aware of the severity of the asbestos crisis in the advanced industrial world, they have continued to use asbestos in massive quantities. One reason is that they are still in the process of building modernized economic systems of their own. Another is that they are also in the midst of military build-ups, and are using asbestos because of its exceptional material qualities and low cost.

Efforts to control asbestos hazards have not been successful even within modern economic systems that use asbestos in massive quantities. This reality highlights a failure of government administrative and political systems, which should be giving top priority to the protection of human life, health, and fundamental human rights. Had most national governments heeded the alarms sounded by the International Labour Organization (ILO) and implemented asbestos investigations and controls on that basis, the damage from asbestos might not have become as extensive as it now is. The problem, as has been seen with other cases of industrial pollution and disasters, is that preventive measures and curbs against the social losses posed by modern economic systems give priority to economic growth, and are built on a structure of collusion between bureaucrats, politicians, and big business rather than a public–private relationship of checks and balances. Additional blame may be ascribed to the weakness of social movements, led in particular by labor movements, that are unable to remedy these governmental flaws, and to the weakness of news organizations that lack freedom of speech or a firm sense of justice. In developing countries characterized by collusion between bureaucrats, politicians, and big business, these issues have delayed the preparation of effective legal curbs and regulations, and even the academic community has not been able to effectively shed light on the harm caused by asbestos or sound an alarm. Further, owing to the lack of freedom of speech or association, labor movements and citizen-led drives advocating the introduction of asbestos curbs and regulations have not materialized because of the difficulty of alerting the public to the dangers of asbestos or reporting on asbestos-related accidents and illnesses.

Perspectives of this kind facilitate a better understanding of the relationship between asbestos-related accidents and the fundamental nature of modern political and economic systems. Examining the historical background of this relationship is of significant value to an analytical understanding of socialist and capitalist systems, as well as modern civilization. At the same time, it seems imperative that we do not neglect the task of weighing the benefits and risks of asbestos alternatives as long as these systems remain in place.

2.3 Responsibility for, and Relief from, Complex Stock Pollution

2.3.1 Judicial Relief and Administrative Relief: Experiences at the Country Level

Most advanced industrial nations have created public indemnification frameworks to provide relief for the victims of workplace asbestos exposure and contamination. This is a form of no-fault insurance, and as such, in most cases it is limited solely to the provision of economic or financial compensation. However, asbestos pollution is different from flow pollution in that not all victims may be guaranteed relief. The reasons are as follows: (1) certain causal relationships may be unknown; (2) certain victims may not be aware that they have been affected; (3) only certain victims may apply for relief; (4) the companies liable for the pollution may no longer exist. Furthermore, no relief frameworks have been set up to handle cases of asbestos exposure affecting the family members of workers in asbestos-related industries, or of asbestos pollution affecting local residents. Recourse in those cases would involve seeking relief through the courts or having public indemnification frameworks revised and updated. That is basically where the situation currently stands in relation to measures for relief in most industrial countries. Capitalist market societies abide by the principle that businesses operating on the basis of free competition are responsible for their actions. Hence, a victim that stands to benefit through litigation will, as plaintiff, typically take the liable company to court, have the defendant's liability established, and seek damage compensation accordingly. However, given the aforementioned nature of asbestos exposure and contamination, there is a limit to the effectiveness of the courts. For that reason, more countries have begun to establish new indemnification laws which differ from the laws or frameworks they may already have in place for the provision of compensation to victims of workplace accidents. Seeking resolution through the courts is the mainstream approach in free societies such as the US, whereas in countries such as those in Europe with tradition-based social policies, public indemnification frameworks have been set up. This section briefly examines some of the problems facing both approaches.

2.3.1.1 Asbestos Litigation in the US

In the US, each state has its own workers accident compensation framework. These frameworks differ state by state, but do not always function effectively in terms of gaining relief for victims. In New Jersey, a Polish national employed as an office worker for Johns Manville Corp. received relief money through that state's framework, and then filed a lawsuit through which he received $30 000 in damages. In later testimony, the worker disclosed that from the $30 000 court award, he repaid $10 000 back to the state of New Jersey in compensation, used another $10 000 to pay his attorneys' fees, and kept only $10 000 of the award for himself. Asbestos court cases can be complicated exercises in litigation that involve multiple defendants. What is more, plaintiffs face difficulties in establishing a causal burden of proof owing to the long periods of dormancy that usually intervene between the initial exposure and the initial presentation of medical symptoms. However, in the US, litigation is deemed a desirable means of obtaining relief money for victims of asbestos exposure (DiMuzio 2007).

The attorney Robert Horkovich has earned $2 billion from compensation claims in his work as one of the most prolific attorneys involved with asbestos issues and litigation having to do with the Super Fund Law. Drawing from his own experience, Horkovich notes that litigation does not always furnish victims with an acceptable resolution. As mentioned in other work by this author, a huge number of asbestos cases have gone to court; as of the year 2000, 59 000 cases had been filed, involving a total of 8 400 defendant companies and some 600 000 plaintiffs. Horkovich has predicted that the existence of multiple defendants could lead to as many as 300 000 cases of litigation. To date, insurance companies have already paid out approximately $65 billion in damage compensation claims, a sum that exceeds the combined value of losses from the 9/11 terrorist attacks and Hurricane Andrew. Of the total in damage compensation, insurers (half of them based abroad) paid 61%, while defendant companies paid the remaining 39%. About 42% of the total asbestos court awards is paid out as compensation to plaintiffs, 31% is paid as insurance company compensation and legal fees for defense attorneys, and 27% is paid as legal fees for plaintiff attorneys. Although the insurance industry did away with asbestos-related insurance products in 1985, in many court cases, awards for damages were for medical conditions that existed prior to 1985. As this illustrates, plaintiffs, on average, receive no more than 30–40% of the award granted in a successful court case. Nonetheless, due to the high cost of health care coupled with the absence of a public health insurance system in their country, many victims in the US hold out strong hopes regarding litigation.

Over 70 companies, including the aforementioned Johns Manville Corp., have already been bankrupted by court litigation (Carroll et al. 2007). The Federal Government thus decided to revise its bankruptcy laws for corporate protection, make failed firms establish a joint trust, and allow damage compensation issues to be handled through that trust, thus sparing parent companies from litigation and allowing them to stay in business. Using that approach, Johns Manville Corp. entrusted its damage compensation burden to the established trust, and is now thriving and doing better business than before. The trust receives contributions from companies subject to litigation proceedings. However, the amount of funding paid into the trust thus far seems trivial compared with the amount of damage compensation paid out by parent firms before their filings for bankruptcy.[2]

Asbestos litigation has not been limited to trials seeking compensation for workplace accidents or environmental pollution. According to Gary M. DiMuzio, some cases involve the application of product liability and building liability laws. When filing suit under the product liability laws, a plaintiff only needs to demonstrate that damage was incurred as a result of product flaws. Defendants are held strictly liable for product flaws in design, manufacture, and even marketing, as in the case of defective or missing warning labels. Additionally, under building liability laws, builders are legally liable for accidents or damage incurred by the building's owner or tenants due to building flaws.

Accordingly, litigation proceedings can be fairly complicated and involve multiple defendants owing to issues with product or building liability. For this reason, while in some cases defendants may not have acted illegally on their own, they cannot escape liability just because their actions were compounded by the illegal actions of other defendants, causing injury or damage, and it cannot be demonstrated that those actions were committed by the other defendants alone. Because plaintiff assertions of strict product liability or joint liability are also often applied, defendants are more likely to lose their case. According to DiMuzio, defense attorneys in such cases often actively plead for strengthened conditions on the plaintiffs' burden of proof, or for limiting the scope of product liability to products manufactured only within the past 10–15 years (DiMuzio 2007).

In view of the formidable litigation climate that many companies face, the US Senate has proposed legislation calling for asbestos relief. This would establish a \$140 billion fund comprising reserves of \$90 billion for defendant companies, \$46 billion for insurers, and \$4 billion for a pre-established indemnification fund. One goal would be to offer administrative sources of relief in lieu of litigation through the courts. The Senate also enacted legislation imposing a ban on asbestos sales, to take effect in 2 years. That law is relatively loose by Japanese standards, as it will initially apply only to products with an asbestos content of at least 1%. Asbestos victims and environmental groups (e.g., the Environmental Working Group) supporting them have expressed opposition to these new administrative regulations and relief frameworks.

2.3.1.2 The Public Relief Framework in France

As mentioned earlier, France established special funds for compensation for asbestos victims (FIVA) in December 2000. This is a social insurance framework designed to provide relief compensation to people recognized to be suffering from asbestos-induced occupational diseases, as well as people who have been directly exposed to asbestos. FIVA is funded by the national budget and the industrial accident and occupational disease division of the social insurance program. Of the 1 238 billion euro in funding provided to date (2001–2005), 130 million euro (10.5%) came out of the national budget and 1.108 billion euro (89.5%) was provided through the social insurance program. Eligibility for relief compensation from FIVA extends to citizens with a range of conditions, including mesothelioma, lung cancer, pulmonary asbestosis, pleural plaque, and bilateral diffuse pleural thickening. As noted earlier, it is estimated that the fund will need from 26.8 to 37.2 billion euro to cover compensation payments made over the next 20 years.

In 2000, France also established the Asbestos Workers Early Retirement Fund (FCAATA), which mandates retirement for victims and previously exposed workers. It comprises a fund that pays out lifelong benefits to allow asbestos-exposed workers to retire early. In their final 12 months of employment, workers receive an average 65% of their salary in the form of FCAATA benefits. Forty thousand workers used this program in 2004.

FIVA shares the same objectives as the asbestos relief act in Japan. However, together with the FCAATA program, it functions more like a social insurance framework. Like the workers accident compensation insurance framework in Japan, FIVA covers a broad scope of disease conditions. France differs from Japan in that the French government already concedes that there have been policy failures in dealing with the asbestos problem. Furthermore, French relief compensation payments are comparatively large in scale.

Although the French framework excels over its Japanese counterpart in these areas, the current framework drew the following criticisms from Prof. Annie Thebaud-Mony, a researcher representing the Institut National de la Sante et de la Recherche Medicale (Thebaud-Mony 2007).

1. The polluter-pays principle has not been applied.
2. The key persons responsible for this vast health tragedy have eluded civil and criminal liability.
3. Compensation payments to patients have been inequitable.
4. No consideration has been given to prevention.
5. The authorities are resistant to the ideal of seeking criminal justice.

Because of these shortcomings, the number of lawsuits dealing with asbestos exposure rose from around 300 in 2002 to around 500 in 2004, and in most of those cases, employers were reportedly found to be guilty of negligence.

2.3.2 *Issues for Future Study*

Although the Ritsumeikan Asbestos Research Project at Ritsumeikan University is pursuing investigative research inside and outside Japan and holds interviews with other researchers on asbestos exposure, its activities have only just begun. Furthermore, countermeasures against asbestos exposure are in their infancy and many issues remain to be addressed.

First, above all, it is imperative that we develop a complete picture of the extent of past exposure, and establish an outlook for future trends in exposure. To that end, we must launch additional epidemiological surveys and strive to identify all individuals suffering from asbestos exposure. Currently, national as well as local government agencies in Japan have been noncommittal about exposure studies, citing the fiscal budget squeeze as their reason. As noted earlier, it is urgent that current and former workers at asbestos-related businesses be identified and registered and provided with access to medical examinations on a continuing basis. This also applies to current and former workers in the structural, demolition, and scrap industries.

For preventive purposes, we need to know where asbestos has accumulated in the environment. This means that information on nearly 3 000 different products must be publicly disclosed. In particular, disclosures on buildings that contain asbestos are needed, along with clear labeling on asbestos-containing construction materials, as these will be the largest sources of asbestos exposure in the years ahead. Large-scale demolition and scrapping projects are planned to get under way soon, but we also need to require that contractors must submit advance notice about smaller projects of less than 1 000 m^2 in scale, ensure that workers are able to perform their operations with maximum safety, and prevent asbestos fibers from impacting the surrounding environment. Although it is highly likely that many public agencies will declare themselves unable to expand their teams of supervisory personnel due to the current fiscal budget crunch, one conceivable alternative would be for local communities and nongovernmental agencies to explore the idea of setting up their own frameworks for monitoring and whistle-blowing.

In view of the findings of surveys and research conducted to date, Japan must establish a public social insurance-type framework for relief compensation which resembles the Western examples described previously. To offset current deficiencies, Japan must also move forward with litigation seeking civil damages from asbestos polluters. As long as holes in the legal structure remain, victims will have no choice but to launch administrative lawsuits (class-action lawsuits).

Revising the provisions of the New Act will be one task that must be completed in the near future. As already demanded by Diet resolutions and victims' groups, the scope of relief money must be identical to that given for work-place accidents, and the amounts of compensation must be raised to levels which are comparable to those paid to victims of work-place accidents. If these steps are to be implemented, presumably the fund for relief money must also be drastically expanded.

An unknown number of questions remain to be addressed, including whether Kubota and other companies will continue to pay the same levels of relief money to the victims of their pollution and the families of employees exposed to asbestos, whether patients and families compensated through the workers accident compensation insurance framework will demand increased benefits, and how diagnostic procedures and treatment methods can be improved. Currently, however, the scale of relief funding itself remains far too small. Although the Kubota Corporation and the NICHIAS Corporation share the burden of making special contributions to the relief fund, the reality is that there are also many victims from shipbuilding, automotive, electrical components, construction, and other fields. It therefore follows that the burden of funding contributions must be extended to these other business sectors. Identifying causal factors and current trends behind asbestos exposure will demand additional survey work and research, as well as revisions to the structure of the relief fund itself.

Should more cases of asbestos exposure and illness arise in the future among workers in the demolition and repair industries, we will face a hurdle in providing relief to workers in those fields who are not covered by the workers accident compensation insurance framework. In reality, workers for many small businesses are apparently outside the scope of coverage. Even under the provisions of the New Act, relief payments to victims of pollution amount to little more than consolation payments, and certainly do not deserve the label of compensation. Considering these circumstances, it seems amazing that so few cases of litigation have been filed in Japan thus far. Also, as demonstrated by litigation over asbestos pollution in the Sennan district, plaintiffs have sued the national government because many of the actual polluters were small companies that have already gone out of business. Given the deficiencies observed in the relevant regulatory structure to date, it seems only natural that plaintiffs would choose to sue the government. However, should not plaintiffs also be filing lawsuits against those textile companies and wholesale trading houses that have produced or handled merchandise utilizing asbestos fiber? In my view, holding multiple companies liable, as is common practice in the US, can be effective in determining who the polluters are. Further, unless there is progress in the courts, efforts to amend and improve the New Act will not make any headway either.

As an advocate of asbestos countermeasures in the US, the Environmental Working Group (EWG) has recommended the following solutions to the current state of affairs in that country (Environmental Working Group 2005).

1. Establish a legitimate relief framework for all victims of asbestos exposure. That task must neither be delayed nor allowed to become too complex. The reason is because asbestos is still in use. The framework for relief money must be in place for at least 50 years, if not longer.
2. To ensure that every individual harmed by asbestos exposure receives relief and aid, a large-scale asbestos health screening campaign must be implemented. The goal will be to screen all persons who have a history of working in an

asbestos-affected environment so that the several million individuals harmed by asbestos exposure can be accurately identified.

3. Restrictions must be placed on the practice of making people forfeit their rights to legal representation. The government trust fund may be only part of the right solution to the asbestos problem. However, participation in the fund must be voluntary. The reason is that we already have a fund to help the families of those lost or injured by asbestos exposure from the collapse of the World Trade Center buildings.
4. Asbestos must be banned. There is no reason to engage in further futile debate over the needless suffering and deaths caused by the use of asbestos. Asbestos substitutes exist. It is time to ban asbestos now.
5. The government institutions and agencies responsible for public health and the environment must strictly monitor the health-care services (relief) provided to the victims of asbestos-induced illnesses. The reason for this is that according to news reports, certain institutions for asbestos relief have been engaged in flawed and dubious operations.

The proposals and recommendations tendered by the EWG seem to apply to national policy in Japan as well as other countries with citizens who have been exposed to asbestos. Asbestos is a global problem. As a proponent of free trade, in 2001 the World Trade Organization (WTO) decided to recognize bans on the importation of asbestos implemented by WTO member countries for health-related reasons. Some years earlier, in 1997, the Canadian government filed a WTO lawsuit against France, claiming that the French ban on asbestos was a technical trade barrier that violated the principle of free trade. Later, in 1999, the EU voted to implement a full-scale ban on asbestos, and requested that by January 1, 2005, all EU member countries should prepare and implement domestic ordinances banning the use of asbestos for all purposes except only as a material for the separation membranes used in chlorine plant electrolysis tanks. This amounted to a measure in support of France's position. Given these circumstances, the WTO decided to recognize the French ban, thus marking its first approval of a measure restricting trade since implementing rules for the resolution of trade disputes.

Given this international backdrop, it is indeed problematic that many developing countries, and in particular China, India, Thailand, and Brazil as well as the transitional economy of Russia, continue to use asbestos (mainly chrysotile) in vast quantities. In the years ahead, our hope is that we can adequately convey the lessons of Japan's own failures to these countries, and urge them to adopt appropriate policies as quickly as possible.

Notes

1. These statistics were taken from BANJAN (2007).
2. Robert Horkovich's Lecture (August, 2006, New York City).

References

BANJAN (2007a) Asbestos problems did not finish. Information of Japan Occupational Safety and Health Resource Center (in Japanese) 338:2–17

BANJAN (2007b) International asbestos conference (in Japanese). BANJAN, Tokyo

Carroll S et al. (2007) Asbestos Litigation: Costs and Compensation. RAND Corporation, Santa Monica

DiMuzio G (2007) Overview of asbestos compensation mechanism in the United State. In: BANJAN (ed) International asbestos conference (in Japanese). BANJAN, Tokyo, p 312

Environmental Working Group (2005) Asbestos think again. http://www.ewg.org/asbestos/. Accessed 31 Oct 2005

Miyamoto K (2006) Toward the sustainable society (in Japanese). Iwanami Shoten, Tokyo

Miyamoto K (2007) Environmental economics new edition (in Japanese). Iwanami Shoten, Tokyo

Osaka Prefectural Government, Osaka city, Sakai city, Takatsuki city, and Higashiosaka city (2007) The report of health damage of Osaka Prefecture in 2006 (in Japanese)

Osaka Prefectural Government (2007) The report of health risk of asbestos of Osaka Prefecture in 2006 (in Japanese)

Thebaud-Mony A (2007) French approach to achieve justice for all asbestos victims. In: International asbestos conference (in Japanese). BANJAN, Tokyo, p 73

The Ministry of the Environment and Amagasaki City (2007) The epidemiological analysis of asbestos exposure in 2006 (in Japanese)

Chapter 3
Why Did the Asbestos Disaster Spread?

Takehiko Murayama and Yuji Natori

3.1 Introduction

After a historical review of the asbestos issue in Japan, we have listed the following four points as factors through which the asbestos disaster was spread in Japan: the social application of medical knowledge, the activities of governments and industries, the economic advantage of asbestos products in the short term, and a knowledge gap among stakeholders.

3.2 Insufficient Application of Current Medical Knowledge of the Toxicity of Asbestos to Public Policies

3.2.1 Prewar

Medical knowledge about asbestosis in foreign countries was introduced to Japan in the early 1930s. After the first international conference on silicosis, held in Johannesburg, South Africa, in 1930 by the League of Nations (ILO 1930), a Japanese government official published a paper in 1931 which reported the conclusion of that conference that asbestos dust definitely induces pneumoconiosis (Ohnishi 1931). In 1932, the International Labour Organization (ILO) published a list of studies on pneumoconiosis which included 70 papers on asbestosis (ILO 1932). As one branch of the ILO was located in Tokyo at that time, it was presumed that the government and related medical experts could easily get updated information

T. Murayama (✉)
Faculty of Science and Engineering, Waseda University, 51-05-07A, 3-4-1 Okubo, Shinjuku-ku, Tokyo 169-8555, Japan
e-mail: tmura@waseda.jp

Y. Natori
Asbestos Center, Z-Building 5F, 7-10-1 Kameido, Koto-ku, Tokyo 136-0071, Japan

K. Miyamoto et al. (eds.), *Asbestos Disaster: Lessons from Japan's Experience*,
DOI 10.1007/978-4-431-53915-5_3, © Springer 2011

on asbestosis from this source. In the late 1930s, this led to the first survey of asbestosis among workers in asbestos factories in the Sennan district of Osaka Prefecture (Sukegawa et al. 1940).

In addition, government officials would share the information about asbestosis through military activity with foreign countries because asbestos use for military purposes was just beginning. For instance, in 1936, Germany, with which Japan had formed an alliance, brought out an order that added asbestosis to the list of occupational diseases to be compensated (Selge 1942:131–135). Germany also included lung cancer with asbestosis for workers compensation in 1943 (Castleman 2005:45).

While several facts suggest that the Japanese government had considerable knowledge of the toxic aspects of asbestos, it did not instigate any public policies for reducing asbestos risk in wartime. It was not until the postwar period that substantial measures against asbestos pollution were implemented.

3.2.2 *Postwar*

After the war, a International Cancer Congress and an International Congress on Occupational Health were held every 3 years, which Japanese experts continuously attended. These conferences provided updated information to Japanese academicians, government officials in embassies in foreign countries, and business workers. In addition, the annual reports of a national institute for occupational health show that members frequently visited foreign countries in the 1950s, and would have shared the current knowledge about advanced research into asbestos pollution. At the beginning of the 1950s, one of the medical experts, who was a member of a national institute, reported that asbestos was a carcinogenic material (Suzuki 1951).

In particular, it would have been important that Japanese experts were involved in conferences such as the seventh Saranac Symposium in 1952, the International Conference on Pneumoconiosis in Johannesburg in 1959, and the first International Conference on the Biological Effects of Asbestos in 1964, which was held by the New York Academy of Science. Related documents published during the late 1950s and 1960s suggest that medical experts recognized that asbestos was one of the causes of lung cancer. In fact, a research report which was published in 1960 by the Ministry of Labor described a case of lung cancer with asbestosis (Ministry of Labor 1960:72–75).

It is believed that two international conferences held in Tokyo during the late 1960s accelerated knowledge about asbestos pollution among medical experts and government officials. At the ninth International Cancer Congress in 1966, American experts reported on mesothelioma victims whose illness had been induced by environmental pollution around asbestos factories (Ninth International Cancer Congress 1966). As the mass media also paid attention to this conference, the information would have been known to the public. Just after the conference, members of a Japanese academic society on occupational health invited famous foreign participants to an informal meeting (Japanese Society for Occupational Health 1967: 11, 36–37). At this meeting, Dr. Wilhelm Heuper, a team leader on environmental cancer at the National Cancer Institute (NCI), introduced the carcinogenic aspects of asbestos based on case studies of mesothelioma. As leading experts, including the president of the society, attended the meeting, it would suggest that updated information on the toxicity of asbestos was shared among members of the society.

The 16th International Conference on Occupational Health in 1969 arranged a special session on asbestos pollution, which was not included in the original plan (Japan Industrial Safety Association 1971). Dr. Ian Webster, a participant from South Africa, reported data on mesothelioma victims not only among mine workers, but also among local residents around mines and factories. Dr. Tatsuo Sano, a member of the Institute for the Science of Labor, was the chair of this session, and mentioned in the concluding remarks that Japan should encourage research into mesothelioma. While information on asbestos toxicity was shared through related documents and conferences in both prewar and postwar periods, specific measures against the risks of asbestos were only begun in the 1970s.

3.3 Commitments of Governments and Industries

3.3.1 Prewar

Several facts indicate that Japanese governments initiated and encouraged the asbestos industry for military use. Nippon Asbestos was founded for the first time in 1896 after a request by the Japan Naval Authority (NICHIAS Corporation 1996:5–6). The demand for ground transport by the Japan Army Authority was a main cause of the foundation of Akebono Brake, which produced brake linings using asbestos (Akebono Brake Corporation 1969). The Ministry of Planning also requested Chichibu Cement to produce asbestos cement pipes (Chichibu Cement 1974). The standardization of asbestos products for railways, the military, and water companies encouraged the development of the asbestos industry. In wartime, Japanese asbestos companies expanded their factories to East Asia, which was under Japanese rule at that time. The factories located in China, Korea, and Taiwan would affect the health of workers and local residents.

3.3.2 Postwar

After the war, the import of asbestos was controlled by the Japanese government. The asbestos industry had to change their products from military to civil use at that time. As many companies were small or medium enterprises, the government encouraged the development of these companies and protected them from the pressure of foreign affiliated companies. Administrative guidance through the standardization of asbestos products would prove to be a substantial influence in spreading asbestos use, although it was not a legal matter. For instance, the standardization of asbestos cement pipes played an important role in their rapid diffusion, because local governments were the main customers and were recommended to use these products.

At that time, it was the Ministry of Labor and the National Institute of Occupational Health that obtained information about asbestos toxicity. An official who was the chief of the occupational health section in the early 1950s published a survey report on asbestosis for the first time in 1930s (Ishidate et al. 1938), and visited the Saranac Laboratory in New York State, USA, in 1951, which was a famous facility for pneumoconiosis research (Ishidate 1951). Another official reported the outline of an international conference on pneumoconiosis which covered asbestosis and the relation between asbestos and lung cancer (Oikawa 1950). Although it is not hard to imagine that they had substantial knowledge about asbestos toxicity through academic documents and their experiences in foreign countries, practical administrative measures against asbestos pollution were not developed.

A monthly journal supervised by the occupational health section of the Ministry of Labor began publication in the 1960s. An article in this journal covered a report from the committee on the chemical industry and labor of the ILO which stated that asbestosis caused lung cancer (Ministry of Labor 1963). Another article reported a round-table talk on occupational health in which the chief of the occupational health section mentioned that several cases of lung cancer with asbestos exposure had been found in the Osaka area (Japan Industrial Safety and Health Association 1970). All this suggests that government officials had knowledge of the carcinogenic aspects of asbestos in the 1960s at the latest, and were aware of domestic cases of lung cancer in the late 1960s.

In addition to the above-mentioned activities of the national government, Japanese asbestos industries also developed a close relationship with foreign companies. Several articles in a journal of the Japan Asbestos Association noted that many Japanese workers visited foreign companies, and executives of foreign companies also visited Japan. In addition, some companies made agreements for technical cooperation. It must be supposed that Japanese companies obtained information about asbestos toxicity as well as knowledge on technical aspects (Watanabe 1967:13–14, and Watanabe 1971:3). In particular, presidents and other executives in foreign companies visited Japan every year in the late 1960s and early 1970s. This was probably because foreign companies were looking for new markets in response to gradually enforced regulations in Western countries. While the senior members of companies might realize the toxicity of asbestos, other workers were not informed of the situation.

3.4 Cost of Asbestos Products

Another reason why asbestos products spread in Japan was their low cost. We examined this point for three cases: high pressure pipes for water supplies, ceiling products, and roofing sheet for buildings.

Until the 1940s, iron pipes were mainly used for water supplies, and then asbestos cement pipes gradually spread during the 1950s and 1970s. Figure 3.1 shows

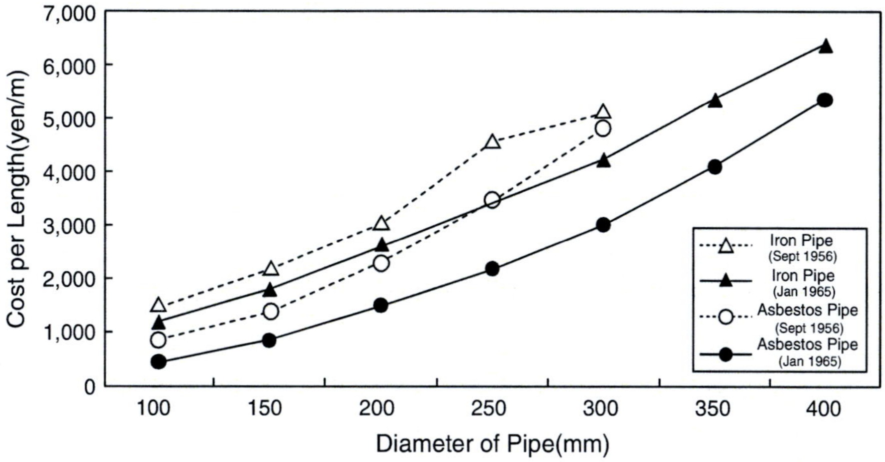

Fig. 3.1 Costs of pipes for water supplies in the mid-1950s and mid-1960s (*broken line*: 1956, *solid line*: 1965). *Source*: Economic Research Association, *Data for Estimate*

the costs of iron pipes and asbestos cement pipes by diameter at that time. In the mid-1950s, narrow asbestos cement pipes were about 40% cheaper than iron pipes, and the price of wide pipes was also about 10% lower. Such differences increased in the mid-1960s, when asbestos cement pipes were 60–20% cheaper. At that time, the price of narrow asbestos cement pipes was less than half that of iron pipes. Another point was that asbestos cement pipes were lighter than iron pipes, which also encouraged the spread of asbestos cement pipes into Japanese society.

Figure 3.2 shows the costs of ceiling products in the early 1970s. Although it would be difficult to compare the differences in insulation properties and durability, the cost for sprayed asbestos was relatively cheaper than for other products. Fast and easy installation also encouraged the use of sprayed asbestos. Figure 3.3 shows the costs of roofing sheets in the 1970s. In this category, asbestos products were also relatively cheaper. These data indicate that asbestos products were less expensive than other products in the 1960s and 1970s, and asbestos spread throughout Japanese society despite the very high social costs.

3.5 Information Manipulation and Abuse of the Knowledge Gap Among Stakeholders

Finally, we show that government authorities and asbestos companies manipulated the information on asbestos toxicity, and abused the knowledge gap in order to spread asbestos products. While executives of asbestos companies and government officials realized the carcinogenic aspects of asbestos in the late 1960s at the latest, workers in asbestos factories, their families, and lay people were not informed in spite of the many ways in which they were subjected to asbestos exposure. Because

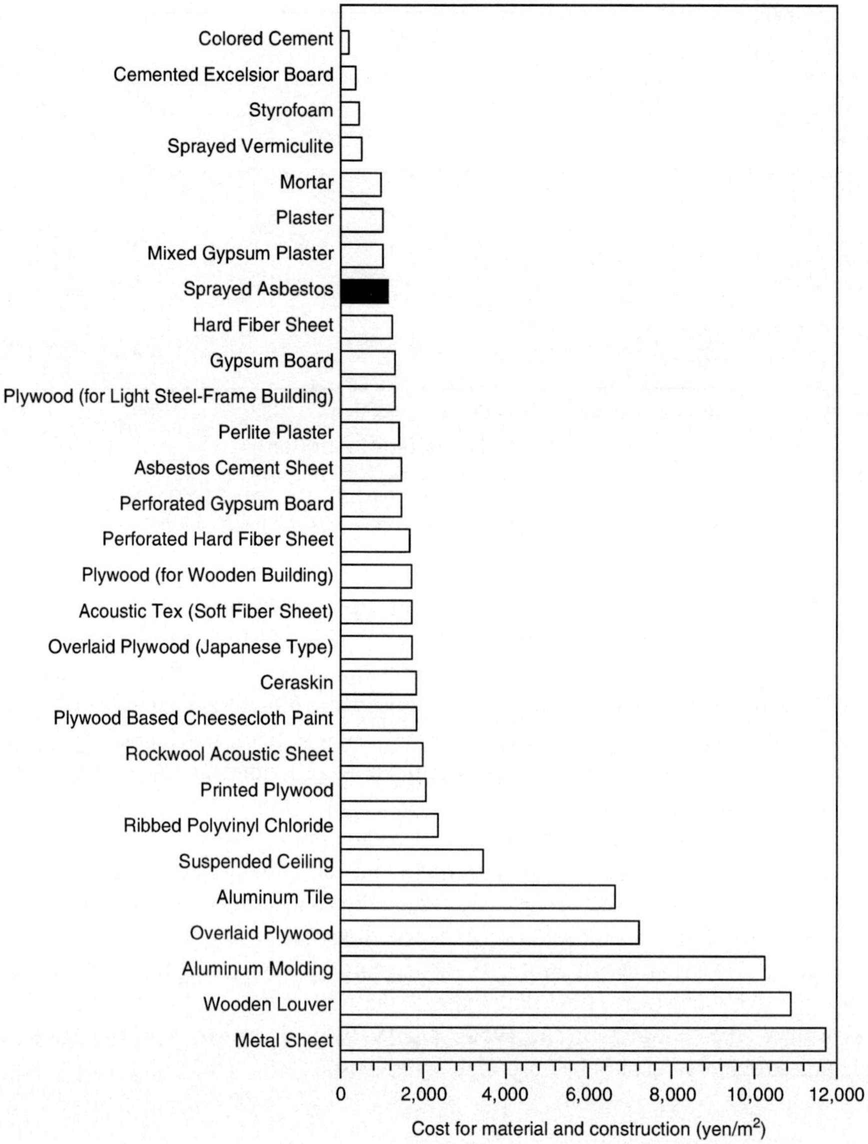

Fig. 3.2 Costs of ceiling products in the early 1970s. *Source*: Japan Association for Construction Estimates, *Construction and Estimates*, No. 22, January, 1972

Japanese society did not completely realize the tremendous impacts of environmental asbestos exposure until the Kubota Shock in 2005, this gap will be continued for 40–50 years.

After the health impacts of asbestos pollution became a serious social issue in Western countries in the 1960s and 1970s, some foreign companies looked for new markets in the Far East, and asbestos imports to Japan rapidly increased. At the same time, some Japanese companies transferred their factories to other countries

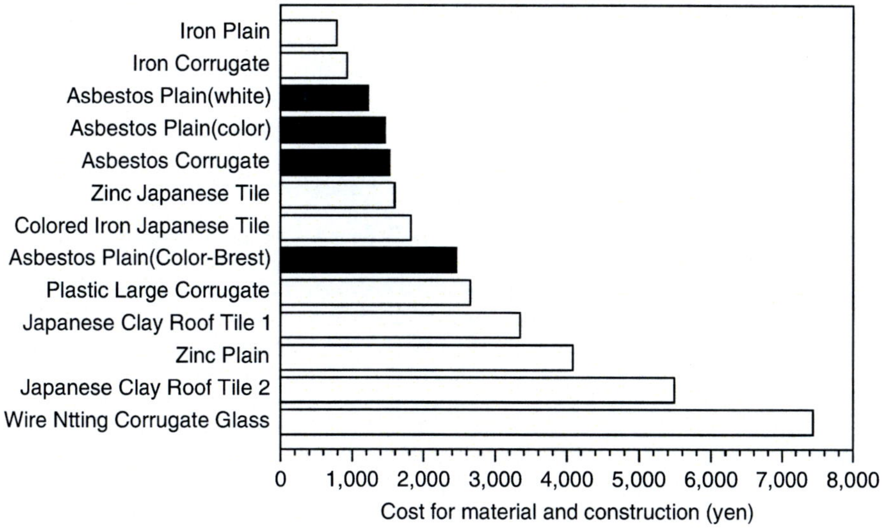

Fig. 3.3 Costs of roofing materials in the 1970s. *Source:* Japan Association for Construction Estimates, *Construction and Estimates* No. 49, April, 1974

in East Asia. Later, one factory was transferred again to Indonesia. Such a negative chain reaction of asbestos use was partially induced by the knowledge gap about the toxicity of asbestos. Some asbestos companies intentionally abused the lack of information among developing countries, and social tragedies from asbestos use are being repeated even now (Murayama 2009).

While some experts insisted that Japanese society needed asbestos products for modernization, it was a fact that nonasbestos products did exist in many fields before asbestos products were developed. It was widely believed that asbestos products were over-used, and that their use should have been limited to particular and closed environments. Asbestos products are still used in developing countries where medical knowledge has not been made public, and some Asian countries, including China, India, and Indonesia, have increased their consumption in recent years. Some companies in Western countries and national governments which own asbestos mines encourage the diffusion of asbestos products among Asian countries. Such behavior, based on the knowledge gap, will be considered criminal by future generations.

References

Akebono Brake Corporation (1969) Forty years progress of Akebono Brake (in Japanese). Diamond Inc, Tokyo

Castleman B (2005) Asbestos: medical and legal aspects, 5th edn. Aspen Law & Business, New York

Chichibu Cement Corporation (1974) Fifty years history of Chichibu cement (in Japanese). Chichibu Cement Corporation, Tokyo

International Labour Office (1930) Silicosis: records of the international conference held at Johannesburg 13–27 August 1930. International Labour Office, Geneva
International Labour Office (1932). Pneumoconiosis: a list of references. Studies and reports, series N (Industrial Hygiene), 15
Ishidate F et al (1938) Hygienic consideration on workers of asbestos factories (the first report). J Sci Labour (in Japanese) 15(3):181–184
Ishidate F (1951) Current situation of occupational health in the United States. Dig Sci Labour (in Japanese) 6(8):21–26
Japan Industrial Safety and Health Association (1970) Round-table talk: a perspective of measures for occupational diseases. Labor Health (in Japanese) 11(1):6–13
Japan Industrial Safety Association (1971) XVI. International Congress on occupational health, September 22–27, 1969, Tokyo. Japan Industrial Safety Association, Tokyo
Japan Society for Industrial Health, Research Committee on Occupational Cancers (1967) Informal group discussion on occupational cancer in the world. J Occup Med (in Japanese) 9(2):11,36–37
Ministry of Labor (1960) Research on pneumoconiosis including asbestosis (in Japanese). (Experimental research on occupational health led by Zenji Horai)
Ministry of Labor (1963) ILO documents; protection of workers from occupational disease and addiction: health effects of chemical substances. Labor Health (in Japanese) 4(1):53–56
Murayama T (2009) Historical review of asbestos industrialization, hazard identification, and legal measures: why asbestos disaster spread in Japan? In: International symposium on preventing emerging occupational and environmental risks in South Asia and beyond, 17–19 December, Delhi, India
NICHIAS Corporation (1996) One hundred year's history of NICHIAS (in Japanese). NICHIAS, Tokyo
Ninth International Cancer Congress (1966). Ninth International Cancer Congress October 23–29, Tokyo
Ohnishi S (1931) Asbestos dust and tuberculosis. J Inside Out Treat (in Japanese) 6(1):39–43
Oikawa F (1950) Fear of pneumoconiosis; conference among world authorities in Sydney. Safe Prod (in Japanese) 8(6/7):14–17
Selge F (Yukiyama K Translation) (1942) Social policy of Nazi. Jitsugyo No Nihon Sha. Ltd., Tokyo
Sukegawa H et al (1940) Research on asbestosis among workers of asbestos factories (in Japanese). Ministry of Home Affairs, 94pp.
Suzuki T (1951) Several issues on occupational health. J Workers Accid (in Japanese) 2(9):29–33
Watanabe T (1967) Overseas travel: London and Oxford, UK. Asbestos (In-house journal of NICHIAS Corporation) (in Japanese) 43:13–14
Watanabe T (1971) New Year greetings. Asbestos (In-house journal of Nichias Corporation) (in Japanese) 82:3

Chapter 4
Asbestos Pollution and Its Health Effects: Asbestos-Related Diseases in Japan

Kenji Morinaga and Yasushi Shinohara

4.1 Introduction

Through 1986 and 1987, the public spotlight was briefly focused on an array of asbestos-related risks and hazards: lung cancer among asbestos industry workers, the uproar over baby powder using talc contaminated with asbestos fibers, the illegal dumping of scrap asbestos materials removed from the US Navy Midway aircraft carrier, a case of mesothelioma in a housewife living in the neighborhood of an asbestos mill, and the degradation of sprayed asbestos in schools. However, these topics were soon forgotten (Morinaga 1989). In time, though, the nation would be stirred back into a furor by the June 2005 news coverage of an outbreak of mesothelioma victims among residents living near the now-defunct Kanzaki factory owned by the Kubota Corporation. (This string of tumultuous events has been dubbed the Kubota Shock in Japan.) In this chapter, we strive to present knowledge and information relating to the health effects of asbestos in Japan to date.

4.2 Postwar Asbestos Use in Japan

Japan lost its access to foreign imports of asbestos in 1942. This material, along with petroleum, was ranked as vital to Japan's defense industry during World War II. Accordingly, Japan had no choice but to develop asbestos mines within its borders, as well as on the Korean Peninsula, the Liaodong Peninsula, and in Manchuria.

K. Morinaga (✉)
Department of Asbestos Health Damage Relief, Environmental Restoration and Conservation Agency of Japan, Myuza Kawasaki Central Tower 9F, 1310 Omiya-cho, Saiwai-ku, Kawasaki, Kanagawa 212-8554, Japan
e-mail: k-morinaga@erca.go.jp

Y. Shinohara
National Institute of Occupational Safety and Health, 6-21-1 Nagao, Tama-ku, Kawasaki, Kanagawa 214-8585, Japan

K. Miyamoto et al. (eds.), *Asbestos Disaster: Lessons from Japan's Experience*,
DOI 10.1007/978-4-431-53915-5_4, © Springer 2011

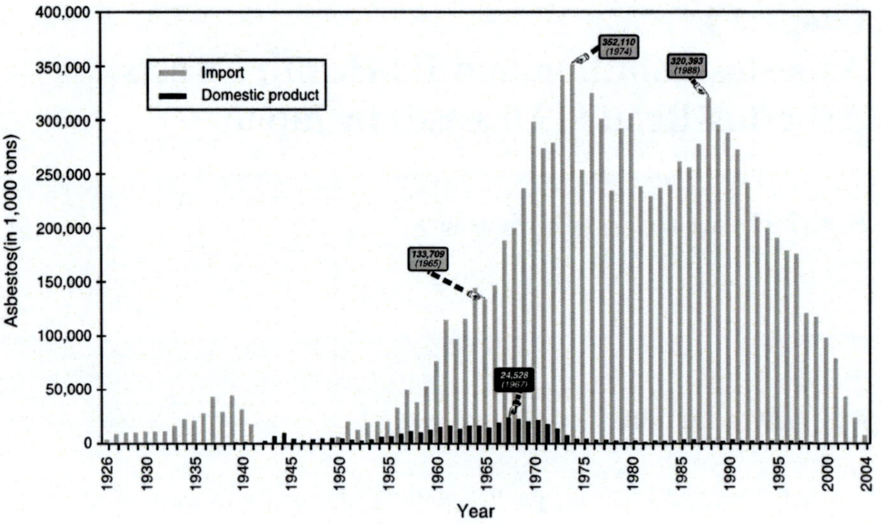

Fig. 4.1 Asbestos imports and domestic product in Japan. *Source*: Ministry of International Trade and Industry and Ministry of Economy, Trade, and Industry, various documents

The output of domestic mining peaked at about 10000 metric tons in 1944, just prior to the end of the war, and then declined until all domestic mines were finally shut down in 1969. During the war, the Furano Mine in Hokkaido handled the processing and stockpiling of Quebec Class-7 and the lower-class chrysotile when a Class-5 grade of product was extracted. (Under the Quebec standard, smaller class numbers referred to chrysotile composed of longer fibers). Tailings of short-fiber chrysotile, which was visible as a powder to the naked eye, were primarily produced from stockpiled reserves. Whereas approximately 4600 tons of tailings were produced in 1975, the annual output tumbled to an estimated 180 tons by 2000, and all production came to a halt in 2002 (Fig. 4.1).

Under the Materials Control Ordinance of 1942, asbestos rationing regulations were issued, and members of the asbestos industries found themselves operating within the framework of a controlled economy. In 1946, the Asbestos Market Adjustment Guidelines were promulgated. The regulation of Class-7 and higher-class asbestos supplies was handled initially by the Asbestos Control Co., Ltd. and then by the Industrial Reconstruction Corporation from the following year. Asbestos supplies (totaling about 12000 tons) stockpiled at Japanese naval dockyards in Kure, Yokosuka, and elsewhere were diverted for civilian use, including the production of asbestos fabric (as an electrolytic filter membrane material) for the electrolysis tanks used in the production of the ammonium sulfate fertilizer which was required to improve rice yields. Further, in 1947 the Government began requesting that General Headquarters (GHQ) should import supplies of the asbestos fabric used in electrolytic filter membranes. It was not until April 1949 that Japan began to import asbestos from Canada and South Africa. In September that year, controls on the relatively lower classes of asbestos products were lifted. In March 1952, rationing controls on raw asbestos were scrapped entirely, but at that time, annual asbestos imports were still around 13000 tons. In FY 1949, the industrial use of asbestos totaled 5610 tons: shipbuilding 1307 tons; chemical engineering operations 845 tons; automotive industry operations 696 tons; land-based transport 427 tons; power generation facilities 373 tons; machinery and equipment 316 tons; steel and coal operations 308 tons; and other sectors combined 1338 tons (Housing Industry, Ceramics, and Construction Materials Division, METI). After the Japanese market for raw asbestos imports was opened up entirely, complete liberalization of the market for asbestos products was carried out in the latter half of 1963. That year, the total import value of asbestos was double that of the previous year.

After GHQ dismantled the asbestos import control framework, Japan was integrated into the international free-market economy and there began a remarkable growth in the use of asbestos. It was around 1955 that asbestos spray-on applications began. This was a distinctive example of work that involved short-term exposure to asbestos in high concentrations. In 1964, asbestos began to be used as soundproofing in a school near the air base, which triggered the widespread use of asbestos as a building material. Revisions to the Building Standards Law in July 1963, and the implementation of revised enforcement regulations in January 1964, prompted revisions to regulations on structural fire resistance and fireproofing. Asbestos products then began to be used heavily in high-rise buildings as a lightweight fire-resistant cladding material. In 1971–1972, the period of economic growth when investment in plant and equipment was at its peak, around 300000 tons of asbestos was imported per year, with decreases or increases in the volume of imports depending on the exchange rate and the price of raw asbestos. From 1970 through 1985, estimated annual asbestos consumption accounted for around 70–80% of the total imported volume. Kubota's defunct Kanzaki factory – the source of the public uproar that broke out in late June 2005 – manufactured high-pressure asbestos pipe products using crocidolite and chrysotile until 1975 (Fig. 4.2) (Morinaga 2002).

4.3 Current Asbestos-Related Disease Trends

4.3.1 Types of Asbestos-Related Diseases

As the term implies, asbestos-related diseases are disorders that are caused by the inhalation of asbestos dust. They include pulmonary asbestosis (one type of pneumoconiosis), lung cancer, mesothelioma (a malignant neoplasm of the pleura or peritoneum), benign (noncancerous) pleural disorders such as circumscribed pleural thickening known as pleural plaques, pleurisy, and diffuse pleural thickening (Table 4.1). Cigarette smoking and a variety of other factors are known causes of lung cancer. (The inhalation of tobacco smoke and asbestos has a multiplicative synergistic effect on the risk of developing lung cancer.) Pleural mesothelioma, however, is almost exclusively associated with asbestos exposure.

Of the asbestos-related diseases, malignant mesothelioma and lung cancer are known to manifest their symptoms only after a long incubation period, which

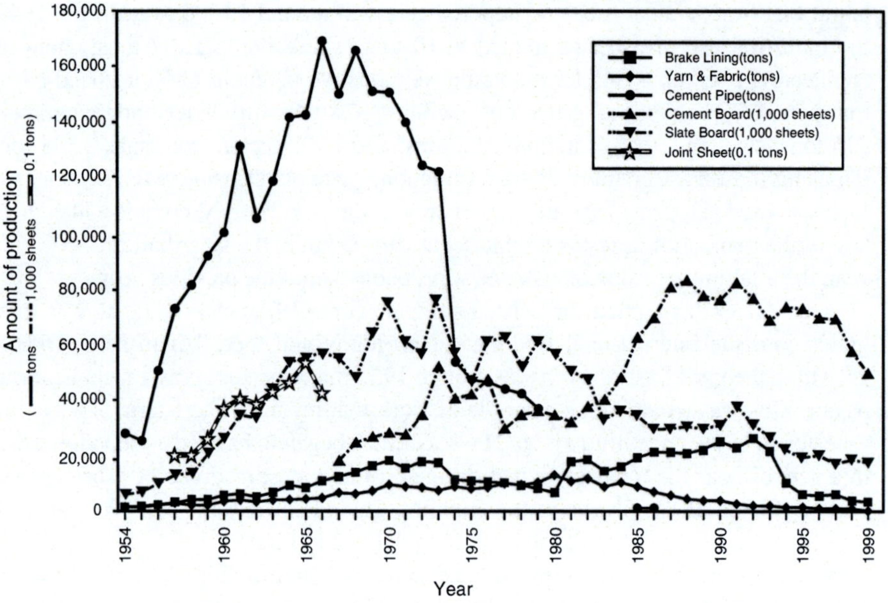

Fig. 4.2 Main asbestos products in Japan. *Source*: Ministry of International Trade and Industry, Census of Manufactures, various years

Table 4.1 Diseases caused by asbestos

Nonneoplastic pleural disease
Pleural effusion
Pleural fibrosis (diffuse pleural thickening)
Pleural plaques
Nonneoplastic parenchymal disease
Asbestosis
Carcinoma of the lung
Malignant mesothelioma of pleura and peritoneum
(Pericardium and tunica vaginalis testis)

ranges from 20 to 50 years from the first exposure to asbestos. The latent period for mesothelioma is even longer than that for lung cancer; cases with more than 50 years' latent period are not at all rare.

4.3.2 Pulmonary Asbestosis

The first case of asbestosis in Japan was reported by Suzuki and Noda (1929), who were physicians affiliated to the Osaka Railway Hospital. Later, from 1937 through 1940, Ishidate et al. (1938) conducted health check-ups for asbestos factory workers, chiefly in the Osaka area (11 factories in the Sennan district, two other factories in Osaka Prefecture, and one factory in Nara Prefecture). Chest X-rays of 403 asbestos workers from the Sennan district detected confirmed pulmonary asbestosis in 38 and suspected asbestosis in 11 (Sera 1983).

After World War II, numerous cases of asbestosis were identified by studies of workers employed at asbestos factories in Osaka's Sennan district and Nara Prefecture, asbestos mines in Hokkaido, and asbestos factories in Tokyo. These studies were performed with the support of the Ministry of Labor by "Research Concerning the Diagnostic Criteria for Pulmonary Asbestosis" (team leader, Z. Horai), which started in 1955. In its prime, more than 70 asbestos industries in the Osaka Sennan district employed around 2000 workers.

Nationwide, only five deaths from asbestosis were recorded on Japanese death certificates in 1968. However, the annual total gradually increased thereafter. Since 1986 it has averaged at least ten asbestosis-induced deaths per year. The decade from 1995 through 2004 had 1.5 times more deaths from asbestosis than the preceding decade (1985–1994) (i.e., 242 compared with 157) (Fig. 4.3) (Morinaga 2009).

One feature of asbestosis is that women account for slightly more than 20% of all deaths due to this disease. Among the likely reasons for this are that a relatively large share of workers in the asbestos textile industry were women, the industry used relatively long-fiber types of asbestos, it was difficult to control the release of asbestos dust in the dry process, and that compared with other industrial sectors utilizing asbestos, workers in this particular sector were more frequently at risk of exposure to high concentrations of asbestos dust. The numbers in the table indicate cases in which asbestosis was diagnosed as the cause of death. We have conducted

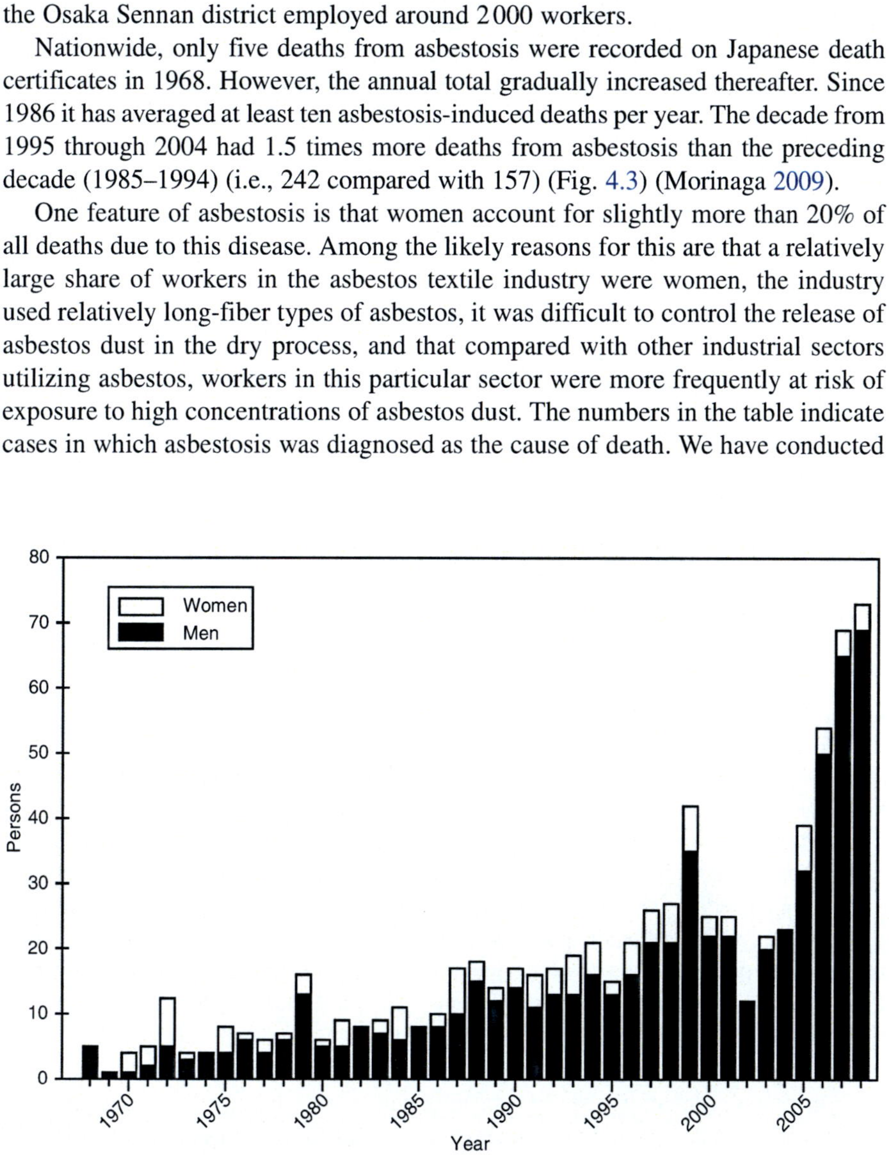

Fig. 4.3 Number of deaths from asbestosis on death certificates in Japan. *Source*: Ministry of Health, Labour, and Welfare, *Vital Statistics* of Japan, various years

a retrospective cohort study of asbestosis patients in the Osaka area who required treatment through the compensation scheme. That study looked at 184 recognized asbestosis patients through the period from 1956 to 1985, and found that 167 of these patients had died by the end of 1995. The causes of death were respiratory illnesses in 97 cases and tuberculosis in 14 cases, giving a total of 111 cases (66.5%). Cancer was the cause of death in 28 patients (16.2%), 25 (15.0%) of whom had died from lung cancer. Asbestosis was listed as the underlying cause for nearly half of the patients who had died from respiratory illnesses or tuberculosis; pulmonary silicosis and pneumoconiosis were among the causes cited for the other deaths. Judging from these data, attention should be given to the fact that the number of deaths from asbestosis as given on death certificates has most certainly been underestimated (Morinaga et al. 1984, 1990; Morinaga 2006).

4.3.3 Lung Cancer

Sera and Tanaka (1960) reported that patients who underwent examinations at the time of the prewar Ishidate study later died of lung cancer. These were the first documented cases of lung cancer caused by exposure to asbestos in Japan (Morinaga 2006).

Not many epidemiological studies of workers exposed to asbestos have been carried out in Japan. Most of the studies which were carried out were small in scale in comparison with those performed in Europe or North America, and one large-scale retrospective cohort study had the a serious fault of providing no follow-up until the workers reached the age of 55 (Morinaga et al. 2001). No large studies have been performed as a matter of national priority in the interests of confirming the evidence and forecasting future cases either prior to, or in the aftermath of, the Kubota Shock (Table 4.2).

Multiplicative synergistic action with tobacco use is one of the important correlations that asbestos has with lung cancer. As a model of this correlation, Hammond et al. demonstrated that the death rate from lung cancer per 100 000 person-years was 601.6 for smokers exposed to asbestos, whereas it was 58.4 for nonsmokers exposed to asbestos. For comparison (no asbestos exposure), the death rate from lung cancer per 100 000 person-years was 122.6 for smokers and 11.3 for nonsmokers. In addition, if the risk of death from lung cancer for nonsmokers who are not exposed to asbestos is set at a baseline value of 1, then the comparable risk is fivefold for individuals exposed to asbestos, tenfold for smokers, and 50-fold for smokers exposed to asbestos (Morinaga et al. 2001; Hammond et al. 1979). In our own follow-up study of asbestosis patients in one hospital, we derived a comparable risk ratio of 12-fold for individuals exposed to asbestos (asbestosis), fourfold for smokers, and 48-fold for smokers exposed to asbestos.

Only two case-control studies on lung cancer in relation to asbestos exposure have been conducted in Japan; one study was in Yokosuka and the other in the Osaka Sennan district. In our study of 62 lung cancer patients, the odds ratio was 9.06 for smokers and 4.53 for patients exposed to asbestos. As for the prevention

Table 4.2 Cohort studies on asbestos-exposed workers in Japan

Study	Year	Observation period	Asbestos exposure	No. of subjects	All causes		Lung cancer		
					No.	SMR	No.	SMR	(95% cl)
Morinaga et al.	1982	1971–1980	Mainly asbestos textiles	789	32	1.15	8	3.88	1.68–7.65
Hosoda et al.	1982	1960–1978	SL repair	32556	NA	NA	27	0.76	–1.57[a]
		1970–1979	SL repair	15260	557	0.58	27	0.82	0.54–1.19
Miyazaki et al.	1983		Asbestos textiles, friction	1210	NA	NA	9	4.81	2.20–9.14
Morinaga et al.	1991	1975–1984	Mainly asbestos textiles	789	96	1.09	12	2.52	1.30–4.39
Morinaga et al.	1990	1964–1983	Asbestos textiles	208	15	1.10	3	6.81	1.40–19.9
Morinaga et al.	1993	1971–1990	Asbestosis patients	M269	131	3.11	43	15.47	11.2–20.8
Sun et al.	1997	1973–1993	Construction	17667	942	0.92	71	1.27	0.99–1.60
Kurumatani et al.	1999	1947–1996	Shipbuilding (lagging/ repair)	249	114	1.22	10	2.20	1.05–4.04

Note: *NA* not available

[a] Risk ratio, follow-up at retirement (age 55)

of lung cancer among people exposed to asbestos, the importance of abstaining from tobacco should be stressed (Morinaga et al. 2001).

Past epidemiological studies have demonstrated a dose–response relationship between cumulative asbestos exposure and the risk of developing lung cancer. It is estimated that exposure to 25–100 fibers/cc of asbestos per year brings a double risk of lung cancer (Tossavainen 1997).

4.3.4 Mesothelioma

The International Classification of Diseases and Related Health Problems (ICD-10) was the first to identify the extraordinary number of deaths from mesothelioma (C42) in Japanese mortality statistics. Based on ICD-10, in 1995 Japan had 275 deaths (201 men and 74 women) from pleural mesothelioma, but in 2008, the comparable total was 852 deaths (710 men and 142women). This was an increase of 3.5 times in men and 1.9 times in women over that 14-year interval (Fig. 4.4). Although the male-to-female multiplier for deaths from mesothelioma over the preceding 14 years was 3.3 for all sites of the disease combined, the comparable multipliers were 4.0 for pleural mesothelioma, 1.6 for peritoneal mesothelioma, and 2.5 for unspecified sites. The principal weakness in the mortality statistics is the large number of deaths attributed to mesothelioma in unspecified sites (no primary site mentioned).

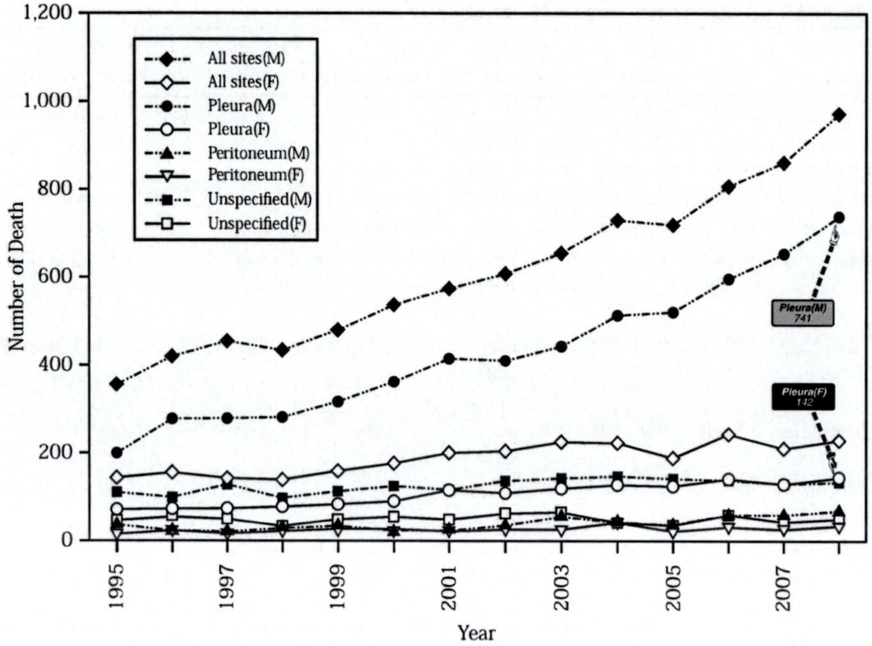

Fig. 4.4 Number of deaths from mesothelioma (1995–2008). *Source*: Ministry of Health, Labour, and Welfare, *Vital Statistics* of Japan, various years

That number was 2426 (1741 men and 685 women) over the preceding 14 years, which accounts for 21.6% (20.3% of men and 26.0% of women) of the total of 11234 deaths (8596 men and 2638 women) during the same period from mesothelioma in all sites combined. Over the same 14-year period, there were 981 deaths (609 men and 372 women) from peritoneal mesothelioma, 81 deaths (52 men and 29 women) from pericardial mesothelioma, and 221 deaths (165 men and 56 women) from mesothelioma in other primary sites. In Japanese mortality statistics covering the preceding 14 years, the pleural to peritoneal mesothelioma death ratio was 7.7 (14.1 for men and 7.1 for women). For comparison, from 1987 to 1999 in Germany, this ratio was 29.2 (1548 to 53) in the Mesothelioma Register (Neumann et al. 2001). It was found that there was a high proportion of deaths from peritoneal mesothelioma, especially among Japanese women. It is strongly suspected that the peritoneal mesothelioma deaths recorded among women might include cases where malignant ovarian tumors became metastatic and spread to the peritoneum. This speculation was confirmed by a recent clinicopathological study, although that might have some selection bias (Takeshima et al. 2009).

We now consider trends in pleural mesothelioma alone. For the period from 1958 to 1967, Japan adopted ICD-7. Malignant pleural mesothelioma was included in the classification for primary pleural cancer, an extremely rare malignancy that at the time accounted for only 3–9 deaths per year, with no evidence of an upward trend. Under the ICD-9 classification adopted in 1978, 62 cases (42 men and 20 women) of pleural mesothelioma were identified in that year, and a gradual upward trend was observed

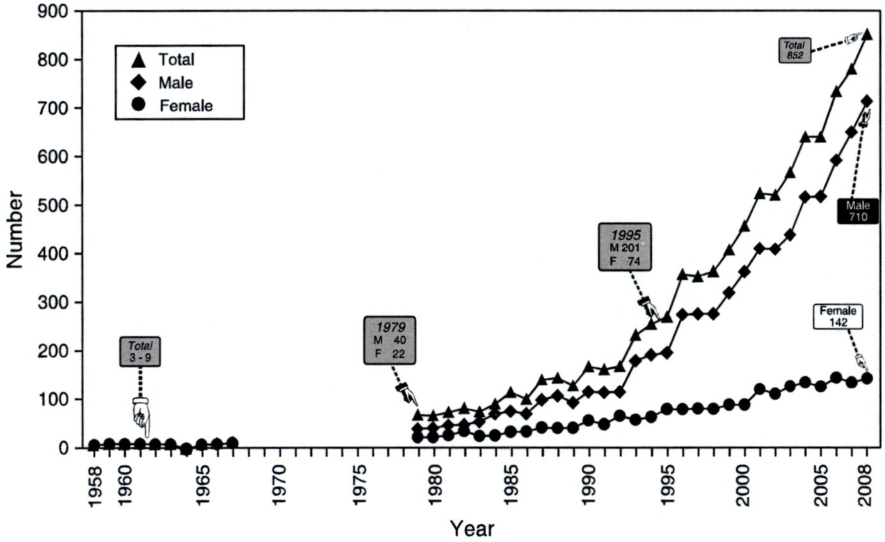

Fig. 4.5 Number of deaths from primary pleural cancer (1958–1967, 1979–1994) and pleural mesothelioma (1995–2008). *Source*: Ministry of Health, Labour, and Welfare, *Vital Statistics* of Japan, various years

thereafter. Unfortunately, owing to classification errors, the trend for the preceding period (1968–1977) is unknown, but it is speculated that the annual number of cases had begun to increase during that time span (Fig. 4.5). The growth trend in the number of deaths from pleural mesothelioma observed from the 1990s coincides closely with trends that started some 30 years earlier (from about 1960) in the annual manufactured output of major asbestos products and the volume of asbestos imports.

An investigation of the death rate from malignant pleural tumor and mesothelioma (all primary sites) by age group and birth cohort from 1980 through 2004 shows that it was higher among the older age group, and tended to be higher the later the year of birth. At a minimum, this leads to a conjecture that the death rate from mesothelioma will be higher for the age groups born prior to 1954 than for any other age group (Fig. 4.6).

Municipalities that have confirmed cases of mesothelioma developing in residents living near asbestos facilities since the summer of 2006 include the City of Chichibu, Ohta City in Tokyo, Tsurumi ward in Yokohama City, the City of Hashima, the Towns of Ohji and Ikaruga in Nara, the City of Amagasaki, and the City of Tosu. Confirmed cases of environmental exposure at home have also been reported (Morinaga and Shinohara 2006). In such cases, particularly where the father rather than the husband is the source of exposure to asbestos dust, the onset or development of mesothelioma (almost all involving the pleura) occurred in patients as young as 20 years of age. The development of mesothelioma in four residents under the age of 40 in the neighborhood surrounding Kubota's defunct Kanzaki factory suggests that these patients were relatively heavily exposed to asbestos soon after birth.

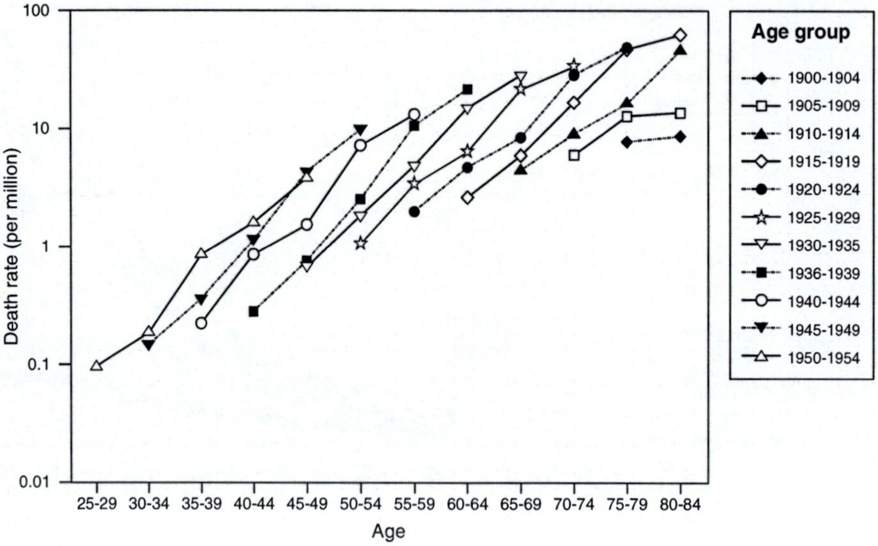

Fig. 4.6 Trends of the death rate (per million) from pleural mesothelioma by birth cohort. *Source*: Ministry of Health, Labour, and Welfare, *Vital Statistics* of Japan, various years

4.3.5 Benign Asbestos Pleurisy

No epidemiological studies of cases with benign asbestos effusion have been carried out in Japan; furthermore, even clinical reports on this condition are rare (Morinaga 1989). "A Survey Study of the Clinical, Pathological, and Epidemiological Aspects of Asbestos-Related Diseases" (chief investigator K. Morinaga,) examined 139 patients (136 men and 3 women) with this condition. Of that total, the average period of exposure for 92 patients (excluding 47 shipbuilding-related cases) was 22.4 years for the high-exposure group (28 patients) and 27.3 years for the lower-exposure group (61 patients). The average age at onset of symptoms was 62.9 years and 67.2 years, respectively. The average period from initial exposure to the onset of symptoms was 38.9 years and 42.5 years, respectively, and the average period from the end of exposure to the onset of symptoms was 16.5 years and 15.0 years, respectively. Although no statistically significant differences were observed, the latent period in the high-exposure group was slightly shorter, ranging from 10 to 69 years (Table 4.3) (Morinaga and Shinohara 2010).

As for benign asbestos effusion, the international literature also contains few epidemiological studies on cases with diffuse pleural thickening caused by asbestos exposure. Dividing the data collected to date on 23 patients in Japan into a high-exposure group (eight patients, two of whom had complications from pulmonary asbestosis) and a lower-exposure group (15 patients), as was done with the benign asbestos pleurisy cases above, we find that the average age at the onset of symptoms was 52.3 and 63.5 years, respectively. The average period from initial

Table 4.3 Characteristics of benign asbestos pleurisy

Exposure level	Case no.	Exposure period (years)	Age at onset (years)	Latent period from first exposure (years)	Duration from end of exposure (years)
High[a]	28	22.4	62.9	38.9	16.5
Not high	61	27.3	67.2	42.5	15

Sources: Morinaga and Shinohara (2010)
[a] Manufacturing asbestos products, asbestos spraying, insulating, demolition of buildings

Table 4.4 Characteristics of diffuse pleural thickening caused by occupational asbestos exposure

Exposure level	Case no.	Age at onset (years)	Latent period from first exposure (years)	Duration from end of exposure (years)
High[a]	8	52.3	29.6	9.4
Not high	15	63.5	38.5	10.2

Sources: Morinaga and Shinohara (2010)
[a] Manufacturing asbestos products, asbestos spraying, insulating, demolition of buildings

exposure to the onset of symptoms was 29.6 and 38.5 years, respectively, and the average period from the end of exposure to the onset of symptoms was 9.4 and 10.2 years, respectively. A statistically significant shortening (95% CI: −16.5 to −1.31) was observed in the average period from initial exposure to the onset of symptoms (the latent period) in the high-exposure group, ranging from 20 to 50 years (Table 4.4) (Morinaga and Shinohara 2010).

Reports concerning the prevalence of pleural plaques in the defined population began to appear in the Japanese literature in the 1980s. However, an international standard for the interpretation of pleural plaque shadows in chest X-rays was lacking, and most studies had false-positives and negatives which could not provide fruitful comparisons between the targeted populations.

In a study focused on Osaka's Sennan district, the authors identified 159 patients with signs of pleural plaques by reviewing approximately 27 000 indirect chest X-ray slides taken during medical examinations of local residents by community health centers from 1979 through 1981. An investigation of their histories of asbestos exposure determined that eight patients had environmental exposure within their homes, while another 19 were exposed as a result of residential proximity to asbestos sources (Yokoyama et al. 1984; Shimizu et al. 1983).

4.4 Workers Accident Insurance Compensation for Patients with Asbestos-Related Diseases

Under Japan's public insurance framework for workers accident compensation, eligibility for compensation benefits was granted to patients with asbestosis starting in 1947, and to patients with asbestos-induced lung cancer and mesothelioma of the pleura and peritoneum starting in 1978. Benign asbestos effusion and diffuse pleural thickening were added to the list of eligible diseases in 2003 (Table 4.5). The compensation eligibility criteria for mesothelioma and asbestos-induced lung cancer have been revised on two occasions: September 2003 and February 2006.

The number of patients given compensation has been climbing sharply year by year since 2005 (Fig. 4.7). As year-of-death data for compensated patients up to 2007 show, signs of an emerging upward trend were already apparent around 2003 (Fig. 4.8). Above all, the sharp surge in compensated patients from 2006 onward is presumed to be the outcome of coverage triggered by the Kubota Shock, and further encouraged by reports in the media, movements driven by patients, their families, and support groups, and information campaigns that government agencies launched in response in order to alert and educate the public.

Table 4.5 Prescribed asbestos-related diseases

Asbestosis (since September 1 1947)
Lung cancer (since October 23 1978)
Pleural and peritoneal mesothelioma (since October 23 1978)
Pericardial mesothelioma of the tunica vaginalis testis (since September 19 2003)
Benign asbestos effusion (since September 19 2003)
Diffuse pleural thickening (since September 19 2003)

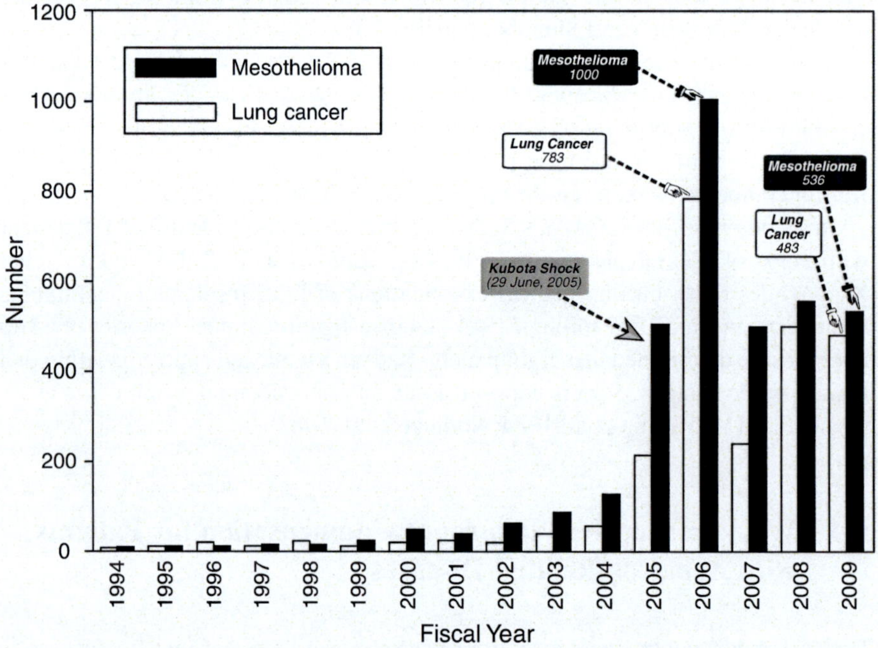

Fig. 4.7 Number of compensated asbestos-related lung cancer and mesothelioma patients by fiscal year (1994–2009). *Source*: Ministry of Health, Labour, and Welfare

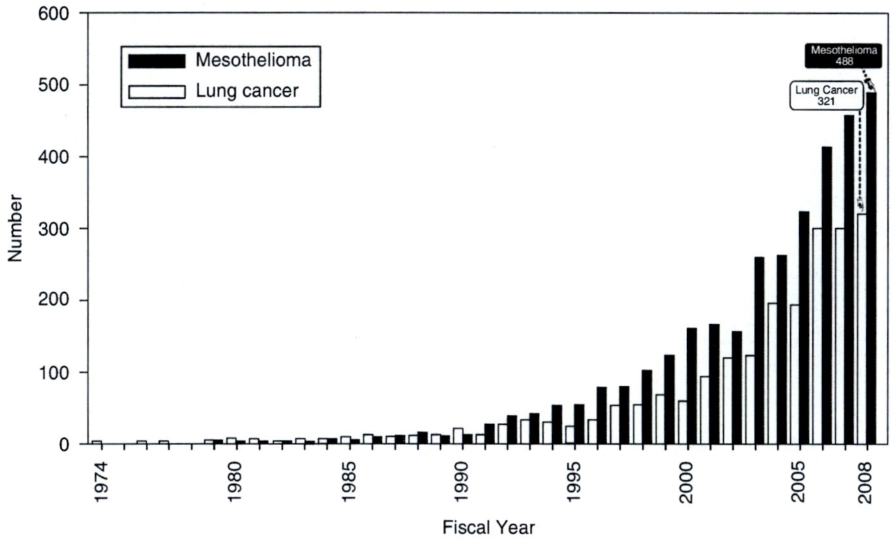

Fig. 4.8 Number of compensated asbestos-related lung cancer and mesothelioma patients by year of death (1974–2008). *Source*: Ministry of Health, Labour, and Welfare

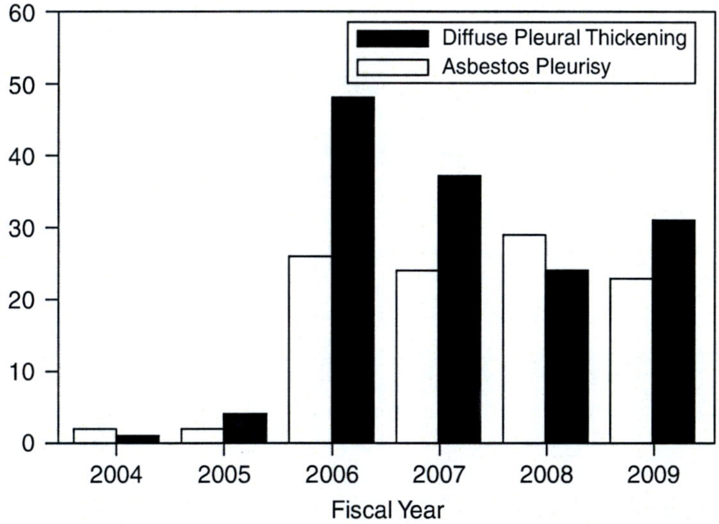

Fig. 4.9 Number of compensated asbestos pleurisy and diffuse pleural thickening patients by fiscal year (2004–2009). *Source*: Ministry of Health, Labour, and Welfare

As of March 31, 2009, compensation under the workers accident insurance framework had been granted to a cumulative total of 106 patients with benign asbestos effusion and 145 patients with asbestos-induced diffuse pleural thickening (Fig. 4.9).

4.5 Relief Measures for Mesothelioma and Asbestos-Induced Lung Cancer Patients

Following the Kubota Shock, reports in the news media shed light on some of the harmful effects that asbestos has had on human health in Japan. In addition, the asbestos issue has evolved into a political one because government agency officials (Ministry of Health, Labour, and Welfare, Ministry of the Environment) ignored the recommendations and advice of experts in the field, and failed to take appropriate action. On July 29, 2005, the Cabinet held its first meeting to deliberate the asbestos issues, and accordingly announced several steps which were to be taken, including measures to (1) prevent the future spread of asbestos exposure and contamination, (2) ease public anxiety over the matter, (3) deal with past cases of asbestos exposure and contamination, and (4) assess the effectiveness of earlier government action. In addition, it declared that efforts would also be made to establish the facts relating to asbestos problems all over Japan. At its second meeting on August 26, 2005, the Cabinet revised certain countermeasures that had been announced in late July, which declared that a new legal framework would be established to provide relief to past victims of asbestos contamination, including workers who had died without receiving any accident insurance compensation, their families, and local residents. The Cabinet also announced that it would continue working primarily with the Ministry of Health, Labour, and Welfare as well as the Ministry of the Environment to gather more factual information on the hazards of asbestos and reach an in-depth conclusion on the matter by September, with the goal of submitting a bill for deliberation to the next ordinary session of the Diet. It was at that time that deliberations began on a structure of relief for a class of citizens who otherwise were not eligible for assistance under the traditional workers accident compensation insurance framework, i.e., mesothelioma victims living in the neighborhood of asbestos facilities. Although the government later promulgated its Act on Asbestos Health Damage Relief on February 10, 2006, that step arguably seems to be nothing more than an attempt to gloss over the policy blunders that the government has committed to date.

This new legislation comprises two key elements. One is to establish a relief scheme for citizens who have been excluded from the compensation scheme under the workers accident compensation program (namely, residents living near asbestos factories, sole proprietors, and employers that have not enrolled in special workers accident compensation insurance plans). The other is to provide compensation to the bereaved families of workers who did not receive any compensation through the traditional workers accident compensation program by prescription (Fig. 4.10-1, -2).

4.5.1 Relief Benefits

The Act on Asbestos Health Damage Relief, enacted on March 27, 2006, granted relief eligibility to victims of asbestos-induced lung cancer and mesothelioma. As of May 30, 2010, a cumulative total of 1275 lung-cancer patients and 3310

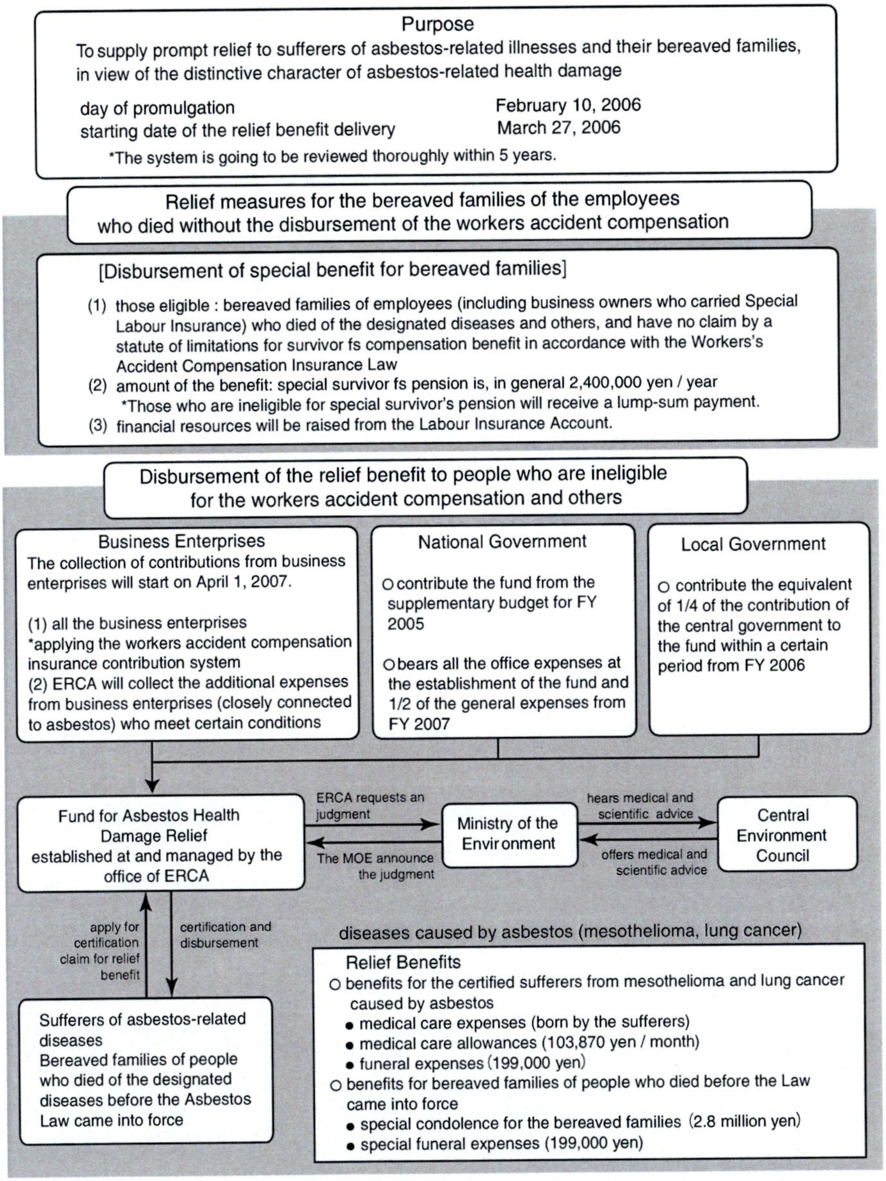

Fig. 4.10 Outline of the act on asbestos health damage relief. *Source*: Environmental Restoration and Conservation Agency of Japan

mesothelioma patients had filed for relief to offset their health-care expenses, and of those, 558 lung-cancer patients and 2253 mesothelioma patients had actually been approved for this relief (Fig. 4.11). In the year from August 1, 2007, through July 31, 2008, 118 lung-cancer patients and 510 mesothelioma patients were approved for relief ; this is an average monthly approval rate of 9.8 lung-cancer patients and 42.5 mesothelioma patients. The photographs in Fig. 4.12 are transmission electron

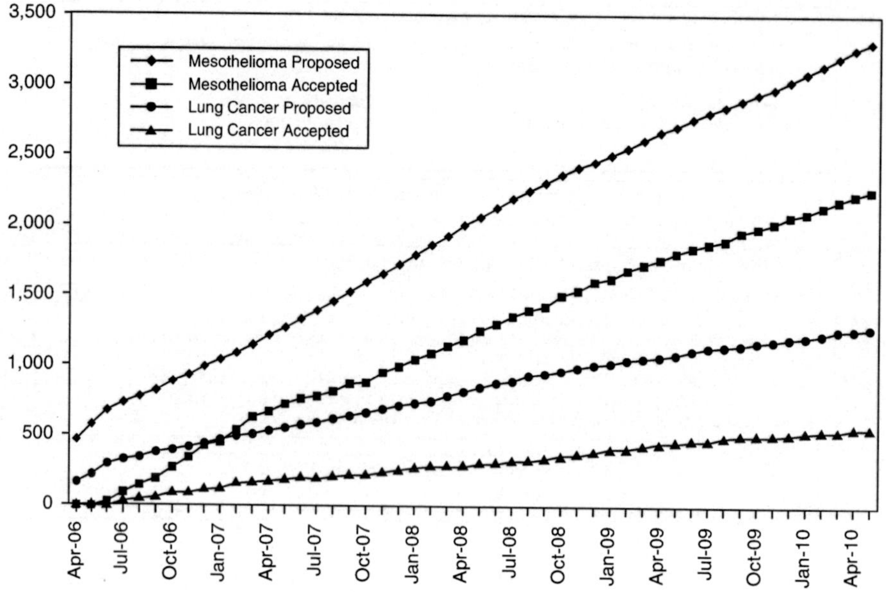

Fig. 4.11 Cumulative numbers of proposals and receivers for asbestos health damage relief (April 2006 – May 2010). *Source*: Environmental Restoration and Conservation Agency of Japan

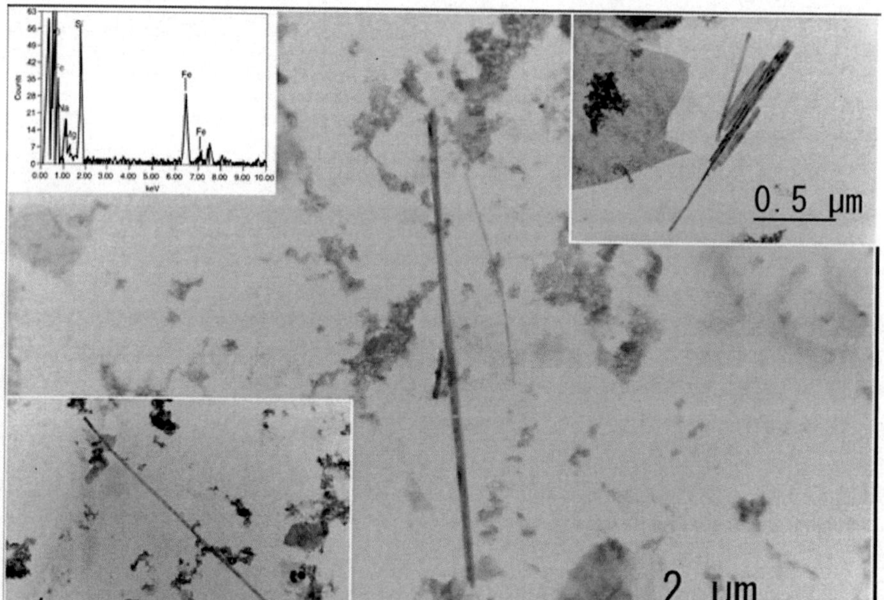

Fig. 4.12 Crocidolite fibers in the lung of a woman whose husband had been working at the Kanzaki asbestos factory in Amagasaki. *Source*: Photo by Shinohara, Y

Table 4.6 Compensation and relief scheme in Japan

Item	Workers accident insurance law (compensation)	Act on asbestos health damage relief
Enactment	April 7, 1947	March 27, 2006, December 1, 2008 (revised)
Subject: all employees	All employees	All (self-employed, family members, inhabitants living near asbestos mines and factories)
Source of revenue	Employers pay installments of 0.03–10.3% of average wages of each employee	(1) All offices and ship's owners pay 0.005% of the compensation fee (2) Local Government (3) National Government (running costs)
Benefit (sufferers)	(1) Medical compensation benefit (2) Nursing care compensation benefit (3) Disability compensation benefit (80% of average wages) (4) Funeral expense	(1) Medical expenses paid by patient (2) Medical treatment benefit: ¥103 870/month (3) Funeral expense: ¥199 000
Diseases	(1) Asbestosis (1947–) (2) Lung cancer (1978–) (3) Mesothelioma (1978–) (4) Diffuse pleural thickening (2003–) (5) Benign asbestos pleurisy (2003–)	(1) Mesothelioma (all sites) (2006–) (2) Lung cancer (2006–) (3) Asbestosis (2010–) (4) Diffuse pleural thickening (2010–)
Necessities of asbestos exposure	Notification of the engagement in the working process with asbestos exposure (February 9, 2006)	Designated diseases caused by asbestos exposure in Japan
Windows of the proposal	Local Labor Office which covers the factory with the last occupational exposure	Health Center, local office of Ministry of Environment, ERCA
Role of physician in charge	Presentation of medical information	Presentation of designated documents and X-ray films, etc
Role of physician in diagnosis	Presentation of pathological documents (all cases), presentation of pathological specimens (some cases)	Presentation of the designated documents (all cases), presentation of pathological specimens (some cases)

microscope (TEM) images of crocidolite fibers in a sample of lung tissue from a woman whose husband worked for a subcontractor providing services on the premises of Kubota Corporation's now-defunct Kanzaki factory. Table 4.6 details the key differences between the New Act and workers accident insurance in Japan.

4.5.2 Special Benefit for Bereaved Families

Until the end of March 2010, the surviving families of 432 lung-cancer patients, 712 mesothelioma patients, 60 asbestosis patients, and two cases with diffuse pleural thickening were approved for compensation under the Act on Asbestos Health Damage Relief on a cumulative basis since the law took effect.

4.6 Concluding Remarks

Although a nationwide mesothelioma register was advocated as a necessity in the early 1980s, the Government still has no framework in place for that purpose. Thus far, action has been limited to a nationwide "mesothelioma panel" campaign that has been held twice on a voluntary-participation basis since 2004. No epidemiological studies on the health effects on either residents living near Kubota's former Kanzaki factory or on the factory workers have been conducted. In a marginal gesture, the city of Amagasaki provided chest X-ray examinations only at the end of 2005. A detailed survey using chest computed tomography (CT) started in the following year, sponsored by the Ministry of the Environment. However, the current situation is totally unsatisfactory because no steps have been taken whatsoever, not even to set up a review committee, including specialists who could interpret the results of the examinations, including chest X-rays and CT. This situation makes these examinations irrelevant (Morinaga et al. 1981; Morinaga 2004:448–452).

Although the first insight into the dangers of asbestos was available in Japan by 1986–1987, it was not until 2005 that some of those risks became crystal clear. In the meantime, many individuals were victimized by asbestos, and many more were put at risk of suffering in the future with asbestos-related diseases. Despite these realities, the complete picture remains hidden to this day. A recent report shows that Japan had imported the largest amount of amphiboles from South Africa between 1980 and 2003. This figure gives us an indication of the terrible future in store for further victims in Japan (Harington et al. 2009).

Acknowledgment The authors wish to thank Shoko Hayasaki for preparation of the manuscript.

References

Hammond C, Selikoff I, Seidman H (1979) Asbestos exposure, cigarette smoking and death rates. Ann N Y Acad Sci 330:473–490

Harington J, Mcglashan N, Chelkowska E (2009) South Africa's export trade in asbestos: demise of an industry. Am J Ind Med 53:524–534

Ishidate F, Tanaka C, Yamauchi S, Tamaki K, Otsuka W, Sukegawa H (1938) Hygiene study on asbestos workers. Science of Labour (in Japanese) 15(3):181–185

Morinaga K (1989) Epidemiological findings of asbestos-related diseases in Japan. Pathol Clin Med (in Japanese) 7(6):686–694

Morinaga K (2002) Legacy of asbestos in the 21st century. Res Environ Disruption (in Japanese) 32(2):6–14

Morinaga K (2004) Research on the progress of the diagnosis and therapy for malignant pleural mesothelioma. In: National Cancer Center (ed) Report of the aid of cancer research fund by Ministry of Health, Labor and Welfare 2003 (in Japanese). National Cancer Center, Tokyo

Morinaga K (2006) Epidemiology of asbestos-related diseases. Mebio (in Japanese) 23(3):12–20

Morinaga K (ed) (2008) Report on the clinical, pathological and epidemiological investigation on diseases caused by occupational asbestos exposure (in Japanese). National Institute of Occupational Safety and Health, Kawasaki, pp 5–34

Morinaga K (2009) Overview of epidemiology on asbestos-related diseases in Japan. Jpn J Lung Cancer (in Japanese) 49:39–47

Morinaga K, Shinohara Y (2006) Occupational and environmental asbestos exposure. In: Morinaga K (ed) Asbestos pollution and its health effect (in Japanese), 2nd edn. Nippon-Hyoron-sha Co., Ltd, Tokyo, pp 47–74

Morinaga K, Shinohara Y (2010) Health effects of asbestos exposure: asbestos-related diseases in Japan. J Clin Exp Med (in Japanese) 232(9):940–946

Morinaga K, Hiyama T, Ohshima A, Hanai A, Fujimoto I (1981) Epidemiology of mesothelioma in Osaka. In: The Center for Adult Disease, Osaka (ed) Report of the prevention and evaluation of medical care by cancer registration (in Japanese). The Center for Adult Disease, Osaka, pp 274–289

Morinaga K, Yokoyama K, Yamamoto S, Sera Y (1984) Mortality and survival of workers receiving compensation for asbestosis in Osaka, Japan. In: Proceedings of the 6th international pneumoconiosis conference, ILO, Geneva, 1984, pp 768–774

Morinaga K, Yasui I, Hara I, Yokoyama K, Yamamoto S, Sakatani M, Sera Y (1990) Mortality of asbestosis patients certified by Osaka pneumoconiosis medical panel. Jpn J Traumatol Occup Med (in Japanese) 38:647–652

Morinaga K, Kishimoto T, Sakatani M, Akira M, Yokoyama K, Sera Y (2001) Asbestos-related lung cancer and mesothelioma in Japan. Ind Health 39:65–74

Neumann V, Gunther S, Muller K, Fischer M (2001) Malignant mesothelioma: German mesothelioma register 1987–1999. Int Arch Occup Environ Health 74:383–395

Sera Y (1983) Asbestosis in Osaka. In: Research Group on the History of Occupational Health in Osaka (ed) Report on the history of occupational health in Osaka (in Japanese). Research Group on the History of Occupational Health in Osaka, Osaka, pp 88–96

Sera Y, Tanaka S (1960) An autopsy case of asbestosis complicated with lung cancer. Sangyo Igaku (in Japanese) 2:326–327

Shimizu H, Morinaga K, Yokoyama K, Hara I, Sasaki M, Fukimoto I, Miura T, Sera Y (1983) Health effects of asbestos exposure with special reference on pleural plaques. J Occup Health (in Japanese) 6(5):24–32

Suzuki K, Noda M (1929) Pneumoconisosis on the railroad workers. J Jpn Assoc Railway Med (in Japanese) 15:10–20

Takeshima Y, Inai K, Amatya V, Gemba K, Aoe K, Fujimoto N, Kato K, Kishimoto T (2009) Accuracy of pathological diagnosis of mesothelioma cases in Japan: clinicopathological analysis of 382 cases. Lung Cancer 66:191–197

Tossavainen A (1997) Asbestos, asbestosis and cancer. Exposure criteria for clinical diagnosis. In: Finnish Institute of Occupational Health (ed) Proceedings of an international expert meeting on asbestos, asbestosis and cancer. People and work research reports 14. Finnish Institute of Occupational Health, Helsinki, pp 8–27

Yokoyama K, Morinaga K, Ohtsuka J, Matsumura T, Hillerdal G, Sera Y (1984) Occupational and non-occupational pleural plaques in Sennan district. In: ILO (ed) Proceedings of the 6th international pneumoconiosis conference, ILO, Geneva, 1983, pp 790–798

Chapter 5
Mesothelioma Due to Neighborhood Asbestos Exposure: A Large-Scale, Ongoing Disaster Among Residents Living Near a Former Kubota Plant in Amagasaki, Japan

Norio Kurumatani and Shinji Kumagai

5.1 Introduction

Immediately after a report in a Japanese newspaper (Mainichi Shimbun Newspaper, June 29, 2005), the clustering of mesothelioma cases in Amagasaki City, Hyogo Prefecture, west of Osaka, was recognized by the public as a large-scale asbestos disaster caused by environmental exposure. The outbreak occurred among residents living near a former asbestos cement pipe manufacturing plant (hereafter "the plant") located in Amagasaki City. The plant was run by the Kubota Corporation, which is now one of the world's leading machinery manufacturing companies. New cases of mesothelioma have continued to arise among local residents, with the total number of mesothelioma patients currently almost 200. This chapter provides a general overview of this ongoing disaster, including our epidemiological findings.

5.2 The Beginning

On January 5, 2005, two members of nongovernmental organizations (NGOs), Akihiko Kataoka (KANSAI Occupational Safety and Health Center) and Kazuko Furukawa (Japan Association of Mesothelioma and Asbestos-Related Disease Victims and Their Families), consulted us about three mesothelioma cases not associated with occupational asbestos exposure. These NGOs assist victims of accidents and diseases encountered in the workplace. From the residential histories of these patients, they strongly believed that all three cases could be ascribed to

N. Kurumatani (✉)
Department of Community Health and Epidemiology, School of Medicine,
Nara Medical University, 840 Shijyo-cho, Kashihara, Nara 634-8521, Japan
e-mail: knorio@naramed-u.ac.jp

S. Kumagai
Department of Environmental Management, School of Health Science,
University of Occupational and Environmental Health, 1-1 Iseigaoka, Yahatanishi-ku,
Kitakyushu, Fukuoka 807-8555, Japan

K. Miyamoto et al. (eds.), *Asbestos Disaster: Lessons from Japan's Experience*,
DOI 10.1007/978-4-431-53915-5_5, © Springer 2011

neighborhood asbestos exposure from the plant. Hiroshi Iida, a member of another NGO (Amagasaki Occupational Safety and Health Center for Workers) which was doing similar work in the same district as the plant, also joined the meeting. These three became the key people in our epidemiological survey. The meeting comprised a roughly 3-h exchange of views on the occupational and residential histories of the three mesothelioma patients. Although this dialog did not reach any conclusions, we later learned that the NGO members, by some diligent work, had identified some additional mesothelioma patients without occupational asbestos exposure but with residential histories near the plant. Against this background, they joined with the patients to initiate negotiations with Kubota. These negotiations eventually led the Kubota Corporation to hold a press conference on June 29. This marked the beginning of a series of asbestos-related problems that still remain a serious concern in our country today. Kataoka (2006) discusses the background events that led up to the Kubota press conference.

The astonishing facts reported in the newspaper (Mainichi Shimbun Newspaper, June 29, 2005) delivered a tremendous shock to the Japanese public. Fifty-one Kubota employees had died from an asbestos-related disease within the previous 10 years. Furthermore, Kubota had made consolidation payments to three mesothelioma patients undergoing treatment at that time, who all had histories of residence near the plant, but who were not considered to have had occupational asbestos exposure. At the press conference, Kubota spokespersons disclosed insider information, including the time period during which the plant had utilized asbestos, the types of asbestos used, and the amount of asbestos cement pipe product the plant had produced. From 1957 to 1975, the plant used an annual average of 4 670 tons of crocidolite and an annual average of 4 600 tons of chrysotile to produce cement pipes (Kubota Corporation 2005) (Fig. 5.1). During those years, Kubota used approximately 5–10% of all asbestos imported into Japan.

5.3 Proposal for an Epidemiological Survey

Following the newspaper report, the NGOs with which Kataoka and others were affiliated, the Amagasaki City Health Authorities, and the Kubota Corporation were all swamped with inquiries from mesothelioma patients or family members of deceased patients who believed that the sufferers had had no occupational asbestos exposure, but had lived near the plant. We proposed launching an epidemiological survey using the contact information of people filing inquiries with Kataoka's group. A case series study would be persuasive, but establishing causal relationships demanded that we apply uniform criteria to assess asbestos exposure, quantify the mortality risk of mesothelioma, and demonstrate correlations between the risk and the assumed source of exposure.

To that end, we proposed the following six avenues of investigation: (1) implement an interview study to determine whether each patient had a history of asbestos exposure, occupationally or domestically, or other chances of asbestos exposure;

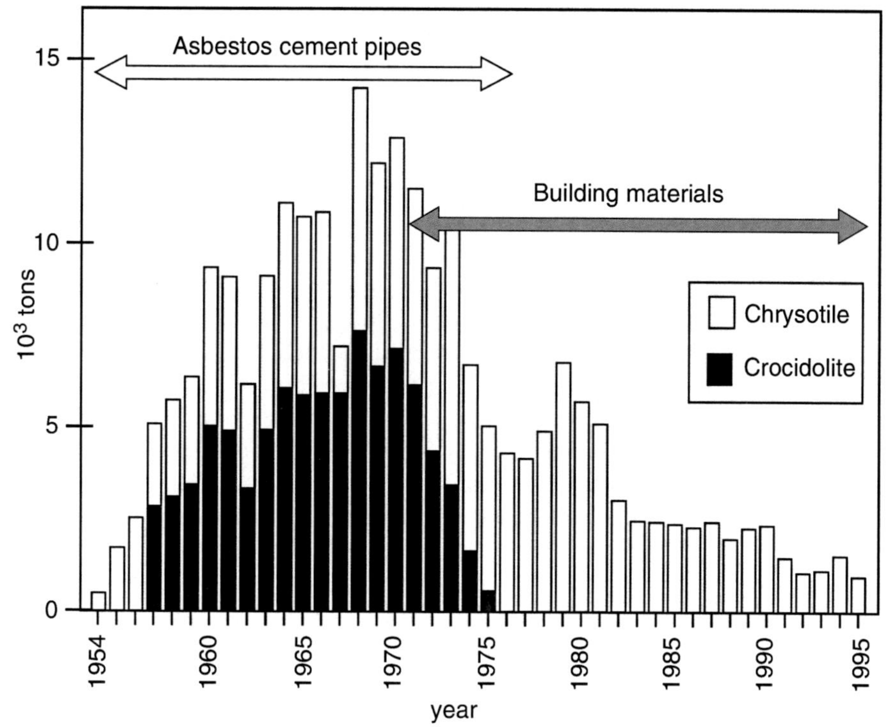

Fig. 5.1 Asbestos consumption in the asbestos cement pipe manufacturing plant run by the Kubota Corporation. *Note*: The plant produced asbestos cement pipes from 1954 until 1975, and building materials containing chrysotile from 1971 until 1995. *Source*: Kubota Corporation

(2) review patient's medical records to confirm the diagnosis of mesothelioma; (3) re-examine pathological tissue as warranted, provided that histological samples of tumor tissue had been retained; (4) calculate the mortality risk of mesothelioma in relation to the residential distance from the plant; (5) assess the mortality risk, incorporating wind direction and other meteorological factors during the time of asbestos use; (6) estimate the asbestos concentrations to which the patients may have been exposed.

These proposals were accepted by the NGOs, and the study was initiated on July 30, 2005. The survey was aimed at the people who had made inquiries.

5.4 Interim Developments

Almost 1 month after initiating the study, we had completed interviews with people related to approximately 50 mesothelioma cases, and performed a preliminary analysis of the existing data. Compared with the baseline national mortality rate for mesothelioma, the area within a radius of 500 m from the center of the plant had a significantly increased standardized mortality ratio (SMR; discussed later) of 9.5.

We also found that calculated values of SMR declined with increasing distance from the plant. We decided to release these preliminary findings through newspaper reports because the production of an interim report was a condition for continued cooperation from the NGOs, and because the incidence of mesothelioma within the vicinity of the plant had become an extremely serious social issue. Our findings appeared in newspapers on August 28, 2005.

The executives of the Kubota Corporation issued a formal apology to patients on December 25, 2005. In April 2006, they established a relief payment program for residents who were not exposed to asbestos occupationally, who lived or worked within a 1-km radius of the plant during the time of asbestos use, and who developed an asbestos-related disease, including mesothelioma (Kubota Corporation 2006). The company paid 25–46 million yen ($312500–$575000) per person to eligible residents or their bereaved families.

Inquiries from living patients and family members of deceased patients continued to pour in, and our epidemiological survey took 2 years to finish. The results of the completed study supported our preliminary findings and produced a much stronger set of conclusions. Our findings were published in two medical journals (Kurumatani and Kumagai 2008; Kumagai and Kurumatani 2009).

In the 2008 article, we (1) described the epidemiological features of mesothelioma due to neighborhood exposure to asbestos on the basis of 162 cases as of December 31, 2006, (2) determined the risk of mesothelioma according to residential distance from the plant, using methods from previous studies (Newhouse and Thompson 1965; Rees et al. 1999; Magnani et al. 2000, 2001), and (3) estimated the area affected by asbestos, taking meteorological conditions into account. This article showed that cases of mesothelioma had occurred in residential areas within a distance of just over 2 km from the plant in the direction of the prevailing wind. This was a rare epidemiological report on the risk of mesothelioma from neighborhood asbestos exposure, and a figure from the paper showing the locations of homes of mesothelioma victims appeared on the front page of the journal in which it was published. In the 2009 article, we (1) estimated the asbestos fiber concentration in the area surrounding the plant during the period of crocidolite and chrysotile use, and (2) predicted the number of future excess mesothelioma deaths in the area.

5.5 Results of Epidemiological Survey

5.5.1 Confirmation of Asbestos Exposure and Underlying Cause of Death

Our initial study population consisted of 162 (96 men and 66 women) of 164 residents with mesothelioma who contacted the NGOs before the end of April 2007 and for whom consent to our interview study was obtained (relatives of 138 deceased subjects and 24 subjects under treatment). The interviewees were requested to bring official documents to verify their mesothelioma diagnosis, occupational history, and place of residence.

We obtained the date of the initial subjective symptom of mesothelioma. We confirmed the diagnosis by reviewing the subject's medical charts and re-examined biopsied pleural tissues pathologically when available. The employment record for receiving a pension after retirement, which we asked interviewees to obtain from the Social Insurance Agency, listed all companies where the subject was employed, along with employment dates. We determined the risk of occupational asbestos exposure by comparing the job descriptions provided by interviewees with a list of 19 occupational groups (with 57 job categories) commonly found in Japan that can lead to asbestos exposure (Osaka Mesothelioma Study Group 2000). The risk of para-occupational exposure was also determined by evaluating the occupations of the spouse, parents, and other household members. We defined each subject's exposure point to asbestos from the plant as the distance from (1) the house where the subject had lived for at least 1 year, or (2) the location where the subject worked, whichever was the closer, between 1957 and 1975, during which time the plant had used crocidolite.

5.5.2 *Subjects for Analysis*

As shown in Fig. 5.2, 121 of the 162 study subjects were judged not to have been exposed to asbestos directly or indirectly at work. Twelve of the 121 subjects worked at a location closer to the plant than their residence. The other 109 had no chance of occupational exposure to asbestos and had lived near the plant for at least 1 year between 1957 and 1975. We then eliminated six subjects for whom pleural cancer was recorded as an underlying cause on the death certificate, but whose diagnosis turned out to be mesothelioma, to maintain comparability with the expected deaths calculated from national vital statistics. Finally, seven deaths which had taken place before the International Statistical Classification of Diseases and Related Health Problems, 10th Revision (ICD-10), went into effect in Japan were excluded. Ultimately, 79 mesothelioma deaths plus 17 patients currently undergoing medical treatment for mesothelioma were included in the final analysis.

All the 17 patients undergoing treatment were diagnosed with pleural mesothelioma histologically. Of the 79 deaths, diagnosis was confirmed by histology in 62 cases (78.5%), by cytology in 12 cases (15.2%), and by clinical findings, including chest computed tomography (CT) scan in five cases (6.3%). The mean age at death was 56.9 years (standard deviation (SD) 10.8 years) for men and 66.3 years (SD 11.8 years) for women. All 79 deceased subjects had pleural mesothelioma except for one patient with peritoneal mesothelioma. In the plant itself, pleural mesothelioma had occurred in 18 workers and peritoneal mesothelioma had occurred in 28 workers as of June 2005 (Kubota Corporation 2005). These findings are consistent with studies reporting that peritoneal mesothelioma occurs more often in people exposed to higher levels of asbestos (Browne and Smither 1983; Reid et al. 2005).

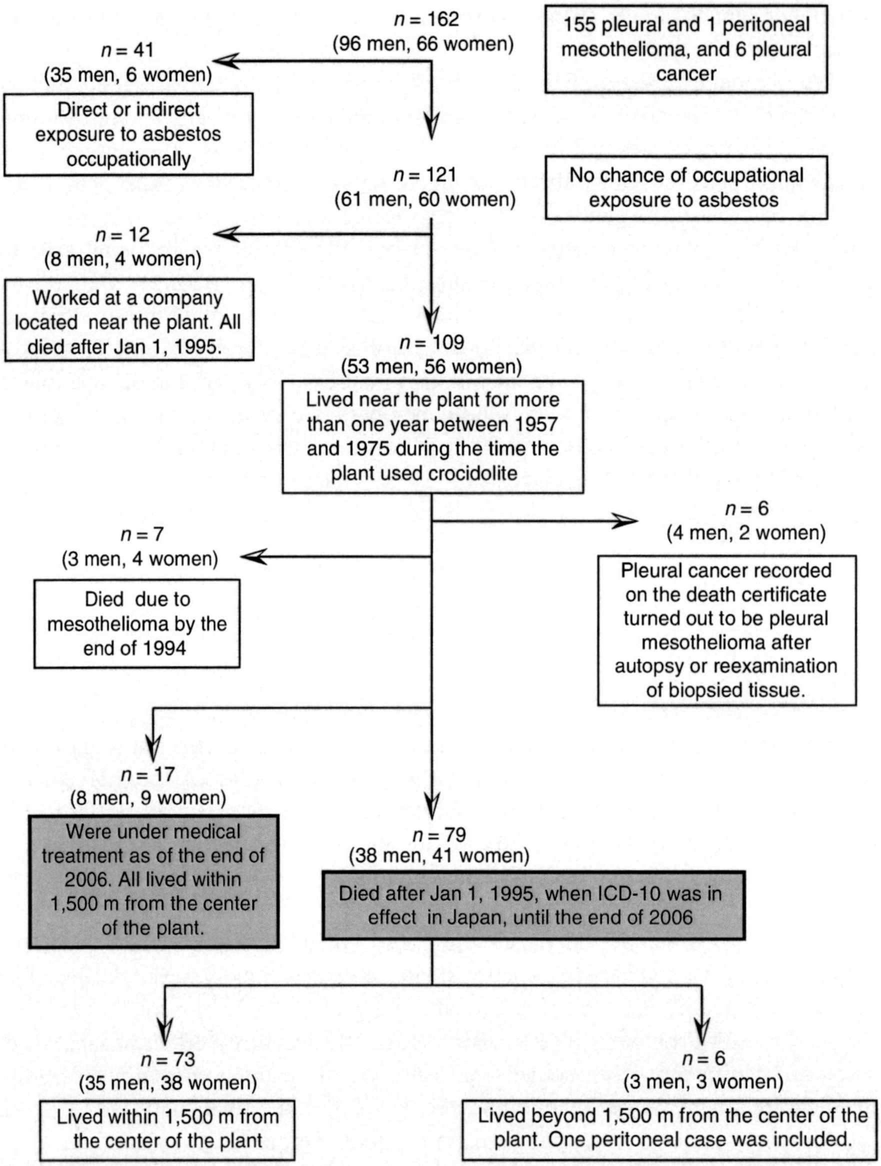

Fig. 5.2 Study subjects by occupational, residential, and vital status. *Notes*: (1) The two *gray boxes* show the subjects selected for further analysis. (2) Seven deaths that occurred before the International Statistical Classification of Diseases and Related Health Problems, 10th Revision (ICD-10), was in use (i.e., before 1995) were excluded from SMR analysis because the expected deaths from mesothelioma can be calculated in relation to the national mortality rate only after the enforcement of ICD-10, which first contained a specific rubric for mesothelioma (C45)

5.5.3 Emitting Source of Crocidolite in the Study Area

The actual concentrations of airborne asbestos fibers in the area surrounding the plant during the period of crocidolite use could not be determined because measurements of asbestos fibers inside and outside the plants were not mandated. However, several factors pointed to the Kubota plant as being the single emitting source of the crocidolite that caused so many mesothelioma cases among residents in the study area.

The Amagasaki City Office identified 136 closed or active companies where asbestos products were made or used (Amagasaki City 2006). However, the office reported that all other companies consumed less than 10 tons of crocidolite per year, which was far less than the 4670 tons (annual mean) used by the plant (see Fig. 5.1).

In the Kubota plant, 46 employees had mesothelioma in the past 20 years, and 40 had asbestosis or asbestos-related lung cancer (Kubota Corporation 2005). Because no epidemiological studies have been performed to date for employees who worked at the plant, the true risk of contracting an asbestos-related disease is still not known. It is said that as many as 1000 workers were exposed to asbestos in the plant. A rough estimate based on that assumption would put the incidence of mesothelioma due to occupational asbestos exposure at the plant at approximately 300 times the national average.

In the Kubota plant, packed raw asbestos was transported to the plant by vehicle, separated by fluffing, and then sent upstairs through pneumatic ducts, where it was collected and mixed with other materials to mold cement pipes, which were dried and cut. We believe that fluffed asbestos was emitted from the duct outlet and dispersed into the surrounding air, where it was inhaled by neighborhood residents. The plant operated around the clock.

Previous studies suggested that (1) transport of raw asbestos with a loosely attached cover (Magnani et al. 2001; Hiraoka et al. 1998), (2) improper use of asbestos residuals (Magnani et al. 2001; Wagner et al. 1960; Peipins et al. 2003), such as for thermal home insulation, gardening, or producing hard pavements, and (3) playing on piles of asbestos (Peipins et al. 2003) were other asbestos exposure routes in the environment around a plant. We did not find these exposure routes in the study area, except for a single narrow 30-m road with asbestos residuals located 500 m southwest of the plant. We did not find naturally occurring asbestos, which can cause mesothelioma (Pan et al. 2005), in the study area, and houses were not constructed with asbestos-contaminated soil mixtures (Metintas et al. 2002).

Another study (Amagasaki City 2008) examining the health status of residents around the plant found that many residents with pleural plaque specific to asbestos exposure had no direct occupational history of asbestos exposure.

We counted asbestos bodies in lung tissue samples available from six subjects (Natori et al. 2006). The mean count from these samples was 410 bodies per gram dried lung tissue, which was clearly below the 5000 bodies per gram threshold for occupational exposure to asbestos, but still far above the mean count of 35 bodies per gram observed in individuals with no history of asbestos exposure. Further, asbestos fiber analysis detected crocidolite in some samples (Natori et al. 2006).

Therefore we investigated the disease occurrence in the study area under the assumption that the plant was the sole emitting source of crocidolite, and focused on residents living near the plant between 1957 and 1975 when crocidolite, which has a much greater carcinogenic potential than chrysotile (Hodgson and Darnton 2000), had been used.

5.5.4 *Risk of Mesothelioma Among the Residents*

5.5.4.1 Mesothelioma Risk by Residential Distance from the Plant

We expressed the risk of death from mesothelioma as the standardized mortality ratio (SMR, ratio of observed deaths to expected deaths). Included in our SMR analysis were deaths that occurred from January 1, 1995, to December 31, 2006, when the ICD-10, which was the first version to contain a rubric specific for mesothelioma (C45), was in effect in Japan. Expected deaths annually during 1995–1997, 1998–2002, and 2003–2006 were obtained by multiplying the sex- and 5-year age-specific population at risk in 1995, 2000, and 2005, respectively, by the corresponding national mortality from mesothelioma during those years.

The population at risk as of 1975 was estimated first using Amagasaki City Census data from 1975 (Amagasaki City 1975). This was 220 809 within a 1 500-m radius from the center of the plant and 280 604 in the area where the relative asbestos concentration (discussed later) was above 2 m^{-3}. Then we estimated the population at risk as of 1995, 2000, and 2005, assuming that the sex- and 5-year age-specific population at risk as of 1975 died according to the corresponding death rates of the Japanese population in every 5 calendar years from 1975 until 1995, 2000, and 2005. Additional details were published previously (Kurumatani and Kumagai 2008).

Table 5.1 summarizes the mesothelioma SMRs and demographic characteristics of the 90 subjects who lived within a 1 500-m radius of the center of the plant. Residential distance was divided into 300-m intervals. The 73 mesothelioma deaths produced a significantly increased overall SMR of 4.3 (95% confidence interval (CI), 3.4–5.4). The highest SMR of 41.4 (95% CI, 15.2–90.1) occurred among women who had lived within a 300-m radius of the center of the plant, and the SMR of women remained significantly elevated in all groups living within 1 500 m from the plant. Similarly, the highest SMR among men was for those living within a 300-m radius, but only one other SMR for men (i.e., the SMR between 300 and 600 m from the plant) was significantly elevated. The SMRs of women were at least double those of men.

To our knowledge, few studies have reported the risk of mesothelioma in relation to the residential distance from an industrial source of asbestos. A significantly increased risk of mesothelioma was observed within half a mile of an asbestos factory in London

(Newhouse and Thompson 1965), within 1 000–2 000 m of asbestos mines, asbestos factories, or shipyards in South Africa (Rees et al. 1999) and some European countries (Magnani et al. 2000), and beyond 2 500 m of an asbestos cement factory in Italy (Magnani et al. 2001). Our results regarding risk relative to residential distance among women are consistent with these earlier findings.

5.5.4.2 Consideration of the SMRs Obtained

Underestimation of the population at risk may lead to overestimated SMRs; conversely, overestimation of the population at risk may lead to underestimated SRMs. However, it is unlikely that large errors occurred in the population estimates, and it is even less likely that such errors varied systematically with exposure.

All our subjects were patients or their families who had voluntarily contacted an NGO. The Kubota Corporation appointed one of the NGOs as an agent for relief payment for asbestos-related death or disease. Not all mesothelioma deaths in the study area during the study period were enrolled because no mesothelioma registry existed. Therefore, deceased people without relatives and affected individuals who were unaware of Kubota's payment did not contact an NGO. This means that the SMRs were underestimated owing to selection bias from unreported cases. We carefully evaluated whether subjects had been exposed to asbestos at their workplace, and excluded those who had been exposed occupationally from the subjects for SMR analysis. However, some misclassification was inevitable. The list of occupations (Osaka Mesothelioma Study Group 2000) to which we referred included all jobs with potential exposure to asbestos, which tends to produce more false positives than false negatives, and also results in an underestimation of the SMRs.

We excluded cases with occupational exposure when counting observed deaths, but could not do so for expected deaths because the relevant data were not available from the national vital statistics office. This bias leads to an underestimation of the SMRs. Such bias may be greater for men than for women, because more men were employed and thus potentially exposed to asbestos occupationally. Consequently, for environmental asbestos exposure from the plant, our results estimate the effects for women more accurately than those for men.

5.5.4.3 Residential Period and Latent Period

When data from the 73 deceased and 17 surviving subjects were combined (Table 5.1), the mean residential period during the use of crocidolite in the plant was 145 months (range 24–227 months), or 12.1 years. This duration of residential exposure is comparable to that of subjects who developed mesothelioma after living within half a mile of an asbestos factory in London (Newhouse and Thompson 1965) (mean 14.4 years), and is more than double that of subjects with mesothelioma who lived in the neighborhood of a crocidolite mine in Australia. However,

Table 5.1 Standardized mortality ratio (SMR) of 73 mesothelioma deaths, and demographic characteristics of the 73 deceased subjects and 17 patients under treatment, by sex and residential distance from the plant within a 1 500-m radius

	Residential distance from the plant (m)	Estimated population at risk in 1975	No. of deceased in 1995–2006	SMR (95% CI)		Age at death (years)	No of patients under treatment	Total no. of deceased and subjects under treatment	Residential months at risk	Latent period (months)
Men	0–300	4 213	7	13.9	5.6–28.7	67.6 ± 10.1	0	7	104.7 ± 58.3	515.1 ± 55.0
	300–600	17 963	12	5.6	2.9–9.8	52.6 ± 10.0	4	16	145.8 ± 58.3	523.3 ± 49.5
	600–900	28 259	7	2.1	0.8–4.3	54.6 ± 9.7	2	9	125.1 ± 52.8	519.1 ± 49.2
	900–1 200	27 891	4	1.2	0.3–3.1	57.9 ± 10.4	1	5	184.1 ± 49.5	523.9 ± 62.0
	1 200–1 500	32 568	5	1.3	0.4–3.0		1	6		
	Subtotal	110 894	35	2.6	1.8–3.7	57.3 ± 11.1	8	43	144.6 ± 59.7	521.3 ± 51.9
Women	0–300	4 175	6	41.4	15.2–90.1	63.3 ± 9.8	3	9	98.4 ± 62.9	498.8 ± 99.4
	300–600	17 804	7	11.3	4.5–23.3	66.7 ± 12.7	0	7	158.1 ± 55.2	516.3 ± 41.9
	600–900	28 010	7	7.2	2.9–14.8	63.7 ± 10.0	3	10	187.1 ± 41.6	516.6 ± 66.3
	900–1 200	27 646	8	8.3	3.6–16.4	64.9 ± 17.8	3	11	142.0 ± 77.8	525.6 ± 48.6
	1 200–1 500	32 280	10	8.9	4.3–16.4	67.7 ± 9.2	0	10	143.6 ± 78.2	533.3 ± 32.3
	Subtotal	109 915	38	9.9	7.0–13.7	65.5 ± 11.8	9	47	146.0 ± 69.1	518.7 ± 60.5
	Total	220 809	73	4.3	3.4–5.4	61.6 ± 12.1	17	90	145.3 ± 64.4	519.9 ± 56.3

Notes:
1. Vital status as of December 31, 2006
2. *$P<0.05$ by one-way-analysis of variance (ANOVA) followed by Dunnett's procedure
3. Age and months are expressed as mean ± SD

89% of the Australian subjects with mesothelioma also lived with an asbestos worker (Hansen et al. 1998). No statistically significant difference in time existed between affected men and women, but for both sexes, the residential exposure in the 300-m radius group was 20–90 months shorter than exposure in the other distance groups.

The latent period, defined as the time from the subject's first year living near the plant to the appearance of the initial mesothelioma-related symptom, varied from 265 to 595 months (mean 520 months; i.e., 43.3 years). We found no significant differences in the latent period between men and women, or among the groups categorized by distance from the plant.

The subjects in the present study died 10 years earlier (men) or 5 years earlier (women) compared with all Japanese who died of mesothelioma (Ministry of Health, Labour, and Welfare 2007). Workers exposed to asbestos occupationally may be older at the time of death because workers are not exposed to asbestos until they are actually working with it. In contrast, residents of areas polluted with asbestos are exposed from the time they start to live there, and sometimes from birth (Reid et al. 2007). Therefore, it is understandable that residential asbestos-related deaths occur earlier.

5.5.4.4 Relative Asbestos Concentrations Around the Plant

None of the previous studies considered the meteorological conditions. Asbestos concentrations in the air surrounding the emission point depend on wind direction and velocity (Laamane et al. 1965), which determine the direction and distance that asbestos fibers travel.

Because the actual airborne asbestos fiber concentrations surrounding the plant were unknown during the period of crocidolite use, we estimated "relative asbestos concentrations" (arbitrary units, m^{-3}) in each of 2500 grid units (each grid unit is 100×100 m). This method assumed that asbestos was emitted from the center of the plant's premises, and used diffusion equations to account for the meteorological conditions (Pasquill 1961; Environmental Management Bureau 2000; Turner 1964; Briggs 1973). Additional details of this method have been published previously (Kurumatani and Kumagai 2008).

Figure 5.3 depicts relative asbestos concentrations in each of the 100×100 m grid units, the places of residence of the 79 deceased and 17 surviving subjects, and the wind-rose of the study area. Relative asbestos concentrations differed considerably in different directions from the plant, even at the same distance from the plant, and concentrations decreased further away from the plant. Higher concentrations of asbestos occurred to the south–southwest of the plant, which is consistent with the prevailing wind direction from the north–northeast. Similarly, the homes of mesothelioma cases were scattered more north–south than east–west, and more were located south of the plant than north of it. We identified three mesothelioma deaths (not depicted) of people who worked in a factory south of the plant; their relative asbestos concentrations were the highest in the study area.

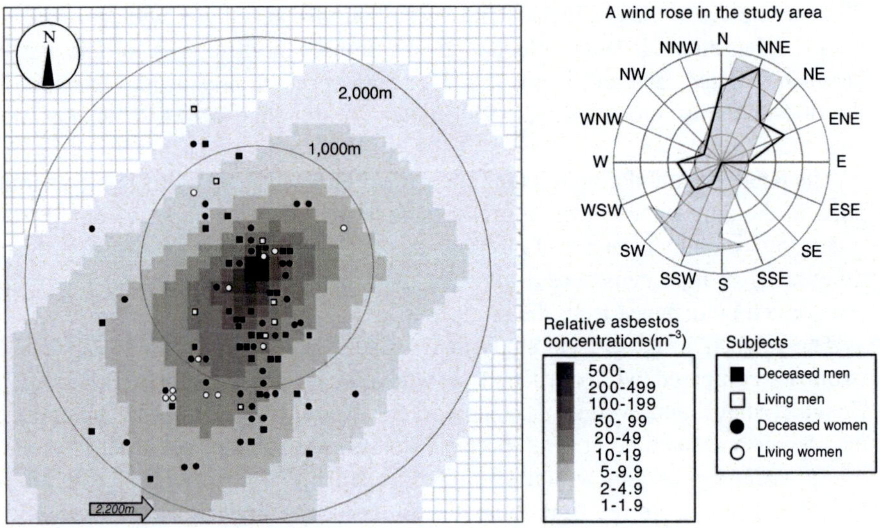

Fig. 5.3 Relative asbestos concentrations in 100 m^2 grid units showing the places of residence of 79 deceased subjects and 17 patients under treatment. *Notes*: (1) Relative asbestos concentrations were determined by diffusion equations incorporating the meteorological conditions of the study area. (2) The *large closed square* in the center of the figure represents the plant. (3) The wind-rose, which was calculated using the meteorological data between 1973 and 1975, shows that wind from the north–northeast was predominant in the study area

5.5.4.5 Mesothelioma Risk by Relative Asbestos Concentrations

The relationship between relative asbestos concentrations and mesothelioma SMRs among deceased subjects is illustrated in Fig. 5.4. An area with a relative concentration of 200 m^{-3} or higher was excluded from the analysis because it contained only the plant and other large factories, but no residences. We also excluded an area with an asbestos concentration less than 2 m^{-3} because the area was too wide to limit its boundary, and we were therefore unable to determine the population at risk. We classified the remaining area into five regions with the following relative asbestos concentration levels: 2–4.9 m^{-3} (mean 3.2 m^{-3}), 5–9.9 m^{-3} (mean 6.9 m^{-3}), 10–19.9 m^{-3} (mean 13.7 m^{-3}), 20–49.9 m^{-3} (mean 30.4 m^{-3}), and 50–199.9 m^{-3} (mean 89.8 m^{-3}). This grouping did not reflect quintiles of asbestos exposure or population exposed, but was defined so that each region included at least five deaths and a large enough at-risk population to allow for statistical analysis. The sex-specific SMR of each area was calculated, and the relation between relative asbestos concentrations and these SMRs was determined by linear regression analysis.

The highest SMR of 47.7 occurred among women in the area with the highest relative asbestos concentration. Significantly increased SMRs were found in areas with asbestos concentrations of more than 5 m^{-3}; the corresponding area extended approximately 2200 m south–southwest and 900 m north–northeast from the plant's center (see Fig. 5.3). Although the SMR regression line for male subjects is significantly less steep than that for female subjects ($P<0.05$), both lines show a linear

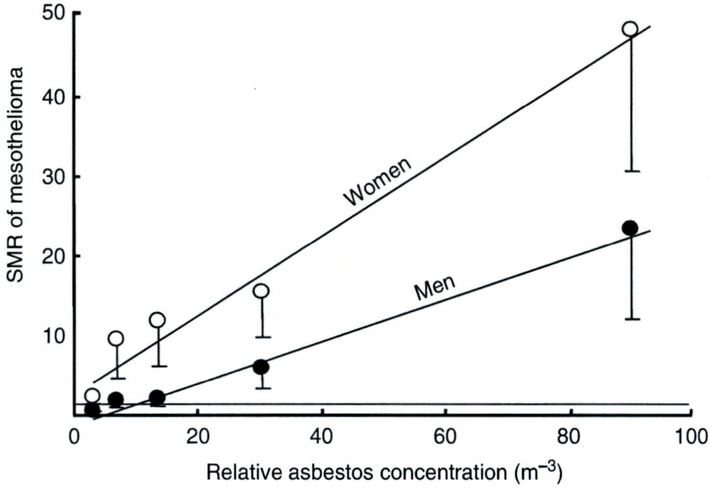

Fig. 5.4 Dose–response relationship between standardized mortality ratios (SMRs) of mesothelioma and relative asbestos concentration levels divided into five exposure groups. *Notes*: (1) The regression lines are $y=0.49(0.36–0.62)x+3.09(-2.44–8.61)$ for 39 women, and $y=0.26(0.22–0.31)x-0.89(-2.83–1.05)$ for 36 men. The figures in parentheses in the equations are 95% C.I.s of the slope and intercept. (2) The *error bar* at each point shows the lower 95% C.I. of the SMR

dose–response relationship between relative asbestos concentration and mesothelioma risk. These results did not change after we excluded 11 deaths of subjects who lived with household members who had a job which could bring about occupational asbestos exposure (Kurumatani and Kumagai 2008).

5.6 Estimation of Asbestos Fiber Concentrations

To estimate asbestos fiber concentrations (fiber/ml) from relative asbestos concentrations (arbitrary units, m^{-3}), we determined a conversion value using both the United States Environmental Protection Agency (USEPA) model (Nicholson 1986) and the Japanese Society for Occupational Health (JSOH) potency factor for mesothelioma, K_M (JSOH 2000:177–186). The EPA model itself is based on the findings of epidemiological studies of mesothelioma deaths due to occupational asbestos exposure. According to this model, the mesothelioma mortality risk (R) is expressed as the cube of the time following initial asbestos exposure (t) multiplied by the length of exposure (d), the asbestos fiber concentration (f/ml), and the coefficient K_M. Therefore, the fiber concentration can be derived from the EPA model if all other variables are determined. The values of t and d were determined operationally, and the K_M value recommended by JSOH (7.75×10^{-9}) was used. In the present study, the SMR for female subjects with mesothelioma, a statistic considered to correlate more closely with the effects of neighborhood asbestos exposure, was

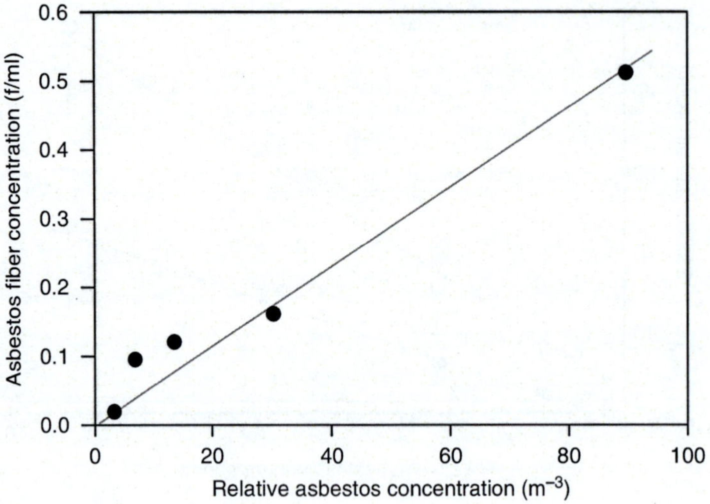

Fig. 5.5 Relationship between relative asbestos concentration and asbestos fiber concentration. *Notes*: (1) The regression line passing through the origin is $y = 0.057x$ [95% C.I., 0.0051–0.0064]. (2) This line was obtained using female mesothelioma mortality because it reflects residential asbestos exposure more accurately than male mesothelioma mortality

used for the value *R*. Figure 5.5 plots asbestos fiber concentrations against relative asbestos concentrations for the five regions covered in Fig. 5.4, and the regression line passing through the point of the origin shows a significantly positive slope (Kumagai and Kurumatani 2009). As noted earlier, the relative asbestos concentration which correlates with a significantly excessive number of mesothelioma deaths is 5 m^{-3}. Substituting this value in the regression formula "fiber concentration (f/ml) = 0.0057 [95% CI, 0.0051–0.0064] × (relative asbestos concentration)" gives a fiber concentration of 0.0285 [95% CI, 0.0255–0.0320] f/ml. Although these findings depend on several assumptions, they suggest that a significantly large number of deaths from mesothelioma are found in areas with an asbestos fiber concentration of approximately 30 f/l or higher. A value of 30 f/l is consistent with the reference value for excess lifetime risk (10^{-3}) of cancer, which was defined by JSOH (2000) for occupational asbestos exposure other than chrysotile.

5.7 Concluding Remarks

In the present study, nearly 100 subjects with mesothelioma were identified as victims of neighborhood asbestos exposure as of the end of 2006. This number is one of the largest reported in studies of mesothelioma among residents around industrial asbestos sources (Newhouse and Thompson 1965; Rees et al. 1999; Magnani et al. 2000, 2001; Wagner et al. 1960; Peipins et al. 2003; Bohlig et al. 1970; Gardner and Saracci 1989; Hansen et al. 1993; Howel et al. 1997; Berry 1997;

Camus et al. 1998). We believe that the mesothelioma outbreak among residents was causally associated with the asbestos, in particular crocidolite, used by the plant. The affected area spread as far as 2200 m from the center of the plant in the direction of the prevailing wind.

Nearly 4 years have elapsed since December 31, 2006, the arbitrary end-point we set for the preparation of our findings. During that period, new cases of mesothelioma have appeared, and we have continued with our investigations. Figure 5.6, compiled by the NGO that also serves as an agent of Kubota, shows the number of applicants for relief payment that have been submitted to Kubota. Although data for 2009 are still being gathered, we are dismayed by the large number of new cases reported. The data reveal a large-scale, ongoing asbestos disaster among neighborhood residents.

In 1960, Wagner and his colleagues (Wagner et al. 1960) first reported a strong relationship between mesothelioma and crocidolite exposure (including occupational, para-occupational, and neighborhood exposure) in a case series study of 33 patients with pleural mesothelioma. Almost half a century has passed, but mesothelioma is still a difficult disease to detect at an early stage or to treat successfully. Asbestos has been used widely since the early twentieth century, and therefore many active and closed factories world-wide that handled asbestos and asbestos-containing products may have spread asbestos into the surrounding communities. Mesothelioma develops more than 40 or 50 years after the first asbestos exposure. Public health policymakers, health professionals, and industry

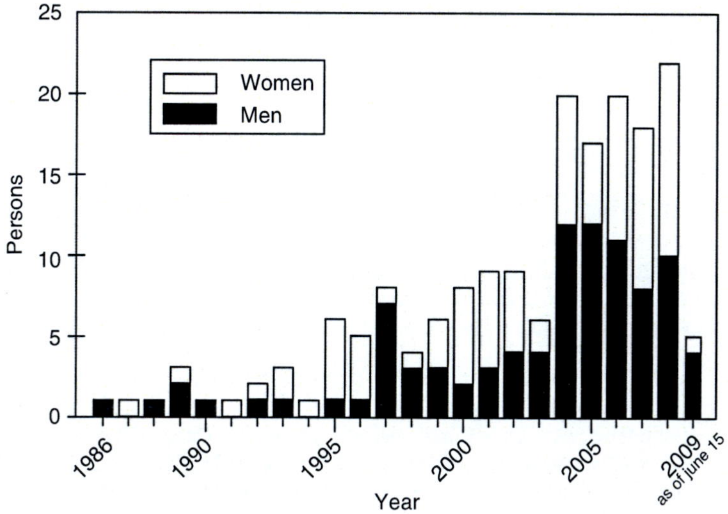

Fig. 5.6 Number of residents with mesothelioma or pleural cancer (before ICD-10 was in effect) who applied for the relief payment program provided by the Kubota Corporation. *Note*: The *X*-axis represents the year of death of applicants. *Source*: Amagasaki Occupational Safety and Health Center for Worker

need to recognize the serious health risk associated with the wide areas of neighborhood asbestos exposure in the present and in the future.

Acknowledgment The present study was supported by a Grand-in-Aid from the Japanese Ministry of Education, Culture, Sports, Science, and Technology (C No, 18590567, 2006–2007) and Nara Medical University.

References

Amagasaki City (1975) Vital statistics of Amagasaki City in 1975: Amagasaki, Hyogo Prefecture, Japan (in Japanese). City Printing Office, Hyogo

Amagasaki City (2006) A report on the amount of asbestos used in the past by 136 companies in Amagasaki City, Amagasaki, Hyogo Prefecture (in Japanese). City Printing Office, Nishinomiya

Amagasaki City (2008) Report on health risk due to asbestos exposure in Amagasaki (in Japanese)

Berry M (1997) Mesothelioma incidence and community asbestos exposure. Environ Res 75:34–40

Bohlig H, Dabbert A, Dalquen P, Hain E, Heinz I (1970) Epidemiology of malignant mesothelioma in Hamburg: a preliminary report. Environ Res 3:356–372

Briggs G (1973) Diffusion estimates for small emissions. ATDL Contribution File 79:83–145

Browne K, Smither W (1983) Asbestos-related mesothelioma: factors discriminating between pleural and peritoneal sites. Br J Ind Med 40:145–152

Camus M, Siemiatycki J, Meek B (1998) No occupational exposure to chrysotile asbestos and the risk of lung cancer. N Engl J Med 338:1656–1671

Environmental Management Bureau (2000) Ministry of the Environment. Manual for regulation of total emission of nitrogen oxide (in Japanese). Environmental Research and Control Center, Tokyo

Gardner M, Saracci R (1989) Effects on health of non-occupational exposure to airborne mineral fibers. In: Bignon J, Peto J, Saracci R (eds) Non-occupational exposure to mineral fibers. International agency for research on cancer, scientific publication No. 90. International Agency for Research on Cancer, Lyon, pp. 375–397

Hansen J, De Klerk N, Eccles J, Musk A, Hobbs M (1993) Malignant mesothelioma after environmental exposure to blue asbestos. Int J Cancer 54:578–581

Hansen J, De Klerk N, Musk A, Hobbs M (1998) Environmental exposure to crocidolite and mesothelioma: exposure–response relationships. Am J Respir Crit Care Med 157:69–75

Hiraoka T, Ohkura M, Morinaga K, Kohyama N, Shimizu K, Ando M (1998) Anthophyllite exposure and endemic pleural plaques in Kumamoto, Japan. Scand J Work Environ Health 24:392–397

Hodgson J, Darnton A (2000) The quantitative risks of mesothelioma and lung cancer in relation to asbestos exposure. Ann Occup Hyg 44:565–601

Howel D, Arblaster L, Swinburne L, Schweiger M, Renvoize E, Hatton P (1997) Routes of asbestos exposure and the development of mesothelioma in an English region. Occup Environ Med 54:403–409

Japanese Society of Occupational Health (JSOH) (2000) Committee for recommendation of occupational exposure limits in Japan Society for Occupational Health. Documentation for proposal of reference values of carcinogens (in Japanese). J Occup Health 42:177–186

Kataoka A (2006) The backdrop to the asbestos disaster at Kubota's defunct Kanzaki Factory in Amagasaki City: facing a reality far worse than imagined. Res Environ Disruption (in Japanese) 35(3):49–54

Kubota Corporation (2005) Demographic data of employees with asbestos related diseases in Kanzaki Plant, a now-defunct asbestos cement plant in Amagasaki City, Hyogo (in Japanese)

Kubota Corporation (2006) Requirements for making relief payment (in Japanese). http://www.kubota.jp/new/2006/s4-17.html. Accessed 19 July 2008

Kumagai S, Kurumatani N (2009) Asbestos fiber concentration in the area surrounding a former asbestos cement plant and excess mesothelioma deaths in residents. Am J Ind Med 52:790–798
Kurumatani N, Kumagai S (2008) Mapping the risk of mesothelioma due to neighborhood asbestos exposure. Am J Respir Crit Care Med 178:624–629
Laamane A, Noro L, Raunio V (1965) Observations on atmospheric air pollution caused by asbestos. Ann N Y Acad Sci 132:240–254
Magnani C, Agudo A, González C, Andrion A, Calleja A, Chellini E, Dalmasso P, Escolar A, Hernandez S, Ivaldi C et al (2000) Multicentric study on malignant pleural mesothelioma and non-occupational exposure to asbestos. Br J Cancer 83:104–111
Magnani C, Dalmasso P, Biggeri A, Ivaldi C, Mirabelli D, Terracini B (2001) Increased risk of malignant mesothelioma of the pleura after residential or domestic exposure to asbestos: a case–control study in Casale Monferrato, Italy. Environ Health Perspect 109:915–919
Metintas S, Metintas M, Ucgun I, Oner U (2002) Malignant mesothelioma due to environmental exposure to asbestos: follow-up a Turkish Cohort living in a rural area. Chest J 122:2224–2229
Ministry of Health, Labor and Welfare (2007) Vital statistics of Japan 2005 (in Japanese). Health and Welfare Statistics Association, Tokyo
Natori Y, Ishiwata H, Sakai K, Kurumatani N, Kumagai S (2006) Asbestos bodies in lung tissues among mesothelioma cases near asbestos production factories. In: Japanese Public Health Association (ed) Abstract of 65th annual meeting of Japanese Public Health Association (in Japanese), p 1008
Newhouse M, Thompson H (1965) Mesothelioma of pleura and peritoneum following exposure to asbestos in the London Area. Br J Ind Med 22:261–269
Nicholson W (1986) Airborne asbestos health assessment update. Report 600/8-84/003F, pp 86–96
Osaka Mesothelioma Study Group (2000) A list of jobs and occupations that can be exposed to asbestos occupationally. In: Morinaga K (ed) Occupational exposure to asbestos and asbestos-related diseases (in Japanese), 1st edn. Sanshin-tosho, Tokyo, pp 364–365
Pan X, Day H, Wang W, Beckett L, Schenker M (2005) Residential proximity to naturally occurring asbestos and mesothelioma risk in California. Am J Respir Crit Care Med 172:1019–1025
Pasquill F (1961) The estimation of the dispersion of windborne material. Meteorol Mag 90:33–49
Peipins L, Lewin M, Campolucci S, Lybarger J, Miller A, Middleton D, Weis C, Spence M, Black B, Kapil V (2003) Radiographic abnormalities and exposure to asbestos-contaminated vermiculite in the community of Libby, Montana, USA. Environ Health Perspect 111:1753–1759
Rees D, Myers J, Goodman K, Fourie E, Blignaut C, Chapman R, Bachmann M (1999) Case-control study of mesothelioma in South Africa. Am J Ind Med 35:213–222
Reid A, De Klerk N, Ambrosini G, Olsen N, Pang S, Musk A (2005) The additional risk of malignant mesothelioma in former workers and residents of Wittenoom with benign pleural disease or asbestosis. Occup Environ Med 62:665–669
Reid A, Berry G, De Klerk N, Hansen J, Heyworth J, Ambrosini G, Fritschi L, Olsen N, Merler E, Musk A (2007) Age and sex differences in malignant mesothelioma after residential exposure to blue asbestos (crocidolite). Chest J 131:376–382
Turner D (1964) A diffusion model for urban area. J Appl Meteorol Climatol 3:83–91
Wagner J, Sleggs C, Marchand P (1960) Diffuse pleural mesothelioma and asbestos exposure in the North Western Cape Province. Br J Ind Med 17:260–271

Chapter 6
Asbestos Disasters and Public Policy: From the Prewar Era Through the Postwar Economic Boom*

Hiroyuki Mori

6.1 The Nexus of the Problem

The Kubota Shock of June 2005 revived political concern over the issue of asbestos-related disasters, an issue that had already been in and out of the public eye for some time. Later, it would be revealed that as many as 128 former employees of Kubota Corporation's Kanzaki factory had died of asbestos-related causes, and that over 100 residents in the surrounding neighborhood had developed mesothelioma. Until the Kubota Shock, Japan had never witnessed an asbestos disaster on such a tragic scale. This event set the stage for the February 2006 passage of the Act on Asbestos Health Damage Relief, which was put into effect the following March. The new legislation was designed to provide coverage for workers and surviving families who had hitherto been ineligible for benefits under the existing workers accident compensation insurance framework, as well as to cover the victims of environmental exposure to asbestos. As such, it was established to ensure relief for all victims of asbestos exposure without any gaps in the scope of eligibility. France is apparently the only other country which currently has a framework of relief compensation for asbestos victims that is this comprehensive. Cast in that context, recent public policy initiatives to deal with the asbestos issue in Japan may be described as nothing less than ground-breaking.[1]

However, that is not to suggest that the history of asbestos and its harmful effects on human health only recently came to light. Asbestos has been in use from the earliest days of antiquity, but if we examine the literature from the modern age forward, after industry began using asbestos on a large scale, its health hazards were already being reported in many countries by the beginning of the twentieth

---

* Translated and revised from the Japanese original, Mori, H (2008). Asbestos Disasters and Public Policy from Pre-World War II to the High Growth Era. *Policy Science* (in Japanese), 16(1), 7–30, with permission of the Policy Science Association of Ritsumeikan University.

H. Mori (✉)
College of Policy Science, Ritsumeikan University, 56-1 Tojiin Kitamachi, Kita-ku, Kyoto 603-8577, Japan
e-mail: hmt23243@sps.ritsumei.ac.jp

K. Miyamoto et al. (eds.), *Asbestos Disaster: Lessons from Japan's Experience*,
DOI 10.1007/978-4-431-53915-5_6, © Springer 2011

century, and Japan was no exception in that regard. Nevertheless, following World War II, many countries transitioned into a phase of intensive asbestos consumption, and the seriously harmful effects of asbestos on human health during that process have also been documented.

Why was it not possible to prevent the spread of asbestos-related pollution and harm even though the hazardous properties of asbestos itself were clearly known? Finding the answer to that simple question demands that we explore the social value of asbestos as a material resource, and the economic and political clout that such value wields. That approach will give us a better understanding of the economic and government relationships surrounding asbestos, and beyond that, the essential features of the public administrative background.

This chapter examines the economic and public policy-related features of asbestos, and explores the implications of the asbestos problem for economic policy, with a focus on Japan. However, given that the utilization of asbestos has a long history, even in Japan, any discussion that deals primarily with generalizations on this theme cannot be expected to have much analytical precision. Accordingly, this chapter will focus on specifics, and primarily on a case study of the Sennan district of Osaka, which was the first community in Japan to experience the negative manifestations of asbestos pollution on a large scale.

6.2 A Shadow Player Behind the Development of Japan's Modern Industry

6.2.1 The Dawn of the Asbestos Industry

In Japan, asbestos first gained status as an industrial product in the 1890s. The development and sale of "Monobe asbestos insulating material" by the Japanese physician Shoei Monobe, in 1891, marked the beginning of the age of asbestos utilization by industry in Japan. Monobe had just produced and patented an asbestos-based insulating material, effectively opening the door to the development of a new asbestos industry (NICHIAS Corporation 1996:5). Monobe opened a business establishment in Osaka and built a record of success with the support of Nobuyoshi Kawahara, the president of the Osaka Mercantile Steamship Co., Ltd., then one of Japan's leading corporate giants. Following Monobe's death, Kawahara's son set up Kawahara Shoten, a business establishment in Kobe to manufacture and market asbestos insulation products. In 1894, Mitsugu Kubo, a Kawahara Shoten employee, teamed up with Seiki Sakaeya to establish another business in Osaka: Kuei Shoten, the forerunner of the NICHAS Corp. This would be followed in April 1896 with the founding of the Nippon Asbestos Co., Ltd., a company created to assume the business operations of Kuei Shoten and satisfy the high demands of the Japanese Navy, which had ordered the establishment of a major domestic asbestos product manufacturer to facilitate the development and construction of one of the most advanced battleship fleets of that era.

In 1908, Seiki Sakaeya went on to develop a fledgling asbestos textile industry in the Osaka district of Senboku-gun, Shindachi-mura (known as the Sennan district today). This marked the debut of a domestic industry for the manufacture of asbestos yarn products. In 1912, Sakaeya established a joint-stock company (later known as Sakaeya Asbestos Spinning & Weaving Co., Ltd.). That move supported expectations of an increase in the manufactured output of gasket products that incorporated asbestos as a base material. Efforts to expand production capacity were marked, for example, by the merger of the Oi Packing Co., Ltd., an Osaka-based company established in 1903, to become the Kanae Packing Co., Ltd. in 1916. The aforementioned Nippon Asbestos also began expanding its output of gasket products while maintaining a focus on asbestos insulation as its product mainstay. As asbestos slate came into increasingly widespread use as a building material, additional firms sprang up, including the Nippon Asbestos Slate Production Co., Ltd. in 1913, the Asano Slate Co. in 1914, and the Asahi Slate Co. in 1924.

Aside from the aforementioned industrialization trends, the instrumental role played by the demands of the defense industry was another noteworthy factor that helped set the stage for the full-scale industrial utilization of asbestos in Japan. Around the close of the nineteenth century, Japanese industry had begun building its international prestige in an array of key sectors, from textiles and shipbuilding to the maritime trades and rail transport. However, the nation had also adopted policies of expanded defense spending, which had the effect of driving asbestos demand even higher. During the first Sino-Japanese war, which started in 1894, the Japanese Navy captured the Chen Yuen, a Chinese gunboat with a steel hull. Advanced battleships of this kind in the Chinese naval fleet constituted a major threat to the Japanese Navy, which lacked any steel-hulled vessels of its own. During a structural inspection, Japanese naval personnel discovered for the first time that asbestos was required as a key material for engine gaskets and insulation against heat in the engine room and other sections of the captured gunboat. This was the reason why the Japanese Navy, yearning to have asbestos harnessed for industrial applications, had called for the creation of Nippon Asbestos.

The market for asbestos products began expanding sharply after the outbreak of World War I in 1914. At Sakaeya Asbestos, trends at home and abroad, including growth in the shipbuilding and construction industry output and the expansion of the defense industry and the industrial infrastructure, fueled a sharp surge in demand from all quarters for a variety of asbestos products with applications in capital goods as an essential material for machinery and equipment. As such, asbestos would be used heavily as an insulating material for boilers, steam piping, and cooling condensers, as well as in the brake linings for a wide range of vehicles. Furthermore, as a material with general fire-proofing, heat-proofing, and insulating properties, it also enjoyed increasing demand as a material for consumer goods applications. Correlations between asbestos products and the sharp growth at this time in private-sector and military demand have been graphically illustrated in a historical document (Editing Committee of Sennan City History 1987:633).

As this illustrates, it was industrialization, fueled by both private-sector and defense industry demands, that provided the backdrop for the increasingly intensive use of asbestos in Japan.

6.2.2 The Asbestos Industry and Wartime Economic Controls

In 1932, the Japanese military seized and occupied four northeastern Chinese provinces, and established the break-away state of Manchukuo under a puppet government regime. Condemned by the international community for the creation of Manchukuo, Japan withdrew from the League of Nations in 1933 and felt compelled to tread a path of increasing isolation. Facing this state of affairs and fearing that access to the importation of materials required for its defense industry would be cut off, the Japanese government moved to accelerate resource development and extraction operations within Japan as well as in colonized areas abroad, and urged members of the asbestos industry to develop the resources to sustain their own operations.

In the rush to expand and modernize its own military arsenal, the Japanese Navy sought to equip its fleet with the high-temperature and high-voltage engines and machinery required for the development of larger and more powerful battleships. However, the quality components demanded by such machinery would have to be supplied entirely by domestic manufacturers. To that end, the government established policies to cultivate and protect new parts manufacturers. In addition, manufacturers that produced high-quality components were listed in a naval procurements register of authorized or designated suppliers and supervised by the government. This was a testament to the superior quality of products manufactured by firms in the private sector (Nippon Valqua Industries Ltd. 1977:18). Because of the military buildup in 1933 and 1934, the Japanese economy began recovering from the economic malaise that had started with the Showa Depression. Demand for insulating materials, chemical industry electrolytic filter membrane materials, and other asbestos products intensified as new factories sprang up nation-wide and existing facilities underwent expansion.

As one consequence of the second Sino-Japanese war, which broke out in 1937, the Japanese government reinforced its control over the economy and asbestos became one of the key raw materials subject to the new control regime. The Temporary Funds Adjustment Law and the Law for Temporary Measures on Imported and Exported Goods,[2] that were promulgated in 1937, and the National Mobilization Law[3] that took effect in 1938 placed the nation under a framework of total economic control. In a related action, new regulations controlling the supply of asbestos were implemented in 1942 under the material control ordinance. These controls applied to bulk asbestos materials of every category, and remained in effect until the end of World War II (Japan Asbestos Association 1950).

By this time, asbestos was already in wide use in many industries, with applications ranging from asbestos joint sheeting to packing and gasket materials. For that reason, priorities for the utilization of domestic supplies of asbestos were established through government planning.

Under the measure enacted for special wartime administrative authority in 1943, steel, coal, light alloys, shipbuilding, and aircraft manufacturing were designated the five industries with top priority. In order to concentrate capital, material inputs, and labor in these five industries, the government implemented reforms of its administrative apparatus, consolidated the Ministry of Commerce and Industry and the Board of Planning, and in November 1943, created the Ministry of Munitions.

The Ministry of Munitions sought to have the entire economy brought under government control and assembled a workforce of competent bureaucrats for that purpose. Many of the individuals who transferred from the Ministry of Commerce and Industry to the Ministry of Munitions would later be involved in the formulation of industrial policy for the Ministry of International Trade and Industry (MITI) (Johnson 1982:167–168).

At this time, the Sennan district of Osaka also was booming, with activity aimed at meeting defense industry demands. The asbestos industry thrived as more railway lines were built and shipyards expanded to handle the military buildup. After World War II reached the Pacific, local firms became swamped with orders as subcontractors for the Navy Ministry of Japan, the Ministry of Railways, and companies in the aviation industry (Editing Committee of Sennan City History 1987:635).

6.2.3 The Asbestos Industry and Postwar Reconstruction

After World War II, the Japanese government reinstated the Ministry of Commerce and Industry that had been integrated into the Ministry of Munitions during the prewar years. In 1946, the government created an Economic Stabilization Agency as the successor to the prewar Board of Planning, and presented it with a set of policies for postwar reconstruction through economic control. The initial mission of this agency had to do with the priority production system devised by the Coal Committee, a private study group reporting to then Prime Minister Shigeru Yoshida. The priority production system sought to inject all economic resources into a small minority of strategically important industries without consideration for the potential impact on inflation trends or the life of Japan's citizens. Under this system, the number-one objective was the expanded production of coal, then considered an essential for all industrial sectors. The second highest priority was placed on the increased output of steel. Additional goals of the priority production system included the increased production of chemical fertilizers that would be effective in boosting food crop yields.

The farming sector needed the chemical fertilizer ammonium sulfate to help boost food crop yields, which was essential during the postwar reconstruction phase. As a by-product of the food shortages of the war years, coupled with postwar demobilization and the return of Japanese troops to the homeland, food demand at this time was showing rapid and pronounced growth, and the increased output of chemical fertilizer was essential to the task of ensuring an adequate food supply. Japan put priority on ammonium sulfate as a fertilizer that could be manufactured with domestically available resources.[4] Nonetheless, the domestic output of ammonium sulfate in the immediate postwar years had fallen to around only 20% of the level recorded in 1941, when prewar capacity was at its peak (Ministry of International Trade and Industry 1994:187). At the end of World War II, the annual ammonium sulfate production capacity in Japan totaled 183 000 metric tons, which was less than one-tenth of the 1 887 000 metric tons of capacity the nation had before the war.[5]

Initially, the Ministry of Commerce and Industry and the Ministry of Agriculture and Forestry shared jurisdiction over the national policy on chemical fertilizer. However, in May 1947, that policy came under the sole jurisdiction of the Ministry of Commerce and Industry, which placed priority on the intensive injection of raw materials, funding, and other economic resources into expanded fertilizer production. In effect, ammonium sulfate became another priority industry that, like coal and steel, benefited from the priority allocation of material resources. Another essential for the manufacture of ammonium sulfate was the asbestos fabric used in electrolysis tanks. Asbestos fabric woven from throstle yarn produced by the Sakaeya Asbestos Co. was the first example of a domestically produced membrane filter modeled on fabric imported from Europe (NICHIAS Corporation 1996:65).

Having explored priority initiatives for the manufacture of the electrolytic membrane filters required for the production of ammonium sulfate, in 1946 the Commerce Ministry issued electrolytic membrane filter fabric production orders to members of the asbestos industry through its factory designations. Factories so designated received production quotas and delivered product shipments to designated delivery sites. Table 6.1 lists the companies designated for this purpose, and reveals that at that time all the leading asbestos manufacturers earned designated-factory status from the government.

The increasing demand for electrolytic filter fabric aggravated shortages of the bulk asbestos which was used as a raw material, and of the high-quality, long-fiber classes of asbestos in particular. As a consequence, the Asbestos Control Co., Ltd. assumed control and handled the adjustment of former military stockpiles of asbestos. In May 1947, the Industrial Reconstruction Corporation took control of these stockpiles, began serving as the agency for asbestos logistics operations, and made shipments as instructed by the Commerce Ministry.

Table 6.1 Electrolytic membrane production quotas

Contractor	Production volume (kg)	Recipient
Japan asbestos	55 000	Showa Denko, Ube Kosan
Nippon Valqua Industries, Ltd.	19 000	Showa Denko, Nippon Fertilizer
Goryo asbestos	32 000	Showa Denko, Nisshin Chemical, Nippon Chisso
Asahi asbestos	15 900	Nissan Chemical
Sakaeya asbestos	7 000	Nissan Chemical, Asahi Kasei
Toho asbestos	4 000	Nissan Chemical
Izumi asbestos	3 000	Nissan Chemical
Nippon asbestos packing	2 000	Nissan Chemical
Miyoshi asbestos	5 000	Nissan Chemical
Kanae asbestos	2 000	Showa Denko
Santai asbestos	5 000	Showa Denko
Akebono asbestos	12 000	Showa Denko, Toagosei

Note: For the July–September quarter, 1946
Source: Japan Asbestos Association (1946)

Table 6.2 Production volume for key industrial goods

Year	Coal	Pig iron	Raw steel	Cotton yarn	Raw silk	Ammonium sulfate	Cement	Paper pulp
1945	2234	97.7	196.3	2.3	0.5	24.3	117.6	22.8
1946	2252	20.3	55.7	5.8	0.6	47.0	92.9	19.5
1947	2934	34.7	95.2	12.2	0.7	72.1	123.7	25.8
1948	3479	80.8	171.5	12.4	0.9	91.7	185.9	37.6
1949	3730	154.9	311.1	15.7	1.1	118.2	327.8	49.5
1950	3933	223.2	483.9	23.8	1.1	150.2	446.2	64.8
1951	4629	312.7	650.2	33.7	1.3	159.4	654.8	91.1
1952	4375	347.4	698.8	35.3	1.5	186.0	711.8	105.4

Unit: 10000 metric tons
Source: Ministry of International Trade and Industry (1994:187)

In March 1947, the government announced a set of new bank loan accommodation rules and assigned loan access priorities for all industrial sectors. Coal and ammonium sulfate were included among the industries in the top priority category, and the asbestos product industry was assigned to the category of second highest priority. That same year brought the implementation of the central government's designated factory framework, which allocated quotas of raw materials for designated manufacturing operations, and asbestos-related factories were included in the designations (NICHIAS Corporation 1996:66). Table 6.2 shows the trends in the output of key mineral resources and industrial products from 1945 through 1952. It is clear that the volume of ammonium sulfate increased sharply during this time in keeping with the targets of the priority production system.

To deal with shortages of bulk asbestos around this time, the public and private sectors collaborated to encourage an expansion in asbestos imports. Led by key trading houses, the Association of Japanese Asbestos Importers was formed in April 1946. This was followed in November that year by the creation of the Asbestos Import Promotion Council, a consortium of the Japan Inorganic Textiles Industry Control Union, the Asbestos Cement Industry Control Union, the Japan Asbestos Association, and other related organizations. Prompted by these developments, the Japanese government repeatedly requested that GHQ should allow asbestos imports, and as a result, imports of bulk asbestos were resumed (under government trade agreements) in April 1949.

In connection with the resumption of asbestos imports, government insiders also commented on the role of asbestos in the industrial reconstruction drive. Tomizaburo Hirai, then Director of the Chemical Bureau within the Ministry of Commerce and Industry, had the following recollections. "Society has had little interest in, and a poor awareness of, the value of asbestos despite the fact that it is an extremely important resource that has traditionally been used in a wide range of applications, including electrolytic membranes, brake linings, and gasket materials. Whatever the underlying reasons for that, the asbestos industry itself has always quietly and steadfastly pursued its endeavors and we shall never forget the contributions it has made to our industrial community at large. Without access to asbestos imports after the war, it was feared that domestic stockpiles would soon be depleted. However, many small and medium-scale companies had their factories develop asbestos supplies by investing energy in the utilization of recycled products, and deserve to be recognized as cornerstones of the soda, fertilizer, and automobile industries" (Hirai 1997:30).

In addition, Yukio Kato of the Economic Stabilization Agency's Production Bureau remarked, "With each advance made by modern industrial civilization, asbestos has gained value as a key material resource, and demand for asbestos has grown dramatically in the postwar era, particularly with the drive to rebuild our war-ravaged nation." On that understanding, Kato also noted, "The trend now is toward ever more powerful machinery and devices of every kind that place their components under the strains of higher pressures and temperatures. The asbestos industry has a great future ahead. We must recognize this as a turning point, and do our utmost to pursue new innovations and advances in technology" (Kato 1997:31). Judging from these comments, government insiders, in articulating their expectations of Japan's ongoing economic development after the postwar reconstruction phase, obviously felt that asbestos products would have to be utilized on a wide scale as a key industrial resource.

In November 1946, the government enacted new procedural rules for the allocation of important industrial materials. These stipulated allocation procedures for designated industrial materials, including asbestos, in accordance with the provisional laws and statutes that provided the legal basis for the Commerce Ministry, the Economic Stabilization Agency, the Industrial Reconstruction Public Corp., and other public agencies to exercise total powers of commodity control. The objective was to ensure an equitable allocation of material resources in the best interest of Japan's economic reconstruction, and to that end, the new rules called for designated industrial materials to be allocated by designated application or product category to each industrial sector with a demand for them. Under this arrangement, industrial users of a given industrial material would not receive any allocations unless they filed prescribed applications with the relevant administrative authorities stating their needs in advance. For asbestos, the Commerce Ministry prepared allocation plans for individual industrial users on a quarterly basis. It submitted these to the Economic Stabilization Agency, and then issued purchase vouchers to approved industrial users. In 1949, the Economic Stabilization Agency and MITI adopted an order-based production formula, and on that basis asbestos products were temporarily placed under summary controls in five key industrial sectors with asbestos demand: soda ash, maritime transport, electric power, automobiles, and chemical fertilizer (Nippon Valqua Industries Ltd. 1977:170).

During the economic reconstruction phase, demand for asbestos products already spanned all major fields of industrial endeavor. Table 6.3 shows a 1946 fiscal plan for asbestos product allocations to individual industrial sectors of demand.

First, as noted earlier, electrolytic membranes for the production of ammonium sulfate accounted for a significant share of all asbestos products.

Second, by this time, virtually all leading industrial fields exhibited demand for asbestos products. As the table illustrates, asbestos-based textile products and brake linings – two categories with close ties to the Sennan district of Osaka – were being supplied to industrial users in practically every field listed.

Table 6.3 FY 1946 allocation plans by product and source of demand

	Electrolytic membranes	Textiles	Joint sheeting	Processed rubber packing	Graphite processed packing	Asbestos slate	Linings	Lumber
Ammonium sulfate	580 400	77 575	129 190	18 160	41 695	87 310	7 130	2 870
Lime nitrogen	–	29 550	22 182	5 275	2 960	45 953	–	–
Coal	–	6 000	45 000	4 700	4 000	–	54 082	–
Maritime shipping	–	23 280	53 250	26 800	16 250	75 020	–	–
Gas lighting	–	14 141	17 270	1 245	2 568	8 968	3 990	–
Petroleum	–	200	20 000	200	200	5 000	1 000	–
Soda	–	2 485	1 550	–	1 560	500	–	–
Railways	–	1 500	20 000	1 200	300	–	6 500	–
Industrial vehicles	–	8 700	10 300	600	300	17 000	–	10 000
Chemical reagents	–	4 545	20 700	1 100	4 400	10 000	–	–
Steel	–	1 500	8 000	–	–	10 000	–	–
Textiles	–	2 000	10 000	1 500	1 000	17 000	–	–
Electric power	–	6 000	15 000	2 000	3 000	–	2 500	–
Electric cabling	–	17 270	645	37	–	10 190	–	–
Wrought copper	–	2 853	480	62	346	–	–	136
Light metal rolling	–	10 217	4 611	10	–	8 464	–	10 463
Automobiles	–	–	–	–	–	30 000	91 743	–
Glass	–	300	–	–	–	80 000	–	–
Occupation forces	–	2 000	25 000	–	–	584 690	1 452	–
Other sources	–	23 196	56 350	8 981	7 148	260 459	11 025	7 627
Total	580 400	233 312	459 528	71 870	85 727	1 250 554	179 422	31 096

Unit: kg

Notes:

1. Allocation plan prepared by the Mineral Fiber Cooperative Association
2. Some totals are not consistent, but are as listed in the original

Source: Japan Asbestos Association (1947)

By this point, even the government had become well aware of the industrial importance of asbestos. Chozaburo Mizutani, then Minister of Commerce and Industry, expressed the following views, echoing the previously cited comments of other government insiders around the time that asbestos imports were resumed.

"Asbestos is an essential for the manufacture of ammonium sulfate fertilizer, which holds the key to increased food crop yields. It is also an essential material for the electrolytic membranes used by the soda industry that serves as the foundation for the development of our chemical engineering industry. On top of that, it is indispensable for the various components and engine systems used to power our automobiles, as well as the ships and trains that handle our transportation and cargo. It is utilized as a structural fireproofing material, and for the manufacture of assorted pipe products, where it is used as an alternative to steel. It is an important material found in virtually every vital component of every system powering virtually every industry in operation today. Looking further, we observe that the benefits of asbestos products will have an enormous impact on the future of our country, and that members of the asbestos industry can look forward to doing business in a vast array of fields" (Japan Asbestos Association 1948).

6.2.4 *The Asbestos Industry and the Phase of Rapid Economic Growth*

Japan entered its legendary phase of rapid economic growth in 1955. Powered by a flood of private-sector capital spending, industries, principally in the fields of oil and chemical engineering, petroleum refining, shipbuilding, steel, automobiles, consumer electronics, and electric power, eventually occupied a central position in the larger Japanese economy.

During the rapid growth phase, private capital investment in new plant and equipment nation-wide fueled a growing demand for asbestos gasket materials and asbestos slate. Similarly, expanded automobile production had the effect of boosting demand for brake lining materials. Figure 6.1 plots the trends in the volume of imported asbestos relative to real gross domestic product (GDP) during the phase of rapid growth. As this figure clearly illustrates, trends in asbestos consumption coincided closely with Japan's economic growth during this period. From trends in these two indicators alone, it is possible to draw a strong economic correlation between asbestos and economic growth in Japan.

Next, we examine concurrent trends in asbestos products[6] that were closely associated with the Sennan district of Osaka. Figure 6.2 plots the nationwide trend in the manufactured value of asbestos products. As expected, if the focus is limited solely to asbestos, the growth in manufactured output is closely correlated with the nation's rapid growth phase. Table 6.4 elaborates on this trend for specific asbestos products. As can be seen, the manufactured output of asbestos products widened approximately 1.5-fold from 1960 to 1965, highlighting the strong growth in demand. By product category, asbestos insulation and brake linings demonstrated

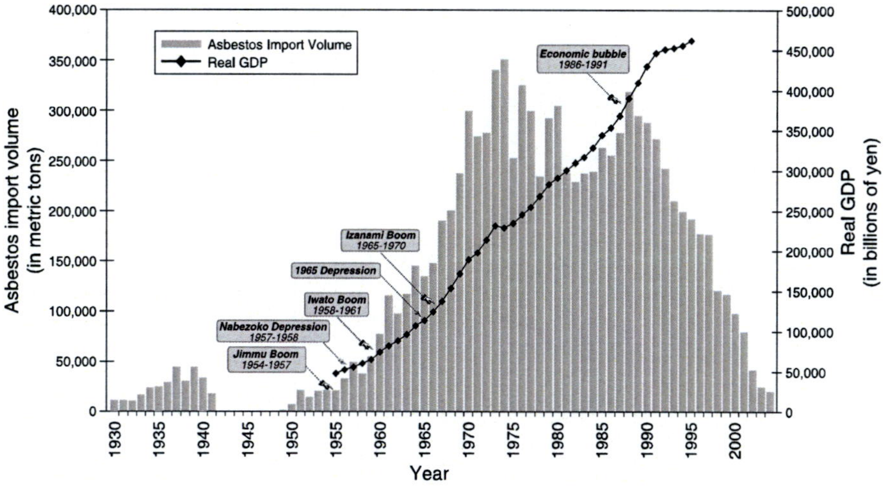

Fig. 6.1 Trends in Japanese asbestos imports and real GDP. *Source*: Compiled from data in the Ministry of Finance, *Trade Statistics,* the Economic Planning Agency, *The Postwar Japanese Economic History: The Economic Planning Agency's First Fifty Years*, and other documents

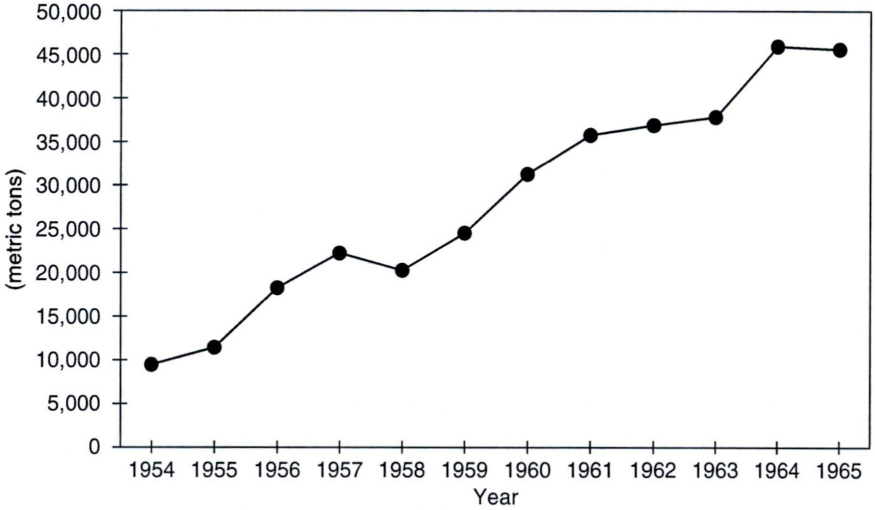

Fig. 6.2 Trends in manufactured output of asbestos products. *Source*: Research Institute in Commerce, Industry, and Economy of Osaka Prefecture (1967:4)

substantial year-on-year gains, with large totals for manufactured output as well. This indicates that the products in these categories effectively helped to sustain levels of corporate capital investment, the ongoing development of the nation's automotive industry, and other facets of Japan's economic structure during the rapid growth phase. Although asbestos lumber output showed a significant downward trend, it is reasonable to conclude that the levels of manufactured output in the other categories were generally sustained. The "other" category includes an assortment

Table 6.4 Trends in manufactured output of assorted Asbestos products

	1960	1961	1962	1963	1964	1965
Asbestos yarn and fabric	3198 (100.0)	3502 (109.5)	3326 (104.0)	2825 (88.3)	3326 (104.0)	2743 (85.8)
Joint sheeting	3869 (100.0)	4379 (113.2)	4027 (104.1)	4555 (117.7)	5259 (135.9)	5027 (129.9)
Asbestos slate	4823 (100.0)	4866 (100.9)	4685 (97.1)	4250 (88.1)	4884 (101.3)	4347 (90.1)
Asbestos paper	378 (100.0)	439 (116.1)	362 (95.8)	339 (89.7)	368 (97.4)	299 (79.1)
Brake linings	4934 (100.0)	6679 (135.4)	6604 (133.8)	7272 (147.4)	9566 (193.9)	8896 (180.3)
Lumber	1800 (100.0)	1804 (100.2)	1594 (88.6)	1127 (62.6)	1534 (85.2)	905 (50.3)
Asbestos insulation	6764 (100.0)	7686 (113.6)	9494 (140.4)	8949 (132.3)	10305 (152.4)	11120 (164.4)
Other products	5412 (100.0)	6380 (117.9)	6728 (124.3)	8578 (158.5)	10656 (196.9)	12294 (227.2)
Total	31178 (100.0)	35736 (114.6)	36820 (118.1)	37893 (121.5)	45900 (147.2)	45633 (146.4)

Unit: metric tons, %

Source: Research Institute in Commerce, Industry, and Economy of Osaka Prefecture (1967:5)

of specialized asbestos products, underscoring a diversification trend in the types of goods aimed at satisfying specialized industrial demand (Research Institute in Commerce, Industry, and Economy of Osaka Prefecture 1967:6). These data portray a situation in which asbestos products supported a broad spectrum of those industrial operations that developed and expanded during the boom years.

It is worth investigating which industrial sectors utilized these asbestos products during the rapid growth era. Table 6.5 lists shipped quantities for a variety of asbestos products in 1956. A perusal of shipment totals by their industrial sector destination shows that many asbestos products were in use, primarily by transport, machinery, and the heavy and chemical industries that served as key engines of growth during the nation's rapid economic expansion phase. Maritime shipping accounts for the largest volume of asbestos products shipped, followed in descending order by automobiles, chemical engineering, electric power, machinery, steel, oil, and land transport. Furthermore, it is clear from the data that many of the textile products produced in the Sennan district were, as might be expected, being utilized by practically all of the industrial sectors that led the economy forward during the boom years: maritime shipping, chemical engineering, land transportation, steel, electric power, machinery, automobiles, oil, and so forth. In effect, this is confirmation of a postwar economic structure in which key industries powering the nation's rapid growth were supported by the asbestos product industry.

We now look more closely at the relationship between asbestos and maritime shipping and other sectors of the transportation industry that have utilized asbestos products in the largest quantities. At the time, shipbuilding was one of the asbestos industry's largest clients, supported by the demand for maritime shipping services. In particular, oil, coal, cereals, and other raw materials required by Japanese industry were transported by tankers and large freighters, and hence demand for these commodities in turn required an increase in shipping capacity. The railroad sector also constituted a large source of demand for the asbestos industry. A wide range of asbestos products were utilized by the rail companies, from insulation and gasket materials to heat-insulating quilting, brake linings, and asbestos slate. As in the maritime shipping industry, the freight handling capacity of the railways had not kept pace with the growth in the volume of freight that needed to be moved during the boom years.

These data illustrate several points: namely, that rapid economic growth and urbanization trends had intensified the demand for transport, and accordingly heightened the necessity of expanding the capacity of modes of transport to support that demand, thus highlighting the existence of an inter-industrial linkage marked by a growth in the demand for asbestos products.

Table 6.6 shows the trends in the relative weighting of selected industrial consumers of asbestos products. On examination, it is clear that during the phase of rapid economic growth, asbestos consumption in the automotive and machinery sectors increased rapidly, while leveling off in the electric power, oil, and maritime shipping sectors.

Reflecting the fact that asbestos products had supported the operations of a diverse range of industries, as shown in Table 6.6, asbestos had grown into an important structural foundation for the entire industrial base. In short, asbestos had

Table 6.5 Shipments of asbestos products (1956)

	Electrolytic fabric	Textiles	Joint sheet	Rubber processed	Asbestos slate	Asbestos paper	Graphite processed	Woven	Resin molding	Rubber molding	Special processed	Lumber	Other	Total
Exports	1.0	105.0	7.4	4.2	24.6	0.0	2.9	9.3	4.2	12.8	0.0	0.0	41.1	212.5
Special procurements	0.0	4.0	0.1	0.0	0.9	1.3	0.2	0.0	–	31.0	0.0	0.0	287.5	335.0
Chemical engineering	61.9	223.8	353.6	13.8	387.1	11.1	39.8	10.2	–	0.0	7.8	135.1	637.0	1881.2
Electric power	0.7	113.0	101.7	27.3	72.2	6.3	14.0	2.8	–	0.1	0.8	555.8	563.1	1477.8
Gas	0.1	30.6	85.6	2.1	18.5	0.8	7.8	6.2	–	0.0	0.1	0.0	88.5	240.3
Mining	–	40.5	54.3	4.4	50.5	2.2	13.2	38.6	–	1.0	1.3	3.1	215.9	424.2
Petroleum	–	46.2	285.3	8.6	95.1	–	10.0	0.9	–	0.0	0.4	0.4	336.6	783.4
Land transport	–	206.0	86.0	13.7	57.1	–	10.2	14.9	–	13.2	6.2	191.8	163.3	762.4
Maritime transport	0.6	701.4	281.8	24.4	388.5	–	15.9	11.9	6.7	46.6	24.6	82.6	1824.0	3419.0
Automobiles	–	91.8	33.5	2.7	338.9	–	6.9	126.9	192.4	1056.2	369.8	0.0	660.0	2285.1
Machinery	2.4	94.0	44.0	10.5	94.7	15.8	23.3	71.4	23.3	98.5	155.8	75.3	298.5	1277.5
Food	–	6.1	27.2	1.4	10.6	–	4.3	3.7	1.1	6.8	8.8	1.0	10.4	81.4
Textiles	0.6	23.0	87.7	3.7	21.1	–	15.8	1.5	–	0.0	0.1	0.0	308.0	521.5
Iron ore	–	119.3	78.6	7.3	128.3	46.3	11.3	20.0	–	3.6	26.9	139.7	295.5	876.8
Nonferrous metals	–	18.6	33.8	2.4	30.5	1.0	6.0	1.4	–	2.9	4.7	2.1	50.7	154.1
Other	0.5	383.2	288.9	14.5	803.4	5.3	17.3	11.2	1.8	1.8	21.9	755.1	594.9	2899.8
Total	67.8	2206.5	2029.5	140.0	2642.0	100.1	198.9	330.9	229.5	1273.6	629.2	1942.0	5841.0	17632.0

Unit: metric tons
Note: Some totals are not consistent, but are as listed in the original source
Source: Japan Asbestos Association (1957, 1958)

Table 6.6 Trends in asbestos product shipments, by source of demand

Year		1956	1957	1958	1959	1960	1961	1962	1963	1964	1965	1966	1967	1968	1969	1970
Chemicals	Fertilizer	785	925	569	871	789	743	723	597	527	519	509	830	855	805	953
	Other	1 096	1 243	1 105	1 406	1 806	2 214	1 835	1 564	2 033	2 018	1 735	2 730	2 682	2 967	5 415
Power	Electric	1 478	2 090	2 661	3 226	4 143	4 523	5 108	4 944	3 301	2 615	3 265	4 034	2 725	4 626	5 519
	Gas	240	218	286	380	574	464	352	602	556	431	391	585	419	647	1 012
Petroleum		783	1 179	1 273	1 389	1 838	2 986	2 000	1 952	1 928	1 688	1 259	1 362	1 799	2 894	3 806
Transport	Land	762	947	584	857	992	1 374	1 361	777	821	599	734	789	779	605	649
	Maritime	3 419	3 830	2 704	2 882	3 205	2 823	2 732	2 923	4 271	4 182	3 620	3 471	4 086	3 978	2 803
Automobiles		2 285	2 672	2 805	3 919	5 652	6 987	7 086	8 286	9 969	9 718	10 787	14 919	16 998	18 612	22 217
Machinery		1 278	1 401	1 275	1 744	2 383	3 192	3 292	3 951	4 014	3 558	3 770	4 609	5 862	7 064	10 287
Alloys	Steel	877	1 243	1 008	1 545	1 932	2 044	1 675	1 418	1 684	1 775	1 335	2 693	2 363	2 680	4 183
Combined totals for other sectors		17 632	21 342	19 808	23 977	30 250	34 902	35 699	37 410	42 459	40 690	42 598	54 582	56 677	64 998	81 946

Unit: metric tons

Note: Data were extracted for key categories only

Source: Japan Asbestos Association, *Asbestos*, various issues, and Asbestos Industry Association, *Industry Association News*, various issues become an essential for any modern economy as a vital industrial material with a wide array of applications in fields ranging from machinery and electric power to construction and chemicals.

6.3 Policy Conflicts Within the Government

Economic growth was the biggest priority of government policy in the postwar years. The government pursued this public policy bias by giving top priority to growth and lower priority to, for example, the environment and labor.

MITI accepted that its role was to act as a locomotive force for the postwar Japanese economy, and as such, implemented certain industrial policies that were considered unusual by international standards. Economic growth was MITI's biggest goal. Suffice it to say that MITI was driven by a doctrine of "growth first," or growth above all else.[7] Because it had popular support as well as the support of the Liberal Democratic Party, which had considerable political power, MITI, as an agency of government bureaucracy, pursued policies that were aimed at carrying through its objectives.

Economy-centric policy-making in the postwar years delayed policy measures that were deemed to pose a risk to economic growth. Policies for workplace safety and environmental protection were typical examples.

The prime reason why corporate operations were accompanied by workplace accidents and environmental pollution was because investment in preventive measures was at odds with the profit motive. In other words, unlike investment in production-related activities, not only did investment in measures to prevent workplace accidents or environmental pollution not translate directly into profits, the cost of implementing them would have to be passed on in product prices, thus undermining the market competitiveness of the company concerned.[8] Because this logic applied to industry as a whole, the idea of boosting the price competitiveness of an industry demanded that investment in workplace safety and environmental protection be minimized, thus helping to keep production costs as low as possible.

As illustrated in the earlier study of specific government ministries and agencies cited above, MITI, having embraced a growth-first stance as being in the "national interest," has had a tendency to be antagonistic to the Ministry of Labor, the Ministry of Health and Welfare, and the Environmental Agency, which all consider workplace safety and environmental protection to be their organizational objectives. Policies that emerged out of these inter-agency conflicts had to be harmonized with the goal of economic growth. As a consequence, policies established for workplace safety and environmental protection have always been inadequate. Furthermore, the government has not shown a strong commitment to actively pursuing remedies in cases of corporate noncompliance with established laws.

For example, in the context of amending or revising the Industrial Health and Safety Law of 1972 or other regulations for worker safety and health, Takeshi Fujimoto points out that the government has consistently failed to keep these laws up to date with innovations in corporate technology, or to fully enforce their provisions despite a pattern of noncompliance among many business owners (Fujimoto 1992:1346). This underscores a tendency for administrative policies on workplace safety to be technically or conditionally outdated relative to trends in corporate manufacturing. Furthermore, Fujimoto notes that large corporations, in their quest to lower costs, are in the practice of using subcontracting and outsourcing arrangements to exploit poor working conditions in smaller companies and subcontracting businesses where the incidence of workplace accidents is relatively high, and that this state of affairs is the product of the government failure to enforce safety regulations at these smaller facilities (Fujimoto 1970:194).

On the subject of environmental protection, since 1955 efforts had been made at the government level to prepare and enact legislation against environmental pollution (the legislative bill for environmental pollution prevention standards). However, resistance from the business community and MITI consistently blocked the passage of such legislation. In 1967, the government did establish the Basic Law for Environmental Pollution Control after the realization that community development at that time required measures against pollution. However, pursuing a balance between industrial growth and environmental protection was still the shared ideal behind this and other environmentally related pieces of legislation up to that time. In other words, a certain level of environmental protection was acceptable provided it was limited to a scope that guaranteed industry profits (Miyamoto 1980:563). In fact, the first article of the Basic Law for Environmental Pollution Control contains provisions stipulating that measures against pollution be harmonized or in balance with healthy economic development. This incorporation of an emphasis on "harmonization" rendered measures against pollution totally ineffective in practice. This was the outcome of changes to the final version of the legislation in the Diet after MITI requested revisions to the original draft introduced by the Ministry of Health and Welfare (Johnson 1982:284). In reference to the harmonization philosophy incorporated into this law, Kenichi Miyamoto has expressed the view that "ensuring that private corporations enjoy healthy growth, that is, profits, is at odds with the goal of protecting the environment. Protecting the environment only to an extent that allows corporations to continue reaping profits is the real motive behind this idea of stipulating harmonization or balance as the objective." Miyamoto is also of the opinion that the spirit of this notion of harmonization remains in the revised Basic Law for Environmental Pollution Control (1970), even though the text of the harmonization-related proviso has been taken out (Miyamoto 1975:110–111). Furthermore, he feels that delays in the introduction of asbestos regulations as safeguards against environmental and workplace disasters is a clear sign that the bureaucracy's loyalties were with the private corporate sector (Miyamoto 2007:284).

Owing to the precedence of economic rationales, workplace safeguards against asbestos disasters were also slow in coming. In 1958, the Director of the Labor Standards Bureau within the Ministry of Labor issued a notice pertaining to the promotion of improvements to occupational environments as a preventive measure against occupational diseases. That notice contained a set of technology guidelines for the prevention of occupational illnesses in the workplace. The text of these guidelines noted that the risk of contracting an occupational illness had shown a marked increase in parallel with the conspicuous advances that the metal industry, the machinery and tool manufacturing industries, the chemical industry, and the civil engineering and construction industries had reported in recent years. On that understanding, the guidelines incorporated provisions for various asbestos-related operations (excavation, pulverizing, sorting, loading, shipping, unbundling, mixing, fabric handling, cutting, and grinding processes) to be included in the 16 hazardous occupations deemed highly important (the 16 occupational routines which were eligible for special health examinations) among the many causes of occupational diseases. The guidelines put an emphasis on harmonization with the economy, noting that "special health examinations of workers with a diagnosis of silicosis, and workers engaged in the 16 designated hazardous occupations, had a detection rate of approximately 12% in 1955 and 1956, and, for other anomalies, 11% in 1956 and 12% in 1957…. It is presumed that an extremely large number of examinations detected the warning signs of occupational diseases…. Occupational environments must be improved by minimizing exposure to hazardous materials as much as possible…. However, from both a technical and economic standpoint, it is difficult to expect that ideal to be achieved in the short term." Furthermore, as explained in the guidelines, "the target limits are provisional targets of effectiveness based on improved measures to curb the emission of hazardous materials at their source." In other words, established concentration limits were nothing more than targets that companies would be expected to make an effort to achieve. Under these guidelines, the target concentration limit for asbestos was 1 000 fibers/cc or 20 mg/m^3. These numbers were extremely relaxed compared with the maximum permissible concentration of 2 mg/m^3 recommended by the Japan Society for Occupational Health in 1965. This was a classic example of the kind of "harmonization" sought: namely, allowing preventive measures against workplace accidents or disasters to be limited to a level that did not impede corporate business operations.

To summarize these points, the emphasis on "harmonization" rendered postwar policies on workplace safety and environmental protection totally ineffective. Suffice it to say that they were implemented within the context of a social structure where most politicians, social groups, and the bureaucracy equated economic growth with the national interest. Furthermore, the economic management policies of MITI – an agency that from the prewar years had been fully aware of the economic importance of asbestos as a material that had supported the "national interest" on a fundamental level – were unquestionably a more powerful force than the social policies of the Ministry of Health and Welfare or the Ministry of Labor. As a consequence, of all the policies under its management, the government had become more sophisticated in administering those with an economic-growth orientation.

6.4 The Sennan District as a Center of the Japanese Asbestos Industry

Thus far in this chapter, the discussion has been shaped by a macro-economic focus on relationships between the asbestos industry and government policy-making from the prewar era through the phase of rapid economic growth. In this section, the fundamental premises concerning the social structure relating to asbestos will apply, but the focus will shift to an analysis of asbestos-related issues provoked by the Japanese government's policy response. The case study for this analysis will be the district of Sennan, an area that from the prewar years had evolved into the largest concentration of asbestos industry operations in postwar Japan.

6.4.1 Asbestos Textile Products: Their Purpose and Role

Sennan's asbestos industry was primarily involved in the manufacture of asbestos yarn and fabric. The processes utilized for the manufacture of these products were basically the same as for other textiles. The most common processing stages are listed in Table 6.7.

The key processes for the manufacture of asbestos textiles include the following stages: blending (crushing the crude asbestos and opening or bagging the cotton bales to form fibers), carding (assembling the fibers into a bundle of loose strands), roving (twisting together fiber strands into thread), and finishing (weaving into fabric, spinning into yarn). Asbestos fabric and yarn products manufactured with these primary processes were utilized directly in gasket packing materials and electrical insulators, and as a processed raw material for asbestos gaskets, asbestos quilting, brake linings, and electrolytic filter membranes. Asbestos yarn, fabric, gaskets, and quilting are essential products for use in locations that demand resistance to heat and chemicals (specifically, in thermal electric power stations, steam locomotive engines, steel mills, petrochemical complexes, and an assortment of factory settings). Brake linings are essential for automobiles and heavy machinery. Electrolytic membranes, moreover, are used in the production of ammonium sulfate fertilizer, as discussed above.

Until around 1970, a time when the extremely hazardous properties of asbestos were still not that widely known, many industries undoubtedly considered these

Table 6.7 Asbestos textile manufacturing processes

Process	Operational content
Blending	(1) Opening of asbestos fibers
	(2) Opening of cotton fibers
	(3) Blending of asbestos and cotton fibers
	(4) Bagging
	(5) Selection of asbestos fibers (ancillary step)
Carding	(1) Asbestos blend opened and turned into a uniform web
	(2) Carding machine needle preparation (ancillary step)
Roving	(1) Spinning of single asbestos strands
	(2) Spinning of multiple asbestos strands
Finishing	(1) Fabric weaving
	(2) Packing cord
	(3) Asbestos cord
	(4) Graphite-coated asbestos cord, packing cord

Source: Hyogo College of Medicine (1981:142–150)

asbestos products to be essential components, in view of their usefulness and relatively low cost. The Sennan district of Osaka (then covering the areas currently known as Sennan and Hannan City) was the principal manufacturing center for asbestos yarns and fabrics in Japan. Asbestos yarns and fabrics produced in Sennan were supplied mainly to Kubota, NICHIAS, Yamaha, and Honda, all of which typically processed these materials into packing materials, gaskets, and brake linings.[9]

Compiled in 1984, *Local Senshu Industries* discusses asbestos products as one of the principle industrial commodities of this district. They are described as follows. "Asbestos products are used in many fields as a fire-resistant fabric, insulating filler, and friction material, thus exploiting the heat-resistant and insulating properties of asbestos. However, in most cases, they are not utilized as discrete, stand-alone products, but are integrated into other products as components in locations that are hidden from external view. Nonetheless, they are essential components and fulfill a vital functional role" (Nankaido Research Institute 1984:80). In effect, as with asbestos products in general, those from the Sennan district are presented as fulfilling an indispensable, behind-the-scenes role for industrial operations in virtually every field of endeavor.

6.4.2 The Prestige of Sennan's Asbestos Textile Factories

Although Sennan was only one of many districts in Osaka Prefecture, it was a district where many asbestos textile factories were concentrated. The Japan Asbestos Product Manufacturers Union was an asbestos industry organization established prior to World War II. As entries in its membership register (around 1943) show, 19 of the union's 44 members in the Osaka area were based in the Sennan district. Figure 6.3 plots postwar trends in the number of asbestos textile factories in the Osaka area by district, using the factory data entered in rosters of factory names and other documentation. This figure highlights a heavy concentration of asbestos textile factories in the Sennan district throughout the postwar period. Although the data include factories involved in the manufacture of joint sheeting and brake linings, asbestos fabric and yarn were the principal product lines for many of the factories located in the Sennan district. The municipalities of Sennan and Hannan together account for less than 5% of Osaka Prefecture's total surface area. That statistic provides another insight into just how heavily concentrated the facilities of the local asbestos textile industry had become within the Sennan district.

Moreover, these statistics are only for factories with four or more employees up to 1968, and ten or more employees from 1970 onward, indicating that only the larger facilities are accounted for in the totals. Naturally, many other factories existed, but were not included in the count. Based on interviews with personnel in the Environmental Management Center, Research Institute of Environment, Agriculture, and Fisheries, Osaka Prefectural Government, and data collected from assorted documentation, phone books, and other records, as many as 330 asbestos-related factories were in operation within Osaka Prefecture at one time. Of that total, around 100 were located in Osaka City and 100 in the Sennan district.[10] Also, according to Kazuyoshi Yuoka, during the boom years of the 1960s and 1970s,

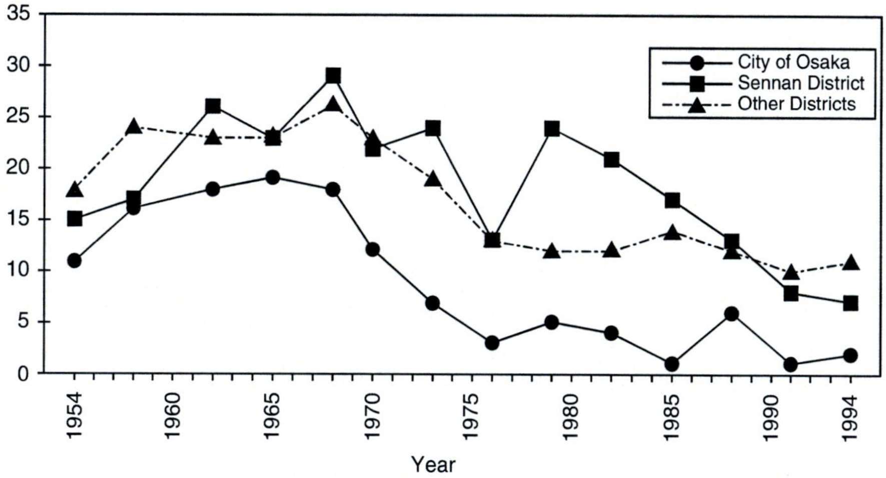

Fig. 6.3 Trends in the number of asbestos textile factories in Osaka Prefecture, by district. *Notes*: (1) The reason for the temporary 1976 dip in the number of factories in the Sennan district is because the collection of data on many small-scale businesses was temporarily halted. (2) Operations with four or more workers were counted up to 1968. From 1970 onward, only operations listed in the *Handbook of Osaka Prefecture Factories* with ten workers or more were counted. *Source*: Data for 1954 are from the *General Directory of Osaka Prefecture Industries*, compiled by the Osaka Prefecture, Osaka Office of Commerce and Industry, Labor Standards Bureau. For 1958 through 1968, the data are from the Osaka Prefectural Government, *Osaka Prefecture Factory Register*, various years. For 1970 and later years, the data are from the Handbook of Osaka Prefecture Factories, various years (compiled by Shinjiro Minami)

the Sennan district had around 60-odd companies with integrated asbestos processing factories. However, if smaller-scale subcontractors and cottage-industry-type businesses are also included, the total comes to 200 or more (Yuoka 2006:22). For the Sennan district, the graph in Fig. 6.3 shows a total of only 29 companies with asbestos textile facilities in 1968, which was the peak year. This number is far smaller than the total for Osaka Prefecture as a whole, or the total cited by Yuoka.

The history of Sennan's asbestos textiles industry extends all the way back to 1912. The industry's founder, Seiki Sakaeya, selected Sennan because it was a booming center of the textile spinning and weaving industries. It started out as a market for the agricultural production and sale of cotton as a cash crop around the middle of the Edo period, and thrived from the Meiji era onward with expanded imports of cotton and the introduction of high-efficiency looms. Additional factors included: (1) access to waterwheel-based hydropower thanks to the development of an extensive irrigation network for the local rice crop production; (2) ready access to a labor pool of workers traditionally engaged in spinning yarn from spinning wheels; (3) access to supplies of inexpensive waste cotton from other textile business operations of various kinds; and (4) a location to which shipments of crude asbestos from dock facilities in Sakai could be delivered within 1 day (Yuoka 2006:21). Sakaeya launched his asbestos textiles business by harnessing these competitive advantages of the Sennan district, and from that point forward, Sennan's concentration of asbestos-related factory operations grew steadily.

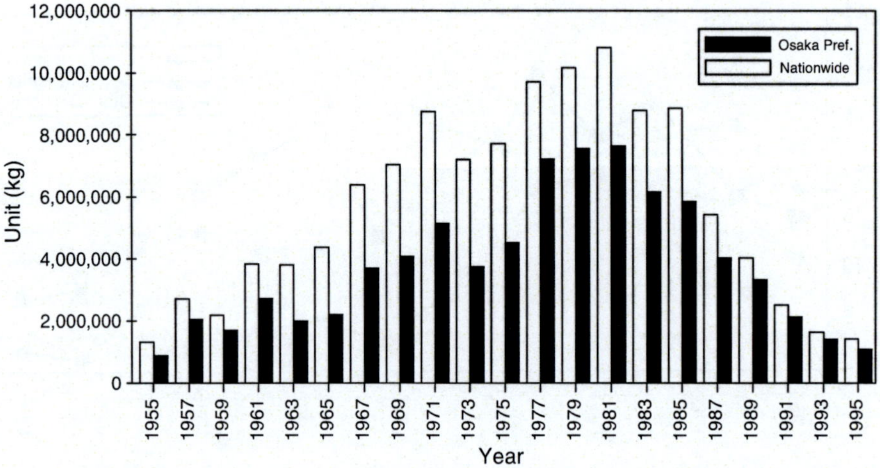

Fig. 6.4 Trends in shipped amounts of asbestos yarn and fabric in Osaka Prefecture and nation-wide. *Note*: *Data for years later than 1995 are not listed*. *Source*: Ministry of International Trade and Industry, *Tables of Industrial Statistics* (*by Product Category*), various years (prepared by Shinjiro Minami)

Table 6.8 Profile of asbestos product manufacturers in Osaka Prefecture (1965)

	Osaka Prefecture	Nation-wide	Osaka share of national total
No. of businesses	66	182	36.3
No. of workers	1 798	8 105	22.2
Value of shipped goods	3 645 million yen	18 280 million yen	19.9
Value-added	1 269 million yen	7 104 million yen	17.9
Workers per business	27.2	44.5	61.6
Value of shipped goods per business	55.2 million yen	100.4 million yen	55.0
Value of shipped goods per worker	2 027 000 yen	2 255 000 yen	89.9
Value-added per worker	706 000 yen	876 000 yen	80.6

Note: Nation-wide data are for 1964
Source: Research Institute in Commerce, Industry, and Economy of Osaka Prefecture (1967:15)

Figure 6.4 shows the trends in the manufactured output of asbestos yarn and fabric both nation-wide and for Osaka Prefecture alone. Osaka has accounted for an overwhelming share of asbestos yarn and fabric output, supplying half the national total even in its weakest year. This effectively illustrates the exceptionally strong position that manufacturers in the Osaka area commanded in the asbestos yarn and fabric industry. That Sennan was the actual center of production becomes obvious if the data are viewed together with a distribution chart of factory locations.

Small-scale operations can be cited as one of the features that asbestos product manufacturers in the Sennan area generally had in common. Table 6.8 is a comparison of selected statistics on asbestos product manufacturers in Osaka Prefecture and nation-wide in 1965. As the comparison shows, the total number of manufacturers in Osaka Prefecture accounted for over 36% of the national total. By contrast, they accounted for only 22.2% of the nation-wide workforce, 20% of nation-wide shipments by value, and 18% of the nation-wide value-added amount. The average number of employees per workplace for Osaka Prefecture manufacturers was only 62% of the national average, and was similarly only 55% of the national average for the value of shipments per workplace. However, this did not mean that all manufacturers in Osaka Prefecture were small-scale operations; rather, it was defined by the heavy concentration of manufacturers in the Sennan district. To put it in perspective, manufacturers in the Osaka area led the nation in terms of key asbestos product output,[11] and included a significant number of factory operations with work forces of 100 employees or more. The Osaka Survey Office explained this as follows: "The reason for the smaller than average scale for Osaka is that the Sennan district has a cluster of operations manufacturing asbestos yarn and fabric. Although they supply their products to clients within a radius covering not only Osaka Prefecture but the entire Kinki region, most are rather small enterprises" (Research Institute in Commerce, Industry, and Economy of Osaka Prefecture 1967:16).

According to K. Yuoka, asbestos product manufacturers in the Sennan district tended to be surprisingly small production units; even the largest comprised no more than 40 employees, 4–5 operations had around 20–30 employees each, and the rest had only 12–15 employees. Subcontractors and home-based shops had even fewer employees. Only a few companies had integrated facilities for the production of yarn, fabric, cord (packing), and ribbon from crude asbestos. The majority handled operations up to the spinning process, fabric production was outsourced to small-scale weaving businesses, and asbestos packing and ribbon production were dependent on small sideline businesses managed by local farmers. These fabric weaving and other small sideline businesses engaged only a few workers, and many were family operations handled by parents, siblings, children, spouses, and so forth. In other words, the asbestos product industry in the Sennan district consisted mostly of small-scale businesses directly run by their owners and other family members, and dependent on other small cottage-industry operations nearby (Yuoka 2006:22).

Taken together, these points confirm that the asbestos product industry in the Sennan district was a vast concentration of small-scale factories and shops that as a group boasted an overwhelming share of the national market for asbestos yarn and fabric products finished through primary stage processing.

The larger asbestos industry was fully aware of the importance of the asbestos products manufactured by this concentration of businesses in the Sennan district. In the February 15, 1964, issue of *Asbestos*, the Japan Asbestos Association had the following to say about the upward trend in prices for asbestos textile products sourced in Sennan: "Although actual details of the situation remain unclear because they have not yet been organized, the small factory operations in this district together form a major force within the asbestos industry, and trends in that force inevitably have a major influence on industry affairs, directly or indirectly. In particular, considering the extent of their involvement as suppliers and subcontractors to the larger manufacturers in this industry, their current inclination to raise prices is not something that can be lightly ignored" (Japan Asbestos Association 1954a). These remarks illustrate the position of importance that asbestos product businesses in the Sennan area had earned within the asbestos industry as subcontractors and suppliers to the larger manufacturers. Another noteworthy point is the concern that the asbestos textile product price increases implemented at that time by businesses in the Sennan area would have a ripple effect on the asbestos product industry as a whole.

6.4.3 Asbestos Product Manufacturing Processes in Sennan and Public Regulations

6.4.3.1 Study by the Insurance Institute, the Department of the Interior (The Sukegawa Study)

A study carried out by the Ministry of Home Affairs' Insurance Institute from 1937 through 1940 showed that the incidence of pulmonary asbestosis had reached serious levels among employees of asbestos product factories in the Sennan district. Led by Dr. Hiroshi Sukegawa of the Health Insurance Counseling Center, which was affiliated with the Insurance Institute's Social Insurance Bureau, this study (hereafter called the "Sukegawa Study") comprised health examinations of 650 workers employed at 14 manufacturing facilities (11 in the Sennan district, 2 in Osaka City, and 1 in Nara Prefecture) engaged in the manufacture of a variety of asbestos products, including yarn, fabric, packing, insulation material, asbestos sheet, and brake linings. As such, it was far larger in scale even than the world's first large asbestosis study, carried out by Edward Merewether and Charles Price in the UK.[12] Of the subjects examined, 80 were determined to have contracted pulmonary asbestosis, giving an incidence rate of 12.3%. The incidence increased in proportion to length of service, from 1.9% for periods of less than 3 years, to 20.8% for 3–5 years, 25.5% for 5–10 years, 60.0% for 10–15 years, 83.3% for 15–20 years, and 100% for lengths of service of more than 20 years. When the different types of work were considered, the highest incidence rate, 30.2%, was detected in workers engaged in blending operations. All the operations in increasing order of incidence were spinning, carding, roving, and finishing, with highly dangerous rates of incidence particularly for workers engaged in the carding and blending operations. By age group, the incidence of asbestosis was highest among those in the 35–39 years age group, and the second highest was among those aged 50 years and over. This is assumed to be attributable to a selection phenomenon contingent on the incidence rate. Of 231 subjects with a length of service of 3 years or more, the study showed that 150 had symptoms in some form or other.

In the light of these findings, the Sukegawa Study recommended that factories should make their employees wear dust masks, implement other steps in workplace management, and install facility-wide dust collection systems. The study also noted that the findings suggested that grave pathological changes are caused by the inhalation of asbestos dust, as pointed out in the study by Merewether and Price.

According to Kenji Morinaga, the Sukegawa Study had already identified a bilateral accessory murmur and asbestos bodies in expectorate as features of the physical findings from chest examinations. The study also showed that nodule shadows were not present even on X-ray radiographic images of pulmonary asbestosis patients with long-term exposure, in contrast to the X-ray findings for patients with pulmonary silicosis. In addition, it clearly showed that most of the clinical findings for asbestosis patients in recent years were already well understood at the time the study was performed (Morinaga 1989:686). In effect, the Sukegawa Study not only clarified the relationships between pulmonary asbestosis and asbestos product factories, but it also excelled in the medical diagnoses of cases of pulmonary asbestosis.

Nonetheless, the findings of the Sukegawa Study did not lead to any new government regulations for the asbestos industry. This stands in stark contrast to the outcome of the Merewether and Price study published in 1930, which prompted the British government to establish asbestos industry regulations and mandatory dust controls in 1931.

6.4.3.2 Pulmonary Asbestosis Studies After World War II

Two participants in the Sukegawa Study, Zenji Horai of Nara Medical University and Yoshisumi Sera of Osaka University, renewed their investigation of the incidence of pulmonary asbestosis in asbestos industry workers in 1952 (Hyogo College of Medicine 1981). The results of examinations performed for that study found symptoms of asbestosis in 10 of 203 (5.0%) asbestos workers. Although inadequate diagnostic methods, conservative interpretations of the X-ray findings, and the participation of many workers with short lengths of service in the industry have been cited as explanatory factors behind that outcome, it has been noted that the detection even of this relatively low number of asbestosis cases is still a matter of importance from the perspective of worker health management in asbestos factories.[13] In 1953 and following years, Horai et al. continued their study of asbestosis in the Nara and Osaka districts. In addition, Shoji Yoshimi et al. launched a long-term asbestosis study in Tokyo in 1953, and reported their findings over a period of several years.

Another noteworthy development was the launch of two full-scale Ministry of Labor studies on diagnostic criteria for asbestosis, in 1956 and 1957. For these studies, Horai served as chief researcher, heading a research team that included Sera, Yoshimi, and other investigators who had been involved in asbestosis research up to that time (Hyogo College of Medicine 1981:14–24). The findings of a concurrent study of asbestos factories in Osaka identified symptoms of asbestosis in 54 out of 330 subjects (16.4%) and suspected symptoms in another 36 (10.9%), and also confirmed the appearance of signs of asbestosis in workers with at least 3 years of service, in roughly half of all workers with at least 10 years of service, and in 100% of all workers with service of 15 years or more.[14] The dust conditions in many asbestos factories in Osaka, Tokyo, and Nara were appalling; almost all locations had indoor dust concentrations that were in excess of tolerable limits. The Labor Ministry study pointed out that anyone working under these conditions for long periods of time would inevitably become a victim of asbestosis.

Yet another concurrent study of five facilities by the Industrial Health Division of the Labor Standards Bureau identified extremely poor workplace conditions comparable to those of the prewar years, with dust concentrations of 145 mg/m^3 in blending areas and 58 mg/m^3 in carding areas (Sera 1983:90). As noted by Yuoka, most of the asbestos dust was generated during the spinning process at factories in the Sennan area that produced asbestos yarn and fabric from crude asbestos, and the risk of illness from exposure was concentrated in the Sennan district.

6.4.3.3 Establishment of the Pneumoconiosis Law (1960)

One factor behind the launch of the Labor Ministry's asbestosis study was the need for more data on the criteria for handling asbestosis as an occupational respiratory disease in the same category as silicosis, a disease covered by the Special Protection Act for Silicosis, which had been enacted in 1955. The Labor Ministry study, along with several other studies conducted in 1958 and 1959 on graphite, carbon black, cement, and other sources of dust, came to fruition with the establishment of the Pneumoconiosis Law in 1960. Preventing all forms of pneumoconiosis due to exposure to mineral dust and providing health management for pneumoconiosis victims were the primary objectives of this law. As such, it was the first piece of legislation in Japan to regulate asbestos. However, the enforcement regulations for the Pneumoconiosis Law contain provisions which have to do with pneumoconiosis in general, as caused by the inhalation of mineral dust, and do not contain any provisions that specifically deal with asbestos. However, the regulations do cover processes in locations involved in the pulverizing, blending, spraying, carding, spinning, weaving, loading, unloading, stacking, sewing, cutting, polishing, finishing, or packaging of asbestos materials.[15]

The Pneumoconiosis Law itself had two significant shortcomings relating to the control of asbestos dust.

First, it set no benchmark concentrations for dust, and thus did not implement substantive, quantitative controls. The law only stipulated that employers and workers had an obligation to strive to curb emissions of dust, use protective gear, and make efforts in other related areas, and that employers should make exhaustive efforts to provide workers with all necessary education and training. It did not set, nor did it enforce, any controls based on specific concentrations. On the matter of mandating efforts to prevent pneumoconiosis, the then Labor Minister Raizo Matsuno, in explaining the reasons for the proposed law to a March 3, 1960, meeting at the House of Representatives, noted that in terms of preventing pneumoconiosis, the law would obligate employers and workers to strive to curb emissions of dust, supply and wear protective masks, and take other precautions as appropriate, and would compel employers to strive to their utmost to provide relevant workers with the necessary education and training (Editing Committee of Measures Against Pneumoconiosis in Japan 1985:89). In fact, this issue was specifically brought up by labor delegates called by the government to testify in an advisory role on the outline of the proposed Pneumoconiosis Law before the Pneumoconiosis Committee in 1959. First, in their general comments on the proposed legislation, the labor delegates noted that while the bill conceivably obligated employers to implement preventive measures, it avoided setting dust prevention benchmarks for workplace conditions, and did not contain basic provisions for the promotion of preventive practices, including the obligation to perform dust measurements that would serve as criteria for decisions on the implementation of rational technical measures.

They also noted that as such, the content of the bill left doubts as to whether employers would take responsible action to improve current conditions in their workplaces if faced with future prevention-related challenges. In their detailed opinion, the labor delegates insisted that rational and effective preventive steps should be implemented, that tolerable limits for dust concentrations should be defined as an inseparable component of health examinations, that a dust management organization should be established, and that a legal framework mandating the reporting of dust measurements should also be created (Ebihara 1976:76–78). Despite these viewpoints, the Pneumoconiosis Law was established without imposing any limits on dust concentrations, and went no further than stipulating that efforts should be made to keep concentrations under control. As such, it was extremely doubtful from the outset that the law would have any effectiveness in terms of providing for preventive measures. In 1958 the Director of the Labor Standards Bureau within the Ministry of Labor issued a notice pertaining to the promotion of improvements to work environments as a preventive measure against occupational illnesses. Based on that notice, a set of technological guidelines had been drawn up for the prevention of occupational illnesses in the workplace. As noted earlier, the asbestos dust concentrations set as control targets by these guidelines were ill-conceived. It was no surprise that workplace safety measures of this kind would have their true effectiveness undermined by the requirement for "harmonization" with economic growth.

Second, the law did not take into account the actual conditions of the small business operations in the Sennan district, and accordingly failed to foster the installation of dust-collection systems or promote other steps in workplace safety management. According to former business owners in the Sennan area, the local consensus was that it would have been enormously expensive at that time to install dust-collection systems, and the high bills for the electricity required to run them would have left many businesses with no margin of profit.[16] In effect, small businesses perceived dust-collection systems to be highly expensive in those days, and business owners had no choice but to remain noncommittal about their installation.

6.4.3.4 The Pneumoconiosis Law and Sennan District Asbestos Factories

As a consequence of developments on the legislative front, the workplaces of small-scale businesses in the Sennan area did not experience any alleviation in the incidence of industrial illnesses due to dust inhalation.

After the Pneumoconiosis Law went into effect, examinations for asbestosis in the Osaka area were handled primarily by the Japan Anti-Tuberculosis Association. Examinations performed in Osaka detected symptoms of asbestosis in 48 of 633 patients (7.6%) in 1960, 27 of 240 patients (11.3%) in 1963, 29 of 260 patients (11.1%) in 1966, and 42 of 231 patients (18.2%) in 1970. As the detection rate increased, the number of patients undergoing examination showed a decline (Sera 1983:90). The following points can be made about this observation.

First, as noted earlier, after the passage of the Pneumoconiosis Law, there was no progress toward improving those workplace environments that posed a risk of causing asbestosis. Judging from examination detection rates alone, the situation actually deteriorated after the law came into effect.

Second, there was an observed pattern of aversion to the health examinations mandated by the law. Yoshisumi Sera notes that the decline in the number of examinees was a by-product of payroll cuts by companies striving to streamline or downsize their operations, coupled with a continuing trend among companies under strengthened administrative guidance to rely more on subcontracting to home-based businesses and other operators that utilized part-time workers (Sera 1983:90).

The Labor Standards Bureau conducted field surveys of actual conditions at 44 Sennan asbestos factories in 1967 and again at 59 factories in 1970. These surveys noted that one-third of all the dust-collection systems installed lacked powered ventilation, that localized exhaust systems suffered from faulty design, and that 20% of the factories surveyed had not implemented health examinations for their workers. These findings led to a report citing the need for powerful measures in guidance and oversight. Another field survey of 101 asbestos manufacturing businesses and 1 291 workers in the Osaka area found that 66 businesses and 899 workers were concentrated in the Sennan district; that of a total of 10 businesses that outsourced processes to smaller home-based shops, nine were located in the Sennan district; that of 29 businesses that had not implemented routine health examinations for their employees, 23 were in the Sennan district; that in 1968, 1969, and 1970, only 34.5%, 32.5%, and 46.9%, respectively, of asbestos workers in the Sennan district received annual pneumoconiosis examinations compared with examination rates of 41.2%, 45.9%, and 67.2%, respectively, for the larger Osaka area (Sera 1983:91).

Over a 25-year period from 1961 – the year after the Pneumoconiosis Law came into effect – to 1985, Kenji Morinaga et al. studied a total of 175 (105 male and 70 female) asbestosis patients in Osaka Prefecture (Morinaga et al 1990:647–652). The study subjects consisted of 10 patients in the period from 1961 to 1965, 19 from 1966 to 1970, 43 from 1971 to 1975, 53 from 1976 to 1980, and 50 from 1981 to 1985, thus highlighting a clear increase in the number of asbestosis patients following enactment of the Pneumoconiosis Law. As of December 31, 1987, 73 (69.5%) of the male subjects and 33 (47.1%) of the female subjects had died, giving a total of 103 deaths (59.8%). Further, all subjects identified as asbestosis patients by 1969 had died. Of 21 male subjects who had died of neoplastic lesions, lung cancer was the cause in 18 cases (zero mesothelioma cases). Another 45 male subjects (61.6%) died of respiratory illnesses, including complications from tuberculosis. Of the female subjects, three died of neoplastic lesions (including one case of lung cancer and one of mesothelioma) and 25 (75.8% of all deaths of female subjects) died of respiratory illnesses, including complications from tuberculosis. Kenji Morinaga et al. also performed a similar study from 1956 to 1985 (Morinaga 2006:12–20). Of 182 subjects in that study known to be still alive or already dead as of 1999, the prognosis was worse for those with more serious X-ray findings and far worse than the comparable prognosis for other patients with pneumoconiosis. Of those subjects who died of respiratory illnesses or tuberculosis, approximately half were listed as asbestosis patients; the remainder were listed as victims of silicosis, pneumoconiosis, pulmonary fibrosis, or other diseases. Judging from these findings, the number of deaths from asbestosis is considered to be significantly underestimated in demographic statistics. One feature of asbestosis is that women account for over 20% of all deaths from the disease. Several causal factors may be cited for this: (1) many workers in the asbestos textile industry are women; (2) the asbestos textile industry uses relatively long-fiber forms of asbestos; (3) it is difficult to control dust emissions in the drying stage; and (4) there were more opportunities for exposure to higher concentrations of dust than was the case in other asbestos product industries.

One may conclude from these points that workplace conditions in the Sennan district remained poor even after the Pneumoconiosis Law was put into effect in 1960, and that the law itself produced very little benefit. Furthermore, this state of affairs was not necessarily something that could be blamed on business owners. The reason is that many, if not most, of the small-scale businesses in the Sennan district were family-run operations in which the business owners and their family members were also victims, and were even at risk of higher exposure to dust than other employees (Yuoka 2006:23). This point is critically important to an understanding of the nature of asbestos exposure and harm in the Sennan district. This is because in the case of Sennan, the reasons for asbestos exposure and harm can be attributed more to a lack of awareness about the hazards of asbestos and effective countermeasures than to decisions by business owners to save money by forgoing investment in worker safety, as was observed in the more common cases of workplace accidents and injury. Had it been widely known that asbestos was such a hazardous material, business owners and/or their family members would at least have taken the initiative to prevent themselves from becoming victims. That reality vividly demonstrates how inadequate government policies on asbestos were at the time.

6.4.4 The Government and the Sennan District

The government at least indirectly acknowledged the value of asbestos products as essential for economic growth, and the importance of the businesses in the Sennan district that specialized in the primary processing of those asbestos products. However, it also extended direct support through production plans that were prepared for the asbestos product manufacturers in this district.

MITI implemented several asbestos production plans in the 1950s. In particular, it drew up a variety of plans for the production of asbestos products, including asbestos slate, asbestos cylinders, high-pressure asbestos pipes, asbestos tiles, and asbestos-lined pipes, as well as plans for the supply of crude asbestos. These plans comprised increases for specific types of asbestos products. For example, in 1954, steps were taken to increase the manufactured output of asbestos lining materials by 20% and of asbestos yarn, fabric, and cord by 10% (Japan Asbestos Association 1954b). Originally published in the Japan Asbestos Association newsletter, *Asbestos*, Table 6.9 is a compilation of data highlighting trends in asbestos product production plans. As this table illustrates, asbestos textiles, the product category with which manufacturers in the Sennan district were most closely associated,

Table 6.9 Year-on-year trends in production plans for asbestos products

Product	1954	1955	1956	1957	1958	1959
Electrolytic fabric	72	120	58	80	90	80
Textiles (yarn, fabric, cord)	1 472	1 590	1 895	2 720	2 300	2 400
Joint sheeting	1 312	1 720	1 847	2 342	2 130	2 240
Processed rubber	180	130	158	192	100	100
Asbestos slate	2 000	2 100	2 195	3 100	2 860	2 760
Asbestos paper	48	56	67	108	200	280
Graphite processed	156	158	197	256	180	160
Woven linings	360	345	310	448	420	440
Resin molding	72	33	90	400	350	400
Rubber molding	1 128	1 140	108	1 240	1 500	1 640
Special processes	532	628	521	780	670	760
Lumber	800	1 226	878	2 906	2 800	1 800
Asbestos-integrated metal packing	–	626	861	1 284	1 390	1 440
Heat-resistant quilting	–	238	323	484	520	560
Chrysotile asbestos insulation	–	710	968	1 448	1 560	1 640
Other categories	3 000	–	1 651	3 132	3 530	3 600
Total	11 132	10 820	12 128	20 920	20 600	20 300

Unit: metric tons
Source: Japan Asbestos Association, *Asbestos, various issues*

consistently accounted for over one-tenth (11–16%) of the total manufactured output measured in tons.

These details demonstrate that the government was both directly and indirectly involved with the cluster of small operations in the Sennan district that handled the primary stages in the manufacture of asbestos products that played a key background supporting role within Japanese industry at large. That was the ultimate outcome of a government policy bias toward a "national interest" defined by economic reconstruction and growth. Eclipsed by MITI-led government policies that embodied this "national interest," government policies on labor and welfare were assigned a subordinate role, and asbestos exposure and harm spread in the Sennan district as a result. Indeed, it is safe to say that thanks to delays in the formulation of labor policies on the installation of dust-collection systems, etc., prices for asbestos products were held at very low levels, enabling Japanese industry to maintain its competitiveness. In that context, it would not be exaggerating matters to portray the Sennan district as having been "sacrificed" for postwar Japan's reconstruction and the subsequent economic boom, and then ignored for many years thereafter.

6.5 Conclusions

As shown in this chapter, during Japan's prewar era and postwar economic boom years, products manufactured from asbestos, the "miracle mineral," were essential whether intended for military or private-sector use. Without those asbestos products, it is highly unlikely that postwar reconstruction and the ensuing economic boom would have been possible in the form that Japan experienced it. Judging from its involvement in the administrative control of asbestos, as reflected in policies for the increased production of asbestos products to satisfy military and civilian demand in the prewar years, policies to boost food crop yields during the postwar reconstruction years, and policies aimed at supporting industrial growth during the economic boom years, we can safely conclude that the government, and MITI in particular, recognized the importance of asbestos within the context of their industrial policy. Delays and deficiencies in government policies on labor and the environment were aided by the national character of MITI and the government in general, which had embraced a growth-first doctrine. This had the end-effect of aggravating the levels of asbestos exposure and harm experienced thus far in Japan. What is more, the government itself was deeply involved in various high-level studies, extending from the prewar to the postwar years, which highlighted the dangers of asbestos. Given this fact, we must inevitably hold the government accountable for the delays in its labor and environmental policies on asbestos countermeasures.

Consistent historical experience of Japanese economic policies of this kind was at its most intense and broadest in the Sennan district of Osaka. Severe asbestos-related diseases caused misery not only for the employees of the small-scale asbestos manufacturers located in the Sennan district, but also for the business owners and their families. This reality underscored not only the deficiencies in the scope and effectiveness of the Pneumoconiosis Law and other elements of government policy on asbestos, but also a lack of familiarity with the severe hazards posed by asbestos. One may make the case that neglect of workplace safety and hygiene in asbestos factories in the Sennan district ultimately contributed to the competitiveness of Japanese industry as well as to the expansion of the Japanese economy. From a broader perspective, the implication is that the small shops in the Sennan district were themselves relegated to a shadow role throughout the economic reconstruction and boom years in the absence of realistic regulatory oversight – despite their status as members of a front-line industry that produced and supplied low-priced primary asbestos products for industrial reconstruction led by giant corporations under the government's "harmonization" policy, and despite their involvement in some of the most dangerous manufacturing processes to that end. Looking back on this history, the Sennan district was treated as a "sacrificial lamb" for the Japanese economy, and it is the government that must bear the heaviest burden of blame for the consequent suffering.

Notes

1. Among the countries known for having comprehensive relief frameworks for victims of asbestos exposure, France has FIVA and Belgium implemented its own asbestos compensation fund (AFA) in April 2007. However, these programs vary country by country. For example, while FIVA is a comprehensive framework extending compensation eligibility even to patients with pleural plaque, AFA excludes lung cancer patients from compensation eligibility owing to the difficulties associated with obtaining a confirmed diagnosis. AFA differs from Japan's

Asbestos-Related Health Damage Relief Fund in that it also includes asbestosis within the scope of eligibility for compensation.

2. Temporary Funds Adjustment Law granted the Minister of Finance the power to divert national and private-sector funds to the defense industry, as necessary. Law for Temporary Measures on Imported and Exported Goods granted the government authority to regulate and ban commodity imports of all kinds and control the processing, distribution, transfer, and consumption of imported commodities. This gave the Ministry of Commerce broad discretion to do as it pleased (Johnson 1982:136–137).
3. The National Mobilization Law had been drawn up by the Board of Planning (later named the Economic Stabilization Agency and then the Economic Planning Agency) and accorded the government totalitarian control over the reorganization of Japan from the bottom up. Johnson 1982, 139.
4. Yamazaki (2007), p. 4. In addition, because ammonium sulfate was considered to be the "king of fertilizers," the recovery in fertilizer production was concentrated in ammonium sulfate (Ohuchi 1957:154).
5. Wartime operations were responsible for approximately 35% of the lost ammonium sulfate production capacity; aging facilities accounting for the other 65% of lost capacity. Ammonium sulfate was considered to be the "king of fertilizers," so the recovery in fertilizer production was concentrated in ammonium sulfate.
6. In this context, asbestos products included "asbestos yarn and fabric, joint sheeting, asbestos slate, asbestos paper, brake linings, asbestos lumber, asbestos heat-insulating material, and other items" (Research Institute in Commerce, Industry, and Economy of Osaka Prefecture 1967:2).
7. Kawakita (1991), 134. Juro Hashimoto also discusses the necessity of identifying these inter-agency policy processes within the government (in Japanese). "In most studies to date, the government has been treated as one integrated entity. However, as I explained in the preceding chapter, there are problems with that concept. The reason is that in Japan, we have the expressions, 'only ministries exist, not government' or 'only bureaus exist, not ministries.' It is also said that there are no conflicting interests within the government even if conflicting interests exist between MITI and the Ministry of Finance, or that there are not even interests within a ministry even if conflicts exist in a specific bureau. Roundabout euphemisms of this kind can be understood as articulating a certain reality. That being the case, it is not realistic to conceive of the Japanese government as being one integrated entity from the outset, or that one ministry effectively represents the entire government" (Hashimoto 2001:194–195).
8. The principle behind this type of corporate investment behavior has been termed "frugality in the interest of a constant supply of capital." Kenichi Miyamoto was the first to apply this concept as an integrated economic principle to everything from workplace accidents to environmental pollution within the factory setting (Miyamoto 1967:163–164).
9. Sawada (2007), pp. 57, and interviews with residents of Sennan.
10. Hearing from Osaka Prefectural Government. December 7, 2007.
11. In 1964, Osaka produced 61.1% of asbestos yarn and fabric, 42.2% of asbestos joint seating, 31.7% of asbestos slate, 5.2% of asbestos brake linings, and 26.8% of other asbestos products in Japan (Research Institute in Commerce, Industry, and Economy of Osaka Prefecture 1967:16–17).
12. Edward Merewether was a factory medical officer and Charles Price was a technology officer. In 1928, they were instructed by the UK government to carry out a study of asbestos textile factories. This was an asbestosis study focused on 363 subjects thought to have been exposed solely to asbestos. For the research conducted by Merewether, E, and Price, C, and the Asbestos Industry Regulation in 1931, see Tweedale (2002).
13. The report by Horai et al. on this occasion noted that while the 1930 Merewether and Price study discovered symptoms in 93 of 363 subjects, for a detection rate of 26.2%, not a single case of asbestosis was identified in workers with a length of service under 5 years (Hyogo College of Medicine 1981:64).
14. The ratios of asbestosis were 22.5% in Tokyo and 19.7% in Nara.

15. The enforcement regulations for the Pneumoconiosis Law, 1960. According to this, four processes (blending, and stacking and sewing) were subject to control under the regulations, but no controls applied to bale opening, a process that released large quantities of dust (Sawada, T. (2007). Chapter 4). However, that interpretation is clearly a misreading of the documentation. The processes listed here had to do with explanations of asbestos-related terminology that the Director of the Labor Ministry's Labor Standards Bureau issued in a notice (Notice No. 331) dated April 25, 1960. The "asbestos pulverizing or untangling" provisions in the enforcement regulations for the Pneumoconiosis Law should be interpreted as naturally including the bale opening process within the scope of control.
16. Based on interviews with former business owners in the Sennan district by the author. Also, according to Teppei Sawada, the dust collection system at one factory in the Sennan district cost 6 million yen (Sawada 2007:61).

References

Ebihara I (1976) History of anti-occupational disease movement pneumoconiosis in Post War Era (in Japanese), vol 1. Iryo Tosho Co., Ltd., Tokyo

Editing Committee of Measures against Pneumoconiosis in Japan (1985) Measures against pneumoconiosis in Japan (in Japanese). Japan Industrial Safety & Health Association, Tokyo

Editing Committee of Sennan City History (1987) History of Sennan City: the historical edition (in Japanese). Sennan City, Sennan

Fujimoto T (1970) Labor accident (in Japanese). Shin Nihon Shuppansha, Tokyo

Fujimoto T (1992) Labor accident. In: Institute for Economic Research Osaka City University (ed) Dictionary of economics the 3rd edition (in Japanese). Iwanami Shoten, Tokyo

Hashimoto J (2001) Growth structure of Post War Japanese economy (in Japanese). Yuhikaku, Tokyo

Hirai T (1997) Celebration on asbestos import Japan. Asbestos Industry Association. Ind Assoc News (in Japanese) 260:4

Hyogo College of Medicine (1981) History of asbestosis study in Japan (in Japanese). Hyogo College of Medicine, Hyogo

Research Institute in Commerce, Industry, and Economy of Osaka Prefecture (1967) Osaka local industries in reality No. 10, asbestos manufacturing (in Japanese). Osaka Prefectural Government, Osaka

Japan Asbestos Association (1946) Asbestos (in Japanese) 3

Japan Asbestos Association (1947) Asbestos (in Japanese) 14

Japan Asbestos Association (1948) Asbestos (in Japanese) 27

Japan Asbestos Association (1950) Asbestos (in Japanese) 52

Japan Asbestos Association (1954a) Asbestos (in Japanese) 97

Japan Asbestos Association (1954b) Asbestos (in Japanese) 98

Japan Asbestos Association (1957) Asbestos (in Japanese) 135

Japan Asbestos Association (1958) Asbestos (in Japanese) 147

Johnson C (1982) MITI and the Japanese miracle. Stanford University Press, Stanford

Kato Y (1997) Meeting a ship with the first imported asbestos. Asbestos Industry Association. Ind Assoc News (in Japanese) 260:4

Kawakita T (1991) Ministry of International Trade, and Industry (in Japanese). Kodansha, Tokyo

Ministry of International Trade, and Industry (ed) (1994) Policy history of International Trade, and Industry (in Japanese), vol 1. Research Institute of International Trade, and Industry, Tokyo

Miyamoto K (1967) Social overhead capital (in Japanese). Yuhikaku, Tokyo

Miyamoto K (1975) Environmental problems in Japan (in Japanese). Yuhikaku, Tokyo

Miyamoto K (1980) Pollution. Dictionary of economics II (in Japanese). Toyo Keizai Inc., Tokyo

Miyamoto K (2006) Toward the Sustainable Society (in Japanese). Iwanami Shoten, Tokyo

Miyamoto K (2007) Environmental economics new edition (in Japanese). Iwanami Shoten, Tokyo
Morinaga K (1989) Epidemiological findings of asbestos-related diseases in Japan. Pathol Clin Med (in Japanese) 7(6):686–694
Morinaga K (2006) Epidemiology of asbestos-related diseases. Mebio (in Japanese) 23(3):12–20
Morinaga K et al (1990) A study on prognosis of asbestosis patients. J Jpn Accid Med Assoc (in Japanese) 38(11):647–652
Nankaido Research Institute (1984) Local industries in Senshu (in Japanese). Nankaido Research Institute, Osaka
NICHIAS Corporation (1996) One hundred year's history of NICHIAS (in Japanese). NICHIAS, Tokyo
Nippon Valqua Industries Ltd (1977) Fifty year's history of Nippon Valqua Industries Ltd (in Japanese)
Ohuchi T (1957) Economics of fertilizer. Hosei University Press, Tokyo
Sawada T (2007) Development of Japanese capitalism and asbestos pollution (in Japanese). Master's thesis, Osaka City University, Osaka
Sera Y (1983) Asbestosis in Osaka. In: Research Group on the History of Occupational Health in Osaka (ed) Report on the history of occupational health in Osaka (in Japanese). Research Group on the History of Occupational Health in Osaka, Osaka, pp 88–96
Tweedale G (2002) Magic mineral to killer dust. Oxford University Press, New York
Yamazaki S (2007) Ammonium sulfate industry in Post War Era (in Japanese). Manufacturing Management Research Center, University of Tokyo. MMRC Discussion Paper 174
Yuoka K (2006) History of asbestos industry and latent damage in Sennan. Roudou Horitsu Jyunpou (in Japanese) 1617:21–24

Chapter 7
Persistent Thorns: Responsibility for Asbestos Disasters

Masafumi Kato

7.1 Introduction

Experiencing sudden respiratory distress, he decided to rush to the hospital. Could this be a cold or the flu? The physician returned with the findings: "Asbestos is the cause of your disorder. Have you ever worked in a factory?"

Holding his hands to his chest, he thought for a moment but could not remember. Mesothelioma, lung cancer, pulmonary asbestosis – all intractable diseases that cause severe suffering for the patient. Somewhere deep in his lungs, that small fibrous thorn had triggered a serious illness that now, after all these years, was finally awakening.

"Where on earth could I have inhaled asbestos?"

The number of people haunted by these thoughts has been multiplying sharply. As many as 6500 individuals with no history whatsoever of handling asbestos materials during their active careers nevertheless have been stricken by mesothelioma and lung cancer, and of that total, half have already died. By contrast, the number of new patients approved for workers accident compensation insurance benefits has topped the 1 000 mark year after year. In 2008, over 1 000 patients died of mesothelioma, a disease with a poor prognosis, but perhaps worst of all is the prospect that the incidence of asbestos-related diseases will rise.

Without question, the number of asbestos victims will certainly surpass the comparable totals attributed even to the four historically largest cases of industrial pollution that Japan experienced during its economic boom years. Japan was fatefully slow to place regulatory controls on asbestos, and did not implement a ban on its use in principle until 2004. By some predictions, the peak in the number of victims of asbestos-related diseases will arrive in 2030.

One event hinted that the asbestos crisis might become known as one of the worst industrial calamities in history. That outlook was driven home in late June 2005 with a press announcement by the Kubota Corporation that there had been a heavy

M. Kato (✉)
The Kobe Shimbun, 1-5-7 Higashikawasaki-cho, Chuo-ku, Kobe, Hyogo 650-8571, Japan
e-mail: bxr05060@nifty.ne.jp

K. Miyamoto et al. (eds.), *Asbestos Disaster: Lessons from Japan's Experience*,
DOI 10.1007/978-4-431-53915-5_7, © Springer 2011

outbreak of asbestos-related diseases among former workers for, and residents living near, its defunct Kanzaki factory in Amagasaki, a city in Hyogo Prefecture.

This was the so-called Kubota Shock, a news event that suggested that the outbreak needed to be treated as a case of industrial pollution on a scale surpassing the conventional bounds of workplace disasters. Now, over 5 years later, a damage compensation lawsuit launched on behalf of a group of former workers from the Sennan community of Osaka has thrown the spotlight of public scrutiny on a pattern of inaction by the national government.

On May 19, 2010, the Osaka District Court issued its ruling, finding the national government liable for failure to take appropriate action against asbestos despite its knowledge of the hazards associated with this material, thereby allowing the extent of victimization from asbestos exposure to spread. This was the first court ruling to find the national government liable for health problems associated with asbestos. The national government rejected the ruling and filed an appeal with the Osaka High Court on June 1. Numerous lingering legal issues and the prospect of a proliferation of litigation in the years ahead have been cited as reasons for its decision to appeal.

In any event, most of the victims are now of advanced age. It is imperative that the national government stop stonewalling in court and come out with sweeping measures for relief as quickly as it can. At the same time, taking the Kubota Shock-inspired Sennan lawsuit as its cue, the government must move forward to bring about a transformation in its asbestos policies and countermeasures.

7.2 Kubota Asbestos Disaster

7.2.1 The Impact of the Kubota Shock on Amagasaki

Amagasaki is a city located in the core of the Hanshin Industrial Region. During the economic boom years, it was designated an "industrial capital" and attracted heavy concentrations of industry in numerous sectors, ranging from steel and chemicals to machinery and ceramics. At the height of the economic boom, local industry thrived and many citizens relocated to Amagasaki in search of employment. This city had lively shopping centers and street markets with mass appeal and a congenial, folksy atmosphere. Its citizens affectionately referred to it as "Ama."

Many corporate industrial operations were located along the city's waterfront zone next to Osaka Bay. However, in 1954, the Kubota Steel Corporation (the current Kubota Corporation), a large machinery manufacturer, put its Kanzaki factory into operation in a heavily populated residential zone located relatively close to the Japan Railways (JR) Amagasaki Station (Fig. 7.1).

From 1954 to 1975, this factory was engaged in the production of asbestos pipes for water company applications, and from 1971 to 1997, it also produced residential building materials. During its years of operation, the facility utilized approximately 88 000 metric tons of crocidolite and 149 000 tons of chrysotile, both known to be

Fig. 7.1 Kubota Hanshin Office (Former Kanzaki Factory of Kubota) (© The Kobe Shimbun)

highly toxic forms of asbestos. No other factory in Japan has ever used crocidolite in such huge quantities.

Today, the district around this factory is the scene of one of the worst asbestos calamities in Asia, if not the entire world.

Of 173 workers at the factory who came down with asbestos-related diseases, 145 have died. An even larger number of local residents – 221 – developed asbestos-related diseases, and of that group, 190 have died. Together, these statistics add up to 394 patients with 335 deaths, and the number of victims continues to grow.

7.2.2 Tears of Grief

"My wife had always put her heart and soul into making our lives better, but suddenly she became ill (with mesothelioma). To everyone with this disease, I say 'Hang in there!'" These were the reflections voiced by the mourning husband of the late Masako Doi at her funeral wake in Itami on October 22, 2007. She was 59. Her husband broke down in tears of grief.

Doi had always been a cheerful and energetic woman. She had taken a leading role at meetings of patients and their families, and had also been involved in negotiations with Yuriko Koike, the Environment Minister at that time. Doi once exclaimed, "I don't care if you offered me 100 million yen, this disease is horrible and I'd rather have my health back." The expression on her face on that occasion was unforgettable.

In Amagasaki, the first citizens to blow the whistle on the Kubota problem have been dying one after another.

Keiko Maeda, who had pointed the finger of corporate liability at Kubota for allowing the scale of harm from asbestos to spread, and who described the company as a sly killer for its failure to take action against emissions of asbestos dust, died at the age of 74 in March 2006.

Sumihiko Hirachi, who early on recognized that this was a serious case of industrial pollution, died aged 60 in October 2007. He had devoted himself completely to the negotiations with Kubota on relief payments.

The surviving families of such victims are of one voice: "Why must we be the bearers of this grief? We only lived in Amagasaki. How we yearn to see the smiling faces of our lost loved ones once again."

7.2.3 The Whistleblowers

June 29, 2005, was to go down as an important day in the historical chronicle of industrial pollution in Japan. Three residents living near the Kanzaki factory site spoke at a press conference (Fig. 7.2). They announced that they were there to blow the whistle on Kubota as the company responsible for an outbreak of mesothelioma and related illnesses not among factory workers, but among local residents who had no experience handling asbestos whatsoever. It was at this instant that the asbestos disaster gained status as a major case of industrial pollution.

Fig. 7.2 The press conference by three residents near the Kanzaki Factory (© The Kobe Shimbun)

Maeda had run a gasoline station near the Kubota factory. Her home was also close by, but in 2003 she was given a diagnosis of mesothelioma. She had never been engaged in any job that involved asbestos. Although her physician informed her she had only 1 year left to live at best, Maeda managed to survive for almost 2 years.

"We don't have much time left. You will see a steady increase in the number of patients from this point on. We want the government and the private business community to provide compensation to victims, to acknowledge this disaster as a case of industrial pollution, and to launch a national campaign to study and develop effective treatments for our illnesses."

Although Maeda suffered from steadily failing health, she expended every effort to help organize groups of patients and their families. On December 31, 2005, she had a meeting with Kubota's then-president, Daisuke Hatakake, and quietly made her case: "This is the last chance to talk. Give us compensation."

Maeda's 74 years in this world came to an end shortly afterward, on the evening of March 27, 2006, the same day the government enacted the Act on Asbestos Health Damage Relief. In the final twilight days of her life, it was as if Maeda was working solely to open the lid on the asbestos issue, and watching to see the shape the new law would take.

Laid to rest in the funeral hall, her face had a conservative touch of makeup that reflected the refined elegance of her complexion in life.

Maeda's eldest son spoke: "Sometimes we never know what the world thinks of a person until the lid on their coffin is shut, but my mother clearly made an impact on this country in her final days."

After the funeral rites, an autopsy was performed on Maeda's body by physicians at the Hyogo College of Medicine. Highly toxic fibers of crocidolite were discovered deep in her lung tissues. Virtually invisible to the naked eye, these thin fibers had triggered a cancerous growth.

Describing herself as a model of good health, Masako Doi lived near the Kanzaki factory until she was around 20 years old. Although she then moved away from the Amagasaki area, the early symptoms of her illness suddenly appeared more than 30 years later. Doi and her husband together ran a *takoyaki* (octopus dumpling) eatery, but in May 2004 she developed a cough with fever. She thought it was unusual and wondered if it were only a cold. A medical examination at the local municipal hospital disclosed a disease condition linked to asbestos: "malignant pleural mesothelioma." After much mental distress, Doi had one of her lungs removed that October at the university hospital. She lamented: "I was such a happy and healthy older woman in the prime of my life as a *takoyaki-ya*. Why did this disease have to strike me?"

Enduring her symptoms, Doi became enthusiastically involved in the activities of victims' organizations. She continued with a chemotherapy regimen of anticancer agents, but complained of the heavy burden of expenses for medical care. "We want the government to move forward quickly with countermeasures, starting with victims like us."

Tissue samples from the lung that Doi had had removed about a year prior to her death revealed large quantities of highly toxic crocidolite fibers. Kubota's Kanzaki factory had been one of the heaviest industrial users of crocidolite by far. Her voice trembling, Doi remarked: "I was shocked to learn that so much crocidolite had been found in my lungs. This was definitive proof of a causal relationship with Kubota's operations."

7.2.4 Inside the Factory

Why was so much asbestos dust released on this scale? What operations was the factory engaged in at the time?

The Kanzaki factory site is located relatively near Amagasaki Station on the JR Kobe Line. I visited the home of a former factory worker still living in Amagasaki, who is now in his 70s.

On a list of former coworkers, many of the names have been stamped "deceased." The former worker concedes he is frightened to see his former coworkers dying one after another.

He began working at the factory in 1959, 5 years after it had gone into operation. In those days, Japan was transitioning into a dramatic economic boom and the factory was gearing up for the full-scale manufacture of asbestos piping used in water company applications.

At that time, the factory had a payroll of approximately 350 employees. The asbestos pipe manufacturing operation comprised the following contiguous stages: raw materials input, pipe formation, curing, and fabrication. Asbestos bales brought from the warehouse were opened and their contents loaded into fiberizers that handled the unraveling operation. The resulting powdered asbestos material was then mixed with cement and formed into the desired pipe shapes. The final product was far lighter than a cast-iron pipe, and above all it facilitated the manufacture of inexpensive piping materials for use by water companies.

Opening the asbestos bales, placing their contents in line machinery, and many other operating stages depended on manual labor.

The former factory worker himself was in charge of the "opener" stage that involved the input of raw materials. That job entailed grabbing the bales of packed asbestos, which were wrapped in dungaree-fabric, with a handheld hook and cutting them open with cutters. The unpacked asbestos was then placed into the fiberizer, which shredded it into finer material. "Fiberizing the raw asbestos gave it a stringy consistency. Thick clouds of asbestos dust were everywhere. I was amazed every time we cleaned out the air ducts. What protective gear did I wear? Only a gauze face mask."

He had to tolerate abysmal levels of dust and harsh working conditions. The factory operated on two daily work shifts, each 12 h long. He came down with a kidney condition and resigned after 7 years. Although from that point on he had led a fairly healthy life, about 10 years ago he developed symptoms of diffuse pleural thickening. Even climbing the stairs at the train station became almost too much of a strain.

"In the summer months we were drenched in sweat that stung our skin.... We didn't have the ventilation or air-conditioning systems that are common now. We often went about our work with the factory windows left open." As may be gathered from the testimony given by the former factory workers, the work environment at the factory was very bad.

The extent of harm to the health of workers in the factory was appalling.

Among the workers engaged in asbestos pipe manufacturing operations for a period of 10 years or more, the incidence of disease was 44.6% and the death rate was 22.7%. That equates to disease in one of every two workers and death in one of every five, which is an astonishing body count for the ranks of these industrial warriors.

7.2.5 *Formal Apology*

On hearing reports in the news, Kubota President Hatakake was at a loss for words. "I couldn't believe that the harm from asbestos had spread beyond the factory's walls."

In April 2005, it was confirmed that health problems caused by asbestos had appeared in Amagasaki residents living around the defunct Kanzaki factory.

That factory had begun utilizing asbestos to manufacture water pipes and other products in 1954. Over two decades later, by 1978, current and former workers began falling ill with mesothelioma, lung cancer, and other asbestos-related diseases. In recent years, the death rate has continued to climb: four deaths in 2001, five in 2002, eight in 2003, and ten in 2004.

"My friends and coworkers, one after the other."

Although shaken up, President Hatakake expressed hope: "The reason the health problems have been concentrated among our workers is because the factory is so well sealed." Probably no asbestos dust had escaped to the surrounding community. That fragile hope was soon shattered.

Hatakake turned to his executive officers. "Disclose everything we know. Releasing information a little at a time will only invite suspicion that we have something to hide. We have no intention of running away from this." In the back of his mind were memories of Mitsubishi Motors' concealment of car recalls, and Snow Brand Food Co.'s beef-mislabeling scam.

December 2005. Meeting asbestos victims and surviving family members at a location in Amagasaki, President Hatakake bowed deeply. "I view the confirmed outbreak of health problems with deep alarm, and as an employer who has utilized asbestos, I extend my sincere apologies with a sense of moral responsibility."

One patient raised the question: "Do you plan to treat us differently from victims who contracted their diseases in the factory?"

Hatakake replied: "Obviously from the standpoint of human life, it should make no difference whether one is an employee or not."

This was the first known instance of an asbestos company issuing a formal apology to residents living near a factory facility, and in this case it had agreed to pay residents monetary Relief on a par with the levels provided to employees. In the history of industrial pollution in Japan, it was an unprecedented response.

Since 1999, the Kubota Corporation had been embroiled in a series of scandals, including involvement in an illegal cartel and the issue of payments to corporate racketeers. President Hatakake had assumed his post with the burden of rebuilding the company's already shaky foundations for business. He said that operating on principles of corporate social responsibility (CSR) characterized his philosophy, and the asbestos problem would put that philosophy to the test.

April 2006. Kubota announced plans to pay relief money in the range from 25.0 to 46.0 million yen per victim. Noting that it would take time to shed light on the causal relationships in view of the long period of dormancy before the symptoms of asbestos-linked diseases become manifest, Kubota had demonstrated its intention to pay large amount of relief money without actually accepting that a causal relationship existed.

President Hatakake explained: "Although the option of battling out questions of liability in court was explored, the truth is that lawsuits involving cases of industrial pollution can take up to 30 years to reach a settlement. Given that the time patients have is limited, can we simply delay a resolution to this problem? We have a responsibility to fulfill from a humanitarian standpoint."

In the case of Minamata disease, which had been identified over half a century earlier, mercury-tainted waste effluent from the Chisso Corporation facilities had been pinpointed as the cause. Nevertheless, action to deal with this disease proved slow in coming because neither the government nor the private business community was willing to accept that explanation. As a consequence, following the formal discovery of Minamata disease, it took close to 20 years to sign a compensation agreement, and then it was only after the plaintiffs had achieved a victory in a court of law.

Given that backdrop, the Kubota response seems all the more striking. Nonetheless, certain shrewd calculations could be perceived through cracks in the facade. At the general meeting of shareholders on June 23, 2006, the questions flew. "This is a case of industrial pollution. How many decades will the company be burdened with this?"

Hatakake replied: "The outlook is not clear. Although we do not have room for optimism, we believe the current heavy burden of relief will not continue over the long term." In other words, the peak had already passed. That was Kubota's position then, and it remains their position today under the helm of its new president, Yasuo Masumoto.

It is widely perceived that Kubota sought to avert the risk of litigation by moving forward with this decision. Company insiders admit: "We cannot state with certainty that no asbestos dust escaped the facility's perimeter." There was no evidence to suggest that the company would be victorious if taken to court on the matter, yet if the litigation process was drawn out, that could do serious harm to its public image.

In the text of its proposed agreement for the payment of relief money, Kubota included a clause stipulating that residents should forfeit their right to future claims against the company. In other words, there was another motive; as its reward for the payment of relief money, Kubota expected to ward off any future efforts at class-action litigation.

7.2.6 Ambiguous Responsibility

Kubota's viewpoint on the asbestos disaster is that as a business that has utilized asbestos, it has a moral responsibility to bear. The text of the declaration distributed at its December 2005 press briefing tells the whole story.

> *The Company has not obtained enough evidence to conclude that the utilization of asbestos materials at its former Kanzaki factory was the cause of health problems experienced by individuals to whom it has already made consolation and condolence payments.*
>
> *However, we cannot state unequivocally that asbestos dust never escaped from the Kanzaki factory, and thus cannot deny the potential risk of dust inhalation to which local residents may have been exposed.*
>
> *We view the facts surrounding the outbreak of asbestos-related health problems among local residents with deep concern, and feel that we have a moral responsibility as a business that has utilized asbestos materials. Hence, on the occasion of this meeting of patients and families, we have extended our sincere apologies and sympathies to everyone undergoing treatment, as well as to their families. Furthermore, we have offered words of condolence to all that have passed away (Underlined sections by author).*

In effect, Kubota refuses to recognize a causal relationship, yet pays asbestos victims large sums of relief money out of sense of moral responsibility. This has been, and remains, its stance to this day.

To date, Kubota has held three important press briefings on the asbestos issue. On those occasions, the questions tendered by attending reporters have centered on whether Kubota recognizes any causal correlations between its use of asbestos and the asbestos-related health problems suffered by workers and local residents.

Reporter: "Can these relief payments be interpreted as a form of compensation?"

Hatakake: "The payments have nothing to do with restitution or compensation as based on any causal relationship, but we do not mind if they are construed as compensation."

Reporter: "You accept that the payments may be viewed as compensation yet you do not clearly acknowledge any causal links. Do you acknowledge any causal relationships to a significant extent? Do you accept that your company is responsible for the incidence of asbestos-related diseases to a significant extent?"

Toshihiro Fukuda (senior managing director): "The word 'compensation' has various meanings. In terms of legal liability, it would be taken to mean damage compensation. At this stage, we do not view it as such. 'Relief compensation' is another expression. Our position at this time is that these payments are of a relief-oriented nature. The word 'compensation' can be understood in a variety of semantic contexts. In our usage, however, it does not carry any of the legal nuances and is more strongly synonymous with 'relief.'"

Reporter: "That being the case, can the new measures discussed today be interpreted as an extension of the condolence and consolation payments that have been made thus far?"

Fukuda: "That is precisely the way we look at it. In terms of moral responsibility, our payments of consolation money were an outcome of asking ourselves what we could do, how we might respond on learning that residents near our factory had contracted asbestos-related diseases. We explored what Kubota could do to demonstrate good faith. These perspectives led to our inauguration of a new framework for the provision of consolation payments."

Hatakake: "There was no precedent for this. To be precise, the causes remain unclear, that is, no definitive conclusions can yet be made. Given this state of affairs, our choice was to offer compensation if you wish to call it that, in other words, take action of this nature as quickly as we could. This is not a typical case that has been dealt with anywhere in the world to date. No question about it, all-out efforts must be made to thoroughly investigate and understand the causal factors, and something must also be done to help the victims actually suffering right now. Again, this is a case without historical precedent. The causal relationships? Certain things remain inconclusive. More precisely, many uncertainties remain. Was Kubota the only possible source of asbestos dust? And of course, we cannot unequivocally say that we were not the cause."

Hatakake: "Crocidolite and Kubota. We utilized enormous quantities of crocidolite at one time. But was Kubota the only source, the only company using it, or were there other companies using it as well? We don't know. Government administrators don't know. One thing we do know is that large quantities of this material were being utilized in Amagasaki."

Fukuda: "We investigated other companies operating in the vicinity of our Kanzaki facility. At the outset, we identified 96 companies that were then handling asbestos materials and the final count rose to 120. We shared that data with the Amagasaki municipal government, and urged that it by all means conduct relevant surveys. Many companies that no longer exist were running factory operations in Amagasaki at the time. Although we do not think they have all been accurately investigated, it is our understanding that an exceptionally large share were demolition companies. We would like to see steps taken to gain a better understanding of the facts."

"Relief" and "compensation" are fundamentally different concepts. Whereas relief refers to efforts in assistance or aid, compensation has to do with acts of monetary or other restitution or indemnification for damage or expenditure, and as such, clearly implies an obligation or liability. Kubota, however, has consistently refused to admit to any causal relationships. Yet, it still provides monetary payments to asbestos victims. This ambiguity conceals the essential nature of the problem. Kubota's asbestos disaster has subsequently gained growing infamy as the worst industrial calamity of its kind in Asia. Against that backdrop, "relief" payments made without any assumption of responsibility cast a shadow of uncertainty over the future outlook for the victims of this disaster.

7.2.7 *Circumstantial Evidence*

"We weren't the only source." "The sources can't be pinpointed yet." These are examples of the assertions Kubota continues to make, but the findings of three different investigations by government administrative agencies and research institutions have highlighted unshakable facts about the source of asbestos contamination.

First, consider the investigation performed by Prof. Norio Kurumatani from Nara Medical University and Researcher Shinji Kumagai from Osaka Prefectural Institute of Public Health (currently an associate professor with the University of Occupational and Environmental Health).

Immediately after the outbreak of asbestos-related diseases had been brought to light, these investigators began interviewing patients and surviving families of deceased patients and have collected data on 162 patient cases to date. Among their findings, they discovered that the risk of death for female and male residents living within a 300-m radius of the Kanzaki factory was 41.4 times and 13.9 times the national average, respectively. They also compiled a concentric ring diagram visually illustrating the distribution of mesothelioma cases from the factory center. In 2005, this graphic had a huge impact. Kubota, however, continued to shrug off these findings as nothing more than anecdotal data.

Second was the health impact assessment performed by the Ministry of the Environment (MOE). Implemented in May 2007, that assessment revealed that the risk of death from mesothelioma for women living in the Oda district of Amagasaki (where asbestos factories were located) was an astonishing 68.6 times the national average, which is much higher than for other districts. When asked about these findings, Kubota President Hatakake declared: "We were not the only source. This was a case of contamination from multiple facilities. The truth is that they don't yet have a solid grasp of all the facts."

As part of an environmental risk assessment performed by the same ministry from FY 2006 through FY 2009, 844 Amagasaki citizens underwent medical examinations, and signs of pleural plaque were detected in 206 (24%). That included 81 (19%) of 417 subjects who had never worked at, or entered, a facility engaged in operations involving the use of asbestos.

Broken down by residential district, the detection rate for pleural plaque was highest, at 36%, in examined subjects from the Hama district where the Kubota Kanzaki factory was located. The rate was 20% in subjects from districts surrounding Hama (Tsugiya, Nishikawa, Jokoji, Nagasu Higashi-dori, Nagasu Naka-dori, and Shioe), and 17% in subjects from other districts.

The third investigation of interest was a survey performed by the City of Amagasaki in response to assertions by Kubota that it was not the only source of contamination. That survey sought to identify businesses in the city that had handled asbestos in the past.

Of 136 businesses listed in the Amagasaki Chamber of Commerce and Industry register and other documentation, 39 were identified as having utilized asbestos at some time in their past. Of that subset, two stood above all the rest in terms of the amounts of asbestos they handled: Kubota and Kansai Slate. Kubota had utilized 237735 metric tons, and Kansai Slate 74750 tons. Two other businesses had utilized 10–100 tons, five had utilized 1–10 tons, and nine had utilized less than 1 ton. The amounts utilized by 21 other firms were unknown.

In other words, Kubota and Kansai Slate stood far above all the rest in terms of quantities of asbestos utilized. Kansai Slate only utilized the chrysotile form of asbestos. Kubota utilized both chrysotile and crocidolite, and of the aforementioned total for Kubota, crocidolite accounted for about 88 000 tons, the most utilized by any single factory in Japan.

Around the time this survey was performed, Sadayuki Tomoe, chief officer of the Department for Countermeasures against Industrial Pollution, the City of Amagasaki, noted that in the light of all available evidence, the source of contamination was none other than Kubota's Kanzaki factory.

What conclusions may be drawn from the findings of the three studies described above? Testimony from Kubota employees also made it clear that the factory operated with its windows open.

As exemplified by the statements that it cannot unequivocally deny that its factory had ever released emissions of asbestos dust, that many other factories in Amagasaki were utilizing asbestos, or that the causes of asbestos-related diseases in local residents cannot yet be pinpointed, Kubota has relied on a line of reasoning aimed at avoiding responsibility.

Kubota should rephrase its statements as follows. "We released asbestos dust into the surrounding environment." "We utilized unprecedented amounts of crocidolite in Amagasaki." "Kubota is the principal cause of most of the health problems observed among local residents." The unshakable truth is that Kubota was the source of pollution, the cause of the outbreak of asbestos-related diseases, and an offending enterprise.

With the number of victims growing at the current pace, Kubota's logic that it is paying "relief" because it has a "moral responsibility" will eventually no longer hold water. It should be using the terms "corporate liability" and "damage compensation" instead.

7.2.8 The First Lawsuit Against Kubota

Kubota has continued to make "relief" payments while arguing that the causal relationships remain uncertain. Against that backdrop, some victims eventually decided to file litigation on claims of corporate liability.

In May 2007, Yasutami Yamauchi, a Kubota employee then age 59, filed a lawsuit against Kubota in the Kobe District Court. Yamauchi's father had worked for 37 years at the factory of another company located 50 m south of Kubota's former Kanzaki facility. Although his job did not involve handling any asbestos, in 1996 he died of mesothelioma at the age of 80.

Yamauchi's lawsuit sought 42 million yen in damage compensation from Kubota and the national government on the charge that the defendants were negligent in curbing emissions of asbestos dust or implementing regulations despite their knowledge that asbestos was hazardous. This was the first lawsuit to be brought against Kubota for asbestos-related health problems.

Yamauchi had deferred acceptance of relief payments and filed his lawsuit because Kubota had failed to explicitly acknowledge any causal relationship between its emissions of asbestos dust and the death of his father.

The written complaint filed by Yamauchi charged that Kubota and the national government had known of the serious impact asbestos had on human life and health, including the impact to local residents, by no later than 1960 based on the findings of studies and reports dating back prior to World War II. In addition, it held Kubota liable for the continued utilization of asbestos in the absence of proper measures against the release of asbestos dust, and accused the government of negligence in the implementation or enforcement of regulations aimed at preventing asbestos-related health problems.

In December 2008, the surviving family of the late Ayako Yasui, an Amagasaki resident who had lived near the former Kanzaki factory and who had died in September 2007 at the age of 85, filed a lawsuit in Kobe District Court seeking damage compensation from the government and Kubota on grounds that Yasui's death was due to mesothelioma caused by exposure to dust from asbestos in use at the Kanzaki facility.

The Yasui family stated that from 1972 to 1995, Ayako lived about 1 km from the Kanzaki factory, and that the bus stop for the fixed-route bus she regularly used was located near the factory entrance. Ayako was given a diagnosis of malignant pleural mesothelioma in 2006.

Her husband, Yasuo Yasui (aged 84), states that at first he felt that the matter was beyond his control, but later decided to file litigation because he as well as his daughter were advancing in age.

Kubota should clearly demonstrate its responsibility by paying damage compensation, not "relief." It can contribute to a full understanding of the scope and scale of the asbestos-related disease outbreak by pursuing investigations and disclosing all relevant information in full. That would be an approach that would demonstrate a true sense of corporate responsibility.

7.3 Sennan Asbestos Disaster

7.3.1 Sennan, Osaka: The Birthplace of Japan's Asbestos Industry

Although the Kubota Shock shook the entire Japanese archipelago, the blow it dealt the Sennan district of Osaka proved especially powerful.

Situated across the bay from Kansai International Airport and comprising the cities of Sennan and Hannan, the Sennan district is known as the birthplace of Japan's asbestos industry. It served as a center for asbestos textile operations that began early in the twentieth century and continued for close to 100 years. The Sennan community had a heavy concentration of small-scale factories and cottage industries. Prior to World War II, they were engaged in the production of goods for the military and defense industry, and after the war, they fulfilled a supporting role as suppliers mainly to the automotive, shipbuilding, and machinery industries.

At its peak, the local asbestos industry reportedly comprised around 200 firms and engaged as many as 2 000 workers. Sennan evolved from the prewar years into the nation's largest center of asbestos textile factory operations.

The chief textile products were yarns that blended cotton with asbestos fiber as the main ingredient, as well as cord and fabric goods that were processed from those yarns. Asbestos boasted excellent heat resistance and insulating properties and was widely used, for example, in an assortment of gasket packing materials, the heat-insulating sleeves in steam locomotives, and the friction materials in car brakes.

Although the Sennan community was once described as "just one giant asbestos factory," its appearance today no longer supports that image.

Working conditions in the asbestos textile factories were harsh. All stages of the operation, including the blending stage that blended asbestos fibers with cotton, and the carding stage that involved breaking up clumps of asbestos and sorting it into fibers of roughly equal length, were conducted within closed areas amid clouds of asbestos dust.

The atmosphere inside the typical asbestos factory seemed especially bad to Mayumi Hara (aged 67) (Fig. 7.3), one of the plaintiffs in a lawsuit. Raised in Tokushima, she was employed at a textile factory in central Hyogo Prefecture, but

Fig. 7.3 Mayumi Hara with a portrait of her husband (© The Kobe Shimbun)

in 1961, at the age of 17, she landed a job with an asbestos factory in Hannan, Osaka, through a connection with a coworker.

"The atmosphere was completely different from the typical textile mill. It was so dusty inside that our arms and the backs of our necks became dry and rough." Hara moved from one job to another at eight different asbestos factories over the ensuing 46 years.

Unsurprisingly, her husband, who also worked in an asbestos factory, contracted a lung disease and died at the young age of 44. On his death bed, he repeatedly implored: "Asbestos is no good. You quit, too!" Hara recalls that in her husband's ashes, some of the bones that would have been close to his lungs were bloodstained brick red.

Hara herself suffered from a productive cough with expectorate, and frequently woke at night in respiratory distress. "Although the government knew about the hazards, it did nothing. I wonder how many victims have died in bitterness and anger."

Factory conditions at the time are portrayed in detail in *Blame the Government for the Asbestos Tragedy* (2009). Kazuyoshi Yuoka, a man who ran an asbestos business started by his father, "This was the destination for many people on the bottom rungs of the social ladder: Korean nationals in Japan and people from rural districts and other areas subject to discrimination."

Although this was a genuine, homegrown industry, it was not properly supervised, and workers who became ill were left to suffer in silent misery. They literally had no one to turn to for assistance.

7.3.2 Asbestos Slum

Ten or more small-scale factories lined the banks of the river that flowed through this district. At the time, they were referred to collectively as *ishiwata-mura* ("the asbestos slum").

Yuoka explains: "This was without question the bottom of the ladder. The workers there were always covered completely in white asbestos dust, knowing full well the hazards they faced. Discrimination was rampant on various levels."

Collecting spent "scrap asbestos" was the job for one man employed by these small factories as a day laborer. After hearing about the Kubota disaster, he visited a hospital for an examination and was given a diagnosis of pulmonary asbestosis. "In those days, the dust in the air was so bad you couldn't see the guy in front of you. All of my coworkers have died; no one is left."

This was Sennan's long-neglected asbestos disaster. As was the case in Amagasaki, the exposure to asbestos dust did not stop with workers, but spread to their families and residents living near the factories.

Yoko Okada (aged 54), a former nurse living in Sennan, Osaka (Fig. 7.4), cannot put down her oxygen inhaler. Her condition has deteriorated over the past few years

Fig. 7.4 Yoko Okada (*left*) with oxygen inhaler and Harumi (© The Kobe Shimbun)

and she is no longer able to work as a nurse, a vocation she always thought of as her dream job. She now has to have an oxygen inhaler tube inserted through her nose 24 h a day.

"I hate going out because people always give me strange looks." The diagnosis by her doctor was pulmonary asbestosis. Should she remove the inhaler tube, she will experience respiratory distress and begin to lose consciousness.

Until the age of 12, Okada lived in company housing adjacent to the asbestos factory where both her parents worked. She was literally raised in a cradle made from a basket for asbestos products. No one at that time had any knowledge of the hazardous nature of asbestos. The president of the company that ran the factory would exclaim, "It's harmless. They even use it for baby powder!" Then he would demonstrate by licking some asbestos in his hand.

However, in 1995, Okada's father died of lung cancer. Her mother, Harumi (aged 74), also suffers from pulmonary asbestosis. Harumi concedes, "Had I even known that asbestos was dangerous, I never would have taken Yoko into that factory." With that remark, she then broke into tears.

Most of the asbestos handled by factories in the Sennan district was the chrysotile variety. Although less toxic than crocidolite, it is the cause of numerous respiratory illnesses, including pulmonary asbestosis and diffuse pleural thickening. Unlike mesothelioma, these diseases do not lead to death within a few years after onset, but most victims agree that the downside is a life of misery caused by long periods of respiratory distress.

For Yoko Okada, that misery started with a persistent cough and expectorate in her mid-to-late 20s. She admits, "I had no idea asbestos was something this dangerous."

After meeting asbestos victims in Amagasaki in November 2005, several months after the Kubota Shock, then Environment Minister Yuriko Koike declared that she would put together a seamless relief plan that left no one out. The Act on Asbestos Health Damage Relief would be enacted a few months later.

However, for Yoko Okada, the safety net provided by law seemed full of holes. The biggest drawback in her case was that asbestosis was left outside the scope of consideration. She has lost her career and has zero income. The approximately 40 000 yen she pays each month for medical treatment and inhaler maintenance has become a heavy burden.

Although asbestosis is a disease which is eligible for coverage under the workers accident compensation insurance framework, Okada did not contract the disease through the inhalation of asbestos dust while working in her job. Consequently, when she visited her local Labor Standards Inspection Office for consultation, officials there informed her that she was not covered.

7.3.3 Damage Lawsuit Against the Government

Sennan's asbestos disaster has been almost ignored by history. However, the Kubota Shock of 2005 awakened society from 100 years of slumber.

Yuoka was startled to hear about the horrific scale of asbestos-related disease in Amagasaki. "Mention asbestos and everyone thinks of Sennan. So Amagasaki was hit, too?" This seems emblematic of the level of local awareness regarding asbestos hazards. Apparently there was virtually no awareness of the disease mesothelioma either. In fact, some asbestos factories were still operating in Sennan.

Yuoka and his peers formed a citizen's association and began searching to identify more asbestos victims.

"When we decided to investigate the level of harm from asbestos, we started with interviews with local residents and counseling activities. At the outset, we learned that many citizens had already died from respiratory diseases, and that quite a few of these were in the prime of life, in their 30s and 40s, when they succumbed. The problem was the attitude of indifference shown by local residents, who looked on all such victims as individuals who had chosen to earn their bread and butter in jobs involving asbestos."

One thing that became evident in this undertaking was the factor of government inaction. "Although the government was in a position to know about the hazards of asbestos and take remedial action, it did nothing."

In May 2006, a damage lawsuit was filed against the national government. This was the first class-action lawsuit of its kind in Japan to charge the government with liability for an asbestos disaster. In that instant, the extent of harm caused by asbestos over many decades was at last out in the open.

Consider the following study from the prewar era.

In 1937, the national government already had an understanding of the extent of asbestos-related disease in the Sennan district. A study carried out by the Insurance Institute in the former Home Ministry had determined that the incidence of asbestos-linked diseases among members of the local working population was as high as 12.3%. The physicians heading that study warned that many afflicted workers were affiliated with hazardous industries, and that legal controls were needed. The incidence of these diseases did not fall below 10% following World War II, prompting Labor Standards Inspection offices to launch additional studies.

However, during the course of the lawsuit, the national government maintained that the aforementioned statistics were nothing more than the findings of limited case studies, and held firmly to its position that it would have been difficult, if not impossible, at that time to recognize the harm caused by asbestos.

Unlike Kubota, which was a corporate giant, almost all of the operations in Sennan were small-scale businesses with inadequate frameworks for the prevention of asbestos-related illnesses. That is precisely the reason why the national government should have been considered to be responsible for measures to prevent the spread of asbestos-linked health problems.

7.3.4 Why the Neglect?

The operations of the asbestos industry in Sennan, Osaka, lasted a full century. However, the asbestos-related health problems of employees in that industry, as well as of local residents living near asbestos facilities, were ignored. Why so?

As an individual born and raised in the area, Kazuyoshi Yuoka explains: "That was because we had the issues of discrimination and poverty."Of the workers in asbestos factories in the Sennan area, a great many were Korean nationals or impoverished people from the lowest classes of society. Although all the asbestos operations have now gone out of business and the factories have been shut down or dismantled, there are still many people who do not want to talk about their former asbestos-related careers.

On one occasion, when I arrived at the site of a former factory with an old map of residential addresses in my hand, one of the residents confronted me with a troubled expression and said, "Just leave us in peace."

Although the health problems caused by asbestos have finally come to the surface, so far we are probably seeing only the tip of the iceberg.

Yuoka explains: "To date our association has probably had around 200 citizens show up for counseling, but we suspect that the actual number of victims is more like 50–100 times that number."

One gentleman I spoke to was an 80-year-old Korean national who had worked in Sennan and contracted asbestosis. He was one of the plaintiffs in the Sennan damage lawsuit that had been filed against the national government. Starting in 1951, he spent 8 years as a day laborer employed by at least ten factories. His work included mixing asbestos and cotton fibers together by hand, and filling bags of asbestos with a shovel. "Everything around me was so white with dust that it was impossible to see the guy working beside me." The work environment was harsh, but he felt he had no choice but to endure it if he wanted to make a living.

Ritsumeikan Asbestos Research Project has been engaged in research on asbestos-related issues in Japan and overseas. With regard to the Sennan case, it notes that one of the reasons that awareness of the asbestos problems there were delayed was that many of the victims were discriminated against and isolated by the local community, and compelled to live quietly with their condition.

One of the most prominent features, not only of the Sennan disaster but also of cases of industrial pollution in general, is that a large proportion of the victims tend to be among the more disadvantaged members of society, who lead lives of desperation compounded by poverty.

7.3.5 Beyond the Limits of Social Tolerance

On May 19, 2010, the Osaka District Court issued a ruling that for the first time recognized the national government's liability for damage compensation in a Sennan lawsuit concerned with asbestos-related health problems. Although it was pouring with rain in front of the courthouse, when several lawyers came running out of the building, splashing through the rain, and waving banners that declared "Victory," or "State Liability Established," several bystanders let out cries of "Banzai!" (Fig. 7.5).

One key feature of the court ruling is that in assigning that liability, it strongly censured the government. Because asbestos-related diseases are chronic in nature and difficult to treat, "countermeasures deserved urgent priority." By the time the old Pneumoconiosis Act was established in 1960, the national government was being strongly urged to mandate that companies must install on-site ventilation systems

Fig. 7.5 Victory in the damage lawsuit against the National Government on May 19, 2010 (© The Kobe Shimbun)

and other dust-control equipment. Despite those recommendations, it did not mandate such measures until the implementation of the Ordinance on Prevention of Hazards Due to Specified Chemical Substances in 1971. As a consequence, concluded the court, "the scale of victimization from exposure to asbestos dust expanded as the asbestos industry posted rapid growth."

In addition, in 1972, when correlations between lung cancer and mesothelioma were disclosed, the government failed to mandate that factories report on, and take steps to reduce, measured concentrations of dust in their workplaces. In other words, it failed to ensure effective implementation. The court thus ruled that the government's inaction was illegal because such policies were deemed unacceptable as a deviation from tolerable limits, and extraordinarily lacking in their rationale.

Until this ruling, the government had consistently denied any liability. Although it had conceded a lack of coordination among its various ministries and agencies, it had repeatedly insisted that its past actions were appropriate, and that they could not be described as any slower than steps taken by other countries.

The principal point of the ruling was that the government and private corporations had engaged in a joint act of tort that set the stage for a spread in the incidence of asbestos-related disease. Further, it threw out the government's assertion that its liability was secondary, or complementary, to any corporate liability. Health problems caused by the corporate use of hazardous materials, and the failure of the government to enforce appropriate regulations, are occupational accidents if they arise only inside the factory, but are treated as the outcomes of industrial pollution if they spread beyond the factory's perimeter. This court ruling signaled that important distinction.

Nonetheless, on June 1, 2010, the government appealed the Osaka District Court ruling – the first to find it liable – thus seeking to have it overturned by the Osaka High Court.

Initially, Health, Labour, and Welfare Minister Akira Nagatsuma and Environment Minister Sakihito Ozawa were ready to drop the idea of appealing the court ruling. However, the decision was left to the Minister of National Policy Unite, Yoshito Sengoku, who chose to move ahead with the appeal. Judging from a series of comments made by cabinet ministers, it appears that an early settlement is the Cabinet's goal. Having chosen to go ahead with the appeal process, the government will conceivably pursue a dual strategy of searching for a resolution while seeking to overturn the first ruling in court. It will be interesting to see how the Kan Cabinet deals with this matter.

7.4 Further Issues

7.4.1 Epidemiological Studies

The problem of industrial pollution starts and ends with victims. This can be described as one of the lessons of history.

Consider the situation in Amagasaki, where the scale of victimization stands out.

Arguably the most important point is that Kubota exposed residents living around its factory to a high risk of contracting asbestos-related diseases during the period (1957–1975) that the facility utilized highly toxic crocidolite as a raw material.

The Ministry of the Environment has conducted asbestos health risk assessments in six communities, including Amagasaki. The scope of those assessments covered the medical records of patients who had undergone health examinations at community clinics and other health-care facilities. According to recently compiled findings for FY 2008, the examinations detected certain signs of illness in 33% of all subjects in Amagasaki who had no history of contact with asbestos in their occupations, etc., with indications of asbestos-induced pleural plaque in 24%.

"This is the tip of the iceberg. We need to try and identify all the hidden victims, including those who have moved away from the Amagasaki area," says Hiroshi Iida, Director of the Amagasaki Occupational Safety and Health Center, an organization engaged in efforts to provide aid to asbestos victims.

Right now, what is really needed is an epidemiological study of residents who lived near asbestos factories during the period that crocidolite was being utilized. Even though operations for the mining and production of asbestos have been brought to a halt, as long as accumulations or stocks of this material remain in human bodily tissues, buildings, and consumer goods, it will continue to cause health problems. Needless to say, given the traits of such "complex stock" disasters, studies aimed at shedding light on the incidence of asbestos-related diseases and their root causes will be essential at the community level.

"If only we had known sooner, I might have had time to have one of my lungs removed." Such laments are extremely common among patients and their families. In one case, a patient who had moved many years earlier from Amagasaki to another city had received a diagnosis of lung cancer. Prompted by news of the Kubota Shock in 2005, he recalled his years of residence in Amagasaki and decided to undergo a comprehensive medical examination, which determined that he had mesothelioma.

No doubt it may be rather difficult to motivate individuals who lived in Amagasaki years ago to be very concerned about a risk that they may have resided in an asbestos-contaminated area.

However, that is precisely why it is so important that efforts should be made to help such individuals recall and rebuild their residential histories step-by-step. They may have lived near, or commuted to a school near, Kubota's Kanzaki factory during the period that it utilized highly toxic crocidolite in very large quantities. They may have worked at the factory itself. Such individuals have to be identified. If certificates of domicile are no longer available, perhaps lists of school graduates can be used. I want to urge the City of Amagasaki and other municipal governments to use all the tools at their disposal in a determined effort ultimately to recreate lists of residents that would cover the period in question.

Many people bearing the stubborn fibrous thorns of asbestos in their lung tissues have scattered to communities nation-wide. That is exactly why we need to uncover the causal relationships through far-reaching epidemiological studies. That, and nothing less, will be the only way to even marginally reduce the number of victims that appear in the years ahead. This is indeed a challenge that must be addressed, especially in Amagasaki, the locus of the largest asbestos disaster in Asia.

7.4.2 From Relief to Compensation

Right now, the trend that started with the Kubota Shock and transitioned to the Sennan lawsuit is gaining momentum. The Tokyo Metropolitan Asbestos Lawsuit, in which a group of 400 plaintiffs, including construction workers and the surviving families thereof, have demanded damage compensation from the national government and several construction materials manufacturers, is playing out in both the Tokyo and Yokohama district courts. Residents living near Kubota's former factory in Amagasaki have filed a lawsuit of their own in the Kobe District Court.

At a general meeting of Kubota's shareholders on June 18, 2010, one attendee described the asbestos problems as a bitter legacy for the company. In retrospect, the World Health Organization (WHO) and the International Labour Organization (ILO) alerted us to the carcinogenic properties of asbestos in the 1970s, and several countries moved to ban the use of this material in the 1980s. However, in Japan, industry continued to utilize asbestos in huge quantities under the administrative guidance of the former Ministry of International Trade and Industry and the Construction Ministry and without adequate regulation amid assertions that asbestos was safe if properly managed. Japan implemented its ban on the use of asbestos in principle in 2004. In total, Japanese industry consumed approximately 10 million metric tons of asbestos.

The symptoms of asbestos-related diseases begin to appear anywhere from 10 to 40 years following the inhalation of asbestos dust. Individuals who were exposed to that dust face the risk of contracting mesothelioma, lung cancer, pulmonary asbestosis, or other respiratory diseases long after they have forgotten the circumstances of their exposure. Asbestos causes these illnesses at every economic stage of the product life cycle, from production and distribution to consumption and final disposal. That is the reason why Kenichi Miyamoto, Professor Emeritus of the Osaka City University, applies the expression "complex-stock disaster."

The world has utilized a combined total of approximately 200 million tons of asbestos since the 1920s. In the early stages of modernization, it was elevated in status to that of a "magic mineral" and used in large quantities. As economic growth began to level off, the microscopic fibers of asbestos – only 1/5 000th the thickness of a human hair – began to demonstrate their true nature as "thorns of death" lodged deep inside the lung tissues of their human hosts. This is the frightening dimension of a complex-stock disaster.

Although some are of the view that we had no choice but to use asbestos because it was inexpensive and lacked substitutes, that in no way justifies the continued use of a life-threatening hazardous material over a period of many years without taking adequate steps to bring its hazards under control. As the ruling by the Osaka District Court indicated, such conduct deserves the description of a joint act of tort by the national government and the asbestos industry.

Compensating residents who are not covered by workers accident compensation insurance will be a huge challenge. Referring to the Act on Asbestos Health Damage Relief that came into force in March 2006, the then Environment Minister

Yuriko Koike declared that relief would be seamless with no one left out. Mesothelioma is a disease with an extremely poor prognosis. Indeed, the 2-year survival rate is only 30%. In terms of compensation, the government's Act on Asbestos Health Damage Relief seems inadequate, since it provides victims with no more than a 100 000 yen medical allowance. If the wage-earning breadwinner of a household is struck by one of the diseases covered, it will be virtually impossible for that household to survive on an allowance that small.

What the government should do is to make the transition from this poor legal framework of relief to a fully fledged system of compensation that is on a par with the workers accident compensation insurance framework. There is a huge difference between ambiguous "relief" and a system of "compensation" built on the admission of government liability. Even though it has appealed, the government can no longer deny liability, and as such, it should acknowledge its administrative responsibilities and develop a comprehensive program of countermeasures comprising the provision of welfare compensation to victims, care for future victims, steps to identify the extent of victimization, and the implementation of controls against dust during building demolition projects. Unless it takes these steps, it will never have the ability to effectively curb the incidence of mesothelioma even 40 years from now.

Reference

Osaka Lawyers for Pneumoconiosis & Asbestos Victims and the Citizens' Group of Asbestos Damage in Sennan (2009) Blame the government for the asbestos tragedy (in Japanese). Kamogawa Press, Kyoto

Chapter 8
Asbestos Industry Transplants from Japan to South Korea

Shinjiro Minami

8.1 Introduction

Using the example of the transplantation of the asbestos industry from Japan to South Korea, this paper explores the "exportation" of the asbestos industry and associated asbestos hazards and harm.

Asbestos industry workers have an exceptionally high risk of exposure to asbestos dust, and in reality the incidence of asbestos-related diseases has been higher among such workers. This has heightened the need for regulations against dust as well as for workers accident compensation for victims, which in turn have confronted asbestos product manufacturing operations with steadily rising costs. Many industries that engage in dangerous manufacturing processes – not only those involved in the manufacture of asbestos products – or that generate hazardous industrial waste have been moving their plants overseas (primarily to developing countries) to take advantage of locally inexpensive labor and weak regulatory frameworks for worker safety, thus operating in an atmosphere of relatively little or no pressure to implement preventive measures against workplace accidents or toxic contamination.

This pattern also applies to the history of the asbestos industry in Japan. When facing stiffer regulations around 1970, members of the Japanese asbestos industry began moving their operations offshore to South Korea, and were therefore involved in the expansion of the asbestos industry as well as incidents of asbestos exposure and harm in that country. This chapter first sheds light on the history of asbestos industry relations between Japan and South Korea, and drawing from that study, then explores the transfer or "exportation" of the industry's manufacturing processes, and the accompanying health problems associated with asbestos exposure.

S. Minami (✉)
Ritsumeikan Global Innovation Research Organization, Ritsumeikan University, 56-1 Tojiin Kitamachi, Kita-ku, Kyoto 603-8577, Japan
e-mail: nann@mbg.ocn.ne.jp

K. Miyamoto et al. (eds.), *Asbestos Disaster: Lessons from Japan's Experience*,
DOI 10.1007/978-4-431-53915-5_8, © Springer 2011

8.2 The History of South Korea's Asbestos Industry

In the early postwar years, South Korea proved to be slower than other industrial countries to register growth in its asbestos industry, or transition into an economic phase marked by mass consumption. As the historical record of the development of the asbestos industry illustrates, the expansion phase in the number of asbestos industry firms and the volume of asbestos imports, production, and consumption coincided with a period extending from the 1970s through the 1990s. This fueled expectations that the country would eventually be hit by asbestos-induced workplace accidents and industrial pollution resembling actual cases in Japan, and in fact, the incidence of asbestos-related diseases has become pronounced on a wide scale in Busan and other communities. The expansion of the asbestos industry in South Korea, the health problems associated with that industry, and measures to counter those problems were all closely related to a trend that originated in Japan: namely, the drive by Japan's asbestos industry to shift its factory operations offshore in a quest to reduce costs and maximize profits. The following section will focus on this feature by tracing the history of the asbestos industry's development in South Korea.

8.2.1 The Development of Asbestos Mining Operations in South Korea

The formative years of the South Korean asbestos industry began during the period of occupation by Japanese forces prior to the start of the Pacific theater of World War II. The world's key centers of asbestos mining were concentrated in only a few countries: Canada, the Soviet Union, and South Africa. Japan itself had become dependent on imports of raw asbestos from these countries because asbestos from these sources was of superior quality. However, during World War II, the flow of foreign imports came to a halt, and Japan found itself pressured by the need to produce and secure the necessary supplies of raw asbestos from domestic sources and territories then under Japanese military control, including the Korean Peninsula, China, and Taiwan.

Japan had explored, and begun developing, asbestos deposits on the Korean Peninsula. By 1945, at the end of the war, 16 asbestos mines were reported to be in operation (The Government–General of Korea 1933). According to a report by the Japanese Agency of Industrial Science and Technology, the *East Asian Co-Prosperity Sphere Asbestos Mining Field Survey* (hereafter the *East Asian Co-Prosperity Sphere Report*), six commercial asbestos mining firms were operating on the Korean Peninsula and all were managed by Japanese business owners. One of the large-scale operations, Teikoku Asbestos Co., Ltd., had been founded as a joint venture by three companies – Kyokuto, Taiyo, and Kansai Slate. The others were the overseas branch of Asano Cement Co., Ltd., Hong Seong Asbestos Industries, Ltd., Korea Asbestos Co., Gi Nae Mining Co., Hong Dong Asbestos Mining Co., and Osaka Asbestos (Koa Mining Co.). Table 8.1 gives asbestos production totals

Table 8.1 Production of asbestos mines in Korea, 1937–1943

	1937	1938	1939	1940	1941	1942	1943
Chrysotile	(–)	(–)	1310	1530	1845	3074	4600
Amphibole forms	(–)	(–)	440	440	450	83	400
Total	7	286	1750	1970	2295	3157	5000

Unit: metric tons
Note: The results for 1941 and 1943 are estimates
Source: Sonokawa (1944:100)

for these operations as a whole (Sonokawa 1944). For comparison, a paper by Choi et al. (1998) on asbestos production totals in South Korea states that chrysotile production in South Korea had peaked at 4815 metric tons by 1945 (Choi et al. 1998:242–253). In the East Asian Co-Prosperity Sphere Report, the total workforce for all six mining operations together was listed at 2707 people.

In terms of operational content, the Korea Asbestos Co. was the only firm that had begun building plants to manufacture asbestos products (asbestos yarn, gasket materials, insulation). The majority of the other operations were involved in the manufacture of asbestos cement products (slate, chimney materials, shingles).

This was how Japan developed and operated asbestos mining facilities on the Korean Peninsula during World War II. Through these actions, Japan basically set the stage for the launch of the South Korean asbestos industry. With the end of the war, Japan withdrew from the Korean Peninsula, and until Japanese companies began returning to do business around 1970, South Korea's asbestos industry operated on a small scale.

8.2.2 South Korean Asbestos Production, Consumption, and Industry Growth and Decline

This section starts with a focus on trends in an assortment of basic data on South Korean asbestos production, consumption, and the number of companies in the industry from the end of World War II to the present day.

First, with regard to South Korean asbestos mining operations after World War II, we will examine some data in the previously cited Choi paper on the volume of asbestos production (Choi et al. 1998:245). No data are available for the period from 1946 through 1949, but for the decade of the 1950s, the volume of asbestos production averaged approximately 55 metric tons per annum, which is an indication that mining operations were continuing, albeit on a marginal scale compared with the prewar years. Although the trend was erratic, with variations from year to year, from 1960 onward the production volume climbed from several hundred to several thousand tons, reaching 6515 tons in 1969. From 1970 until asbestos mining operations ended in 1990, the volume of asbestos production in South Korea averaged at least 1000 tons per year. In particular, during the 6-year period from 1978 through 1983, the annual production volume peaked at an average 13388 tons, with the highest level of output being achieved in 1982, i.e., 15933 tons. (Note, however, that no corresponding data are available for the period from 1971 through 1974.) Access to less expensive foreign imports has been cited as the prime reason for the curtailment of asbestos mining within South Korea.

It appears that the asbestos mined from the end of World War II to around 1970 was utilized chiefly for the production of asbestos cement products (asbestos slate). The 1998 Choi paper contains a statement noting that asbestos slate had been in production for 50 years or more, allowing the conjecture that South Korea retained the slate production technology that Japan had used in the prewar era. The same 1998 paper states that the production of asbestos textiles, as well as brake linings and other friction materials, had begun 27 years earlier, thus dating the inception of South Korea's asbestos textiles sector at around 1970 (Choi et al. 1998:242). Table 8.2 shows data on the number of corporate asbestos factories and employees in selected sectors from 1965 to 1993. As the table shows, in 1965 the construction materials sector had only one asbestos facility (involved in slate production).

From 1970 onward, the number of companies engaged in the manufacture of asbestos products multiplied in various sectors in addition to the construction materials sector. Table 8.2 confirms that this expansion continued until 1993. As domestic demand for raw asbestos rose in parallel, efforts were made to expand domestic mining output, but imports eventually became the main source of supply of raw asbestos. Data on raw asbestos imports from 1976 through 1995 reveal an import volume of 36787 tons in the slowest year and 95476 tons in the peak year, with an aggregate total of 1216505 tons being imported over the entire period (Choi et al. 1998:246).

Construction materials accounted for most of the asbestos consumed within South Korea. This corresponded closely with the pattern of consumption in Japan, with at least 80% of all asbestos being utilized for the manufacture of construction materials. Construction materials accounted for at least 80% of all imported raw asbestos use and 96.1% of the total used in 1976.

The volume of asbestos production and consumption continued to grow year on year, and corporate business operations in the industry thrived. From 1971 onward, domestic demand for asbestos in South Korea ranged between 60000 and 80000 tons in most years, fell to the 40000-ton level in weak years, and increased beyond the 100000-ton level in the peak year, 1992.[1] However, in 1995, the annual totals in domestic demand and imports of raw asbestos began falling sharply, with

Table 8.2 Number of asbestos factories and workers in Korea

	1965	1970	1975	1980	1985	1990	1993
Costruction material	1(207)	2(270)	4(471)	4(446)	6(404)	7(374)	9(376)
Frictions	0(0)	1(112)	3(277)	7(324)	12(414)	23(574)	33(637)
Textiles	0(0)	1(157)	3(187)	4(244)	6(350)	13(330)	15(214)
Others	0(0)	1(7)	4(34)	11(83)	22(128)	37(168)	61(249)
Total	1(207)	5(546)	14(969)	26(1097)	46(1296)	80(1446)	118(1476)

Unit: factories (workers)
Source: Choi et al. (1998:248)

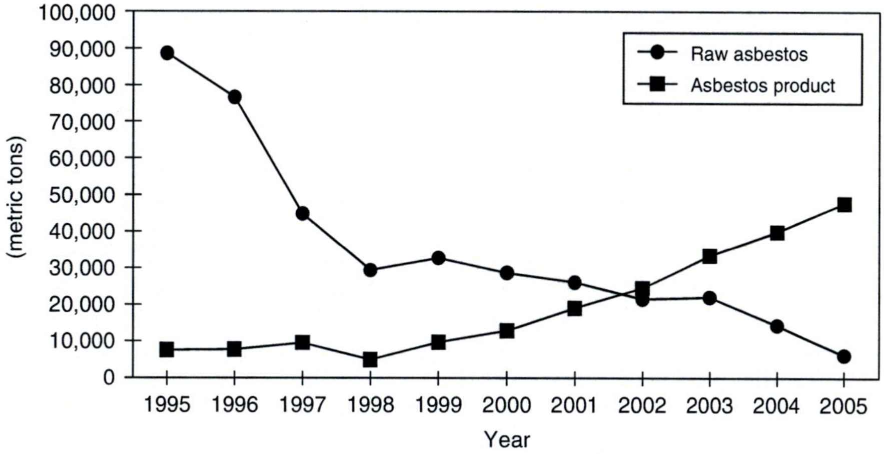

Fig. 8.1 Imports of raw asbestos and products containing asbestos in Korea. *Source*: Ministry of Employment and Labor of Korea (2007). "Measures for Asbestos Control"

the latter hitting a low of 6477 tons in 2005, as shown in Fig. 8.1. In contrast, from 1998 onward, imports of manufactured products containing asbestos showed a steady upward trend, totaling 47967 tons in 2005. One conceivable factor behind these trends was a withdrawal from the industry by elements of the manufacturing sector that faced tighter regulations on domestic asbestos production facilities, and the necessity of implementing measures against asbestos dust. Acting under the provisions of the Industrial Safety and Health Act, in 1988, South Korea established a benchmark exposure ceiling of 2 fibers cm^{-3} (2 f/cc) for asbestos workplaces, and implemented a licensing framework for facilities that handled asbestos. In 1997, a ban was put into effect on the use of the crocidolite and amosite varieties of asbestos, the public gained a certain awareness of the hazards of asbestos, and the image of the asbestos industry itself increasingly became that of a sunset industry in decline. As an outcome of these developments, many companies presumably either shut down, or changed to other lines of business, or relocated their plant operations to countries in Southeast Asia where regulations did not yet exist. The Korean Ministry of Labor register of companies that handled asbestos as of FY 2005 contained the names of only 25 firms.

8.2.3 *South Korea's Asbestos Industry and Factors Contributing to Growth in Domestic Demand*

Beginning around 1970, South Korea experienced an influx of foreign asbestos companies and related technologies, followed by upward trends in the volume of asbestos imported, the total domestic demand, and the number of asbestos companies.

The key factors that contributed to the growth of South Korea's asbestos industry and the quantitative expansion in domestic demand are enumerated below.

First, the Sae-Maul movement may be cited as a factor that promoted the acceptance and use of asbestos construction materials. The Sae-Maul movement ("Sae-Maul" means "new community") was a rural modernization drive that South Korea pursued under the leadership of the then President Park Chung Hee in the 1970s. As one element of that program, rural homeowners were encouraged to use asbestos slate as a roofing material. Considering that the Sae-Maul movement eventually expanded into urban areas of the country, it is reasonable to assume that the consumption of asbestos construction materials increased and spread nation-wide. That trend was reflected by the growth in the domestic South Korean demand for asbestos products from 1971 onward.

South Korean national policies on industrialization also may be cited as a second factor of influence. Two policies in particular were noteworthy: one had to do with the development of an industrial infrastructure, and the other with the provision of tax incentives. In the context of industrialization, from 1962 onward, South Korea hammered out new economic development plans every 5 years and pursued industrialization policies aimed at fostering modernization and economic growth. For that reason, in 1963 the government implemented a new legal framework under the Act on the Comprehensive Plan of National Land Construction, thus facilitating an active industrialization drive at the rural community level. As tools to this end, it adopted plans for the development of industrial parks, and policies that encouraged foreign investment (Miki 1973:17–19). It then followed up with the implementation of the Act for the Subsidized Development of Export-Oriented Industrial Parks in 1964, and the Act for the Establishment of Export-Free Zones in 1970, thus promoting the export of goods manufactured at facilities built with foreign investment capital.

Against this backdrop, new industrial parks were built as part of the industrial infrastructure development program. Among the districts that had new industrial parks, Busan, Masan, Onsan, and other cities on the side of the Korean Peninsula facing Japan were designated as part of an industrial belt for heavy and chemical industrial operations along the southeastern seaboard. The cities in this southeastern industrial belt were well endowed with an array of conditions which were suitable for the location of industrial operations; they had port and harbor facilities, railway lines, expressways, water resources, large labor pools, and energy resources. Most industrial parks, regardless of their purpose, tend to be equipped with a range of facilities for shared use, including canals, electric power plants, and other industry-related infrastructure (Miki 1973:17–19). Hence, to companies setting up operations in South Korea, industrial parks afforded benefits in the form of lower production costs. In particular, those industrial parks located in cities that were part of the southeastern industrial belt and that possessed port and harbor facilities were arguably analogous to oil terminal complexes built with Japanese-style industrial hub development strategies.

In addition to this drive for industrial infrastructure development, the government also established tax breaks for companies (especially foreign-capitalized companies) that set up their operations in industrial parks. Specifically, companies that set up facilities in the free-trade zone established and developed in Masan under the Export-Free Zones Establishment Act of 1970 were eligible for total exemption from income taxes, corporation taxes, property taxes, and acquisition taxes during the first 5 years of operation, and a 50% exemption from these taxes during the ensuing 3 years of operation. Although obligated to export all their manufactured output to destinations abroad, such foreign-capitalized firms were exempt from duties and excise taxes on the capital goods, raw materials, components, and semi-finished goods they imported for their operations in South Korea (Kang 2002:51). Although they may have varied to some extent, direct and indirect tax incentives (including tax breaks for publicly managed corporations that developed industrial parks) were also offered in other districts. The Union Asbestos Co., Ltd. was one of several Japanese firms that set up operations in Masan. Starting around 1970, it became much easier for foreign companies, including firms in the asbestos industry, to set up local operations in South Korea.

A third factor was that wage levels in South Korea were much lower than those in many advanced industrial countries. Foreign firms transplanting their factories to South Korea would be in a better position to cut their personnel-related expenses. The advantage of cutting personnel costs is arguably the classic reason why many companies relocate their factories offshore. The South Korean government itself even stressed this point in the overtures it used to attract companies from abroad. Around 1970, the wage of the average worker in South Korea was reported to be equivalent to only 30% of the average wage in Japan (Kang 2002:51).

As indicated above, domestic consumption of asbestos construction materials climbed as an outcome of the Sae-Maul movement, and a variety of asbestos products came onto the market and into widespread use. In addition, industrialization policies in South Korea contributed significantly to the expansion of that country's asbestos industry. In the 1970s and 1980s, industrialization proceeded at such a rapid pace in Asia (and especially in South Korea and other newly industrializing economies, or "NIEs") that it earned the label "compressed industrialization" among Japanese observers (Akiyama et al. 1992:38–55). However, because the pace was so rapid, it quickly intensified the severity of certain forms of industrial pollution (for example, the outbreak of Onsan disease, a condition caused by exposure to a complex of chemical contaminants). This is the so-called issue of the "exportation of pollution" through industrialization policies in the developing world. According to Teranishi, there are two fundamental points for consideration here: one has to do with a double standard on regulations, and the other with the principles of conduct (in this context, economic growth fostered by government policies on industrialization) followed by the parties involved (corporations and government agencies) (Teranishi 1992:65–66). In effect, a link had been established between the "exportation of pollution" and the pivotal industrialization policies pursued by national governments in the developing world (Teranishi 1992:77).

From this perspective on the features of industrialization in Asia, the aforementioned double standard on regulations may be cited as a fourth factor that influenced trends in the asbestos industry. Whereas the South Korean industrialization policies included efforts to attract foreign companies on the one hand, tighter regulations were being brought into effect on the use of asbestos in Japan. Public awareness of the dangers associated with asbestos had become fairly well established in Japan by about 1970. The 1971 Ordinance on Prevention of Hazards Due to Specified Chemical Substances (hereafter the "Specified Chemicals Ordinance") designated asbestos as a substance requiring control, and on that basis, regulations concerning antidust measures in workplace settings were given more substance and reinforced. Acting in line with the applicable regulatory frameworks, encouraging factories under their jurisdiction to install dust-collection equipment was the primary administrative approach taken by labor standards inspection offices at the community level. Factories subject to this administrative guidance considered the cost associated with installing and operating such equipment to be economically prohibitive. Some South Korean scholars also accept that tighter regulations in Japan were one of the key reasons prompting Japanese companies to move their manufacturing operations to South Korea when its asbestos textiles industry was reaching its prime (Paek 2001). Asbestos did not face serious regulation in South Korea until that country implemented its Industrial Safety and Health Act in 1988. Prior to that time, asbestos-related regulations in Japan and South Korea clearly had double standards.

These were among the key background factors that shaped and influenced the development of the asbestos industry, and spurred growth in the production and consumption of asbestos. However, it was the relocation to South Korea by elements of Japan's asbestos industry that perhaps made the defining impact. With that influence, the South Korean asbestos industry began concentrating in the Busan district, conceivably due in part to the industrial parks that had already been developed in the Busan community, and to the geographic advantages offered by Busan itself (close proximity to Japan and an established port infrastructure for trade).

8.2.4 The Advent of Asbestos Regulations in South Korea

I now briefly review the backdrop leading to the regulation of asbestos in South Korea.

In the 1970s, a period of thriving asbestos industry operations, South Korea did not have any noteworthy regulations in place on asbestos, and did not implement any countermeasures against dust in workplace environments. Although the history of asbestos processing itself had started with the development of mining operations around 1940, owing to the extended dormancy of asbestos-related disease symptoms and a shortage of physicians qualified to diagnose asbestos victims, it was some time before South Korea actually became aware of the health hazards posed by asbestos, and the fact that asbestos disasters had already taken place. However, perhaps motivated by mounting anxiety among workers in the industry, and alerted to growing worldwide awareness of asbestos as a source of health problems, in 1984 the government instructed its Ministry of Labor to begin measuring airborne asbestos dust concentrations in asbestos textile factories, and under the provisions of the Industrial Safety and Health Act in 1988, established a permissible dust concentration ceiling of 2 f/cc for factory workplace settings. In 1994, dust concentration measurements were compiled on the basis of historical survey data. Table 8.3 lists the findings of that exercise (Park et al. 1995:16–39). Based on the data here, it is estimated that asbestos textile factory dust concentrations ranged from 11.0 f/cc to 92.4 f/cc in 1975 (Park et al. 1995:16).

Although the government adopted provisions for a ban on the utilization of amosite and crocidolite in 1981, those provisions could not be effectively enforced. In other words, provisions for a ban had been put into place, but no regulations had been implemented. In 1997, the government went a step further, using provisions of the Industrial Safety and Health Act to institute an all-out ban on the utilization of the amphibole forms of asbestos. By this time, the country's asbestos industry had already transitioned into a period of decline. Hence, as with the total ban implemented by Japan in 1995, the South Korean ban signified little more than an affirmation that the utilization of these forms of asbestos had almost entirely come to a halt.

In the process of strengthening related regulatory provisions year on year, South Korea also moved closer to a policy stance favoring an all-out ban on asbestos, after considering the push for an all-out ban in Japan and the technological prospects for alternative materials. First, on September 13, 2006, the government issued a notification declaring a ban on the use of asbestos in cement construction materials and as a friction material. From then until 2008, it gradually phased in a ban on other asbestos products as well, eventually implementing a total ban in 2009. However, asbestos parts used in submarines and missiles have been excluded from this total ban.

8.3 Asbestos Disasters and Japanese Corporate Operations in South Korea

8.3.1 An Overview of Japanese Corporate Operations in South Korea

Starting around 1970, several asbestos companies from Japan began setting up operations in line with South Korean industrialization policies, which were aimed at spurring the country's economic development. Those Japanese firms known to have set up businesses in South Korea are listed in Table 8.4.

Of the four firms listed in the table, the Korea Rock Wool Co., Ltd. was a joint venture with Japan Asbestos. Given that rock wool was its main line of business, if this company is excluded from the list of Japanese asbestos companies, it may be noted that the remaining three firms had set up factories in Busan or its satellite city, Masan. In other words, the Busan district.

Table 8.3 Annual trends in airborne asbestos concentrations by process

	No. of plants surveyed	Mixing	Carding	Spinning	Twisting	Weaving	Total
1984	6	9.71(0.62–24.80)	3.46(0.65–7.85)	5.29(0.89–16.82)	6.28(0.95–12.80)	8.77(1.17–30.73)	6.28(0.62–30.73)
1987	7	6.26(1.2–31.1)	5.07(1.0–81.7)	6.08(1.0–18.9)	5.00(0.4–28.1)	5.15(0.3–36.9)	5.01(0.20–81.70)
1991	4	(–)(0.38–17.3)	4.63(0.26–15.00)	5.40(0.10–12.60)	3.92(0.28–17.20)	1.72(–)	3.11(0.10–17.30)
1992	7	6.10(6.10)	0.91(0.08–4.71)	0.85(0.11–2.41)	0.94(0.12–4.98)	1.33(0.07–2.80)	1.42(0.07–6.10)
1994	6	0.48(0.22–1.20)	1.98(0.23–10.97)	2.22(0.41–8.93)	1.65(0.21–9.83)	4.29(2.61–11.58)	1.72(0.21–11.58)

Unit: f/cc (range)
Source: Choi et al. (1998:32–33)

Table 8.4 Korean asbestos textile companies that were established and managed by the advance of asbestos companies in Japan

	Name	Company or person from Japan	Year when it advanced	Location	Note
1	Jeil Asbestos Co., Ltd.	Nippon Asbestos Co., Ltd. (now, NICHIAS Corporation)	1971	Busan	Joint venture with Jeil Chemical Ind. Co (established in 1969)
2	Korea Rock Wool Co., Ltd.	Nippon Asbestos Co., Ltd. (now, NICHIAS Corporation)	1971	Incheon	As for the handling of asbestos, it is vague
3	Dong Yang Asbestos Industry. Co., Ltd.	Kanazawa Asbestos Industries (now, Tosho Kogyo) Co., Ltd.	In around 1973	Busan	Joint venture. Kanazawa Asbestos Industries is in Sennan in Osaka
4	Union Asbestos Co., Ltd.	Nemoto Teiki, Watanabe-Sekimen Co., Ltd.		Masan	Foreign affiliated company in Masan Free-Trade Zone

Note: Companies that were able to confirm the information as of February, 2008.
Source: Toyo Keizai Nippo-sha, *Toyo Business Directory (Korea Business Directory)*, various issues and NICHIAS Corporation (1996)

8.3.2 Jeil Asbestos (Jeil Chemical)

Jeil Asbestos was a joint venture between Jeil Chemical and Japan Asbestos. It currently operates under the name, "Jeil E&S." ("Jeil" is the romanized spelling of the Hangul reading of the Chinese characters for "dai-ichi," which translates as "number one."). Its Busan factory has already been relocated, but was once considered to be the largest-scale asbestos textile manufacturing operation in South Korea. In addition to being located on a site surrounded by a residential district, this facility was involved in the processing of crocidolite. Given that background, a local news station, Busan Munhwa Broadcasting Corporation (Busan MBC), working in collaboration with Dr. Dongmug Kang of Pusan National University School of Medicine, conducted an interview-based survey of health problems among former factory workers and local residents living near the defunct factory site. The findings of that survey identified numerous cases of mesothelioma among both residents and workers, thus revealing that there had been an outbreak of health problems stemming from exposure to asbestos dust.[2] Drawing on the approach taken by the epidemiological study performed by a team led by Norio Kurumatani (Kurumatani and Kumagai 2007) in the Amagasaki community hit by the Kubota Shock in Japan, the Kang study[3] surveyed residents living within a 2-km radius of three asbestos facilities: the aforementioned Jeil Asbestos factory and the two next largest plants run by the Dong Yang Asbestos Industry and Han Il Chemical (all of which were sited in the metropolitan Busan area). Of the survey of subjects who had been exposed to asbestos, 13 were identified with symptoms of mesothelioma, and 11 of those were residents living in the vicinity of the Jeil Asbestos facility. The health problems of former Jeil Asbestos factory workers and residents living around that facility are now the best-known case of an asbestos disaster in South Korea. Here we review company handbooks and other documentation to glean insights into the conduct of Jeil Asbestos and its affiliates over time.

Table 8.5 summarizes entries relating to Jeil Asbestos in annual editions of the *Directory of Korea and Japan*.[4] The first confirmed entry was in the 1971 edition. Although it listed "gasket packing materials" among other details suggestive of the asbestos-related focus of the company's business operations, it did not contain any specifics highlighting that asbestos connection. The entry in the 1972 edition added "asbestos" to the statement of the scope of the business. Jeil Asbestos is first mentioned in the 1973 edition as an entry alongside, but separate from, the entry for Jeil Chemical Industries. This was also the format used in the 1974 edition, serving as confirmation that Jeil Asbestos had been placed into operation as a new joint-venture firm. The entries in the 1976 and later editions of the *Directory of Korea and Japan* are for Jeil Asbestos alone, and contain details of the scope of the business that suggest that the two companies had finally merged.

The 1984 edition contains the first confirmed entry listing Japan Asbestos as a joint-venture participant. It is conjectured that joint-venture arrangements with Japanese firms extended up to this time. Another noteworthy observation in the 1984 edition is that Jeil Asbestos changed its name to "Jeil Rex." In subsequent annual editions up to 1998, the company is listed in English as "Jeil & Rex" although in Japanese it is still listed as "Dai-ichi Kagaku (Jeil Chemical)." Although not listed in the *Directory of Korea and Japan*, details of the overseas operations of Japan Asbestos are listed in annual editions of Toyo Keizai Inc.'s *Japanese Multinationals, Facts and Figures*[5] up to 1985. Although the equity stake of Japan Asbestos in Jeil Asbestos was listed at around 40% up to 1982, in the 1983 and later editions it had fallen significantly to 7.7%. It was at this time that the joint-venture's name changed to Jeil Rex. (Incidentally, Japan Asbestos changed its own name to NICHIAS Corporation on October 1, 1981.) Rex Industries Produkte GmbH (hereinafter "Rex") succeeded NICHIAS as the top investor with an equity stake of 41.3%. Given that Rex was a German asbestos company, the joint-venture relationship had apparently shifted from one with a Japanese company to one with a German company. As Table 8.5 shows, "West Germany" or "Germany" is listed as the main country of business in 1982 and later editions.

Rex had been involved in business relationships with Japan Asbestos in earlier years. The *One Hundred Year's History of NICHIAS* even lists technology tie-ups that existed with Rex. Throughout the 1970s, the two firms had several asbestos product-related technology tie-ups with each other (NICHIAS Corporation 1996:122). Their business relationship is thought to have been formed around this time, and is assumed to have set the stage for the shift in the joint-venture arrangement from Jeil Asbestos in the first half of the 1980s to Jeil Rex thereafter.

Table 8.5 Activities of Jeil Chemical and NICHIAS (Nippon Asbestos Co., Ltd.)

Year	Company name	Year when it advanced	Capital	The number of employees	Representative	The head office and factory	Activities	Characteristic
1971	Je Il Chemical Ind. Co.	April, 1968	5 million won	70	Sung Keun, Kim	525, Yunsan-dong, Dongrae-ku, Busan	Sponge rubber, a synthetic resin product, home rubber, and others (slate, packing, and special rubber products)	
1972	Je Il Chemical Ind. Co.	April, 1968	5 million won	70	Sung Keun, Kim	525, Yunsan-dong, Tongrae-ku, Busan, Korea	Sponge rubber, a synthetic resin product, home rubber, and others (slate, packing, and special rubber products), **Asbestos**	
1973	Je Il Chemical Ind. Co.	April, 1968	5 million won	70	Sung Keun, Kim	525, Yeonsan-Dong, Dongrae-Gu, Busan City, Korea	Sponge rubber, home rubber, and others (packing and special ubber products), **Asbestos**	
	Jeil Asbestos Co., Ltd.	March, 1971	40 million won	70	Sung Keun, Kim	525, Yeonsan-Dong, Dongrae-Gu, Busan City, Korea	Asbestos yarn, asbestos cloth, import and export	Joint venture form (Nippon Asbestos Co., Ltd. 45%)

(continued)

Table 8.5 (continued)

Year	Company name	Year when it advanced	Capital	The number of employees	Representative	The head office and factory	Activities	Characteristic
1974	Je Il Chemical Co.	April, 1968	5 million won	70	Sung Keun, Kim	525, Yeonsan-Dong, Dongrae-Gu, Busan, Korea	Sponge rubber, home rubber, packing, special rubber products, **Asbestos**	
	Jeil Asbestos Co., Ltd.	March, 1971	47.7 million won	80	Sung Keun, Kim	525, Yeonsan-Dong, Dongrae-Gu, Busan, Korea	Asbestos yarn, asbestos cloth, import and export	Joint venture form (Nippon Asbestos Co., Ltd. 45%)
1976	Jae Il Asbestos Co., Ltd.	March, 1971	47.7 million won	92	Seung Keun, Kim	525, Yeonsan-Dong, Dongrae-Gu, Busan, Korea	Asbestos yarn, asbestos cloth, asbestos packing, all rubber products, production, import and export [bank] Busan	Joint venture form (Nippon Asbestos Co., Ltd. 45%)
1977	Jae Il Asbestos Co., Ltd.	March, 1971	47.7 million won	92	Seung Keun, Kim	525, Yeonsan-Dong, Dongrae-Gu, Busan, Korea	Asbestos yarn, asbestos cloth, asbestos packing, all rubber products, production, import and export [bank] Busan	Joint venture form (Nippon Asbestos Co., Ltd. 45%)
1979	Jae Il Asbestos Co., Ltd.	March, 1971	135.92 Million won	92	Sung Kun, Kim	525, Yeonsan-Dong, Dongrae-Gu, Busan, Korea	Asbestos yarn, asbestos cloth, asbestos packing, all of rubber products, production, import and export [bank] Busan	Joint venture form (Nippon Asbestos Co., Ltd. 45%)

1980	Jae Il Asbestos Co., Ltd.	March, 1971	135.92 Million won	150	Sung Kun, Kim	525, Yeonsan-Dong, Dongrae-Gu, Busan, Korea	Asbestos yarn, asbestos cloth, asbestos packing, all rubber products, production, import and export [bank] Busan	Joint venture form (Nippon Asbestos Co., Ltd. 45%)
1982	Jae Il Asbestos Co., Ltd.	March, 1971	135.92 Million won	150	Sung Kun, Kim	525, Yeonsan-Dong, Dongrae-Gu, Busan, Korea	Asbestos yarn, asbestos cloth, asbestos packing, all rubber products, production, import and export [bank] Busan, Seoul trust [export destinations] Japan, East Germany	Joint venture form (Nippon Asbestos Co., Ltd. 40%) [corporate group] Jeil Chemical Co., Ltd., Jeil Steel Co., Ltd
1984	Jae Il Rex Co., Ltd.	March, 1971	616 million won	150	Sung Kun, Kim	132, Yusan-ri, Yangsan-up, Yangsan-gun, Kyong-nam, Korea	Asbestos yarn, asbestos cloth, asbestos packing, all rubber products, production, import and export [bank] Busan, Seoul trust [export destinations] Japan, East Germany	Joint venture form (Nippon Asbestos Co., Ltd. 40%) [corporate group] Jeil Chemical Co., Ltd., Jeil Steel Co., Ltd.

(continued)

Table 8.5 (continued)

Year	Company name	Year when it advanced	Capital	The number of employees	Representative	The head office and factory	Activities	Characteristic
1987	Jeil & Rex Co., Ltd.	October, 1969	371.4 million won	250	Jin Tae, Kim	525, Yeonsan-Dong, Dongrae-Gu, Busan, Korea	Asbestos yarn, asbestos cloth, asbestos packing, all rubber products, production, import and export [bank] Busan, Seoul trust [the export destinations] Japan, East Germany	[corporate group] Jeil Steel Co., Ltd.
1988	Jeil & Rex Co., Ltd.	October, 1969	371.4 million won	250	Jin Tae, Kim	525, Yeonsan-Dong, Dongrae-Gu, Busan, Korea	Asbestos yarn, asbestos cloth, asbestos packing, all rubber products, production, import and export [bank] Busan, Seoul trust [export destinations] Japan, East Germany	[Corporate group] Jeil Steel Co., Ltd.
1990	Jeil & Rex Co., Ltd.	October, 1969	371.4 million won	400	Jin Tae, Kim	525, Yeonsan-Dong, Dongrae-Gu, Busan, Korea	Asbestos yarn, asbestos cloth, asbestos packing, all rubber products, production, import and export [bank] Busan, Seoul trust [export destinations] Japan, East Germany	[Corporate group] Jeil Steel Co., Ltd.

1994	Jeil & Rex Co., Ltd.	October, 1969	550 million won	203		Jin Tae, Kim	39-5, Yusan-ri, Yangsan-up, Yangsan-gun, Kyong-nam, Korea	Asbestos yarn, asbestos cloth, asbestos packing, all rubber products, production, import and export [bank] Busan, Seoul trust [export destinations] Japan, East Germany	[Corporate group] Jeil Steel Co., Ltd.
1996	Jeil Chemical Co., Ltd.	February, 1977	550 million won	225		Jin Tae, Kim	39-5, Yusan-ry, Yangsan-up, Yangsan-gun, Kyon-gnam, Korea	Seal products, fluororesin products, production, import and export, real estate business [bank] Busan	Joint venture form (Nippon Valqua Co.,Ltd. 17%)
	Jeil & Rex Co., Ltd.	October, 1969	550 million won	203		Jin Tae, Kim	39-5, Yusan-ri, Yangsan-up, Yangsan-gun, Kyong-nam, Korea	Asbestos yarn, asbestos cloth, asbestos packing, all rubber products, production, import and export [bank] Busan	[Corporate group] Jeil Steel Co., Ltd.
1998	Jeil Chemical Co., Ltd.	February, 1977	550 million won	220		The chairman: Seong Keun, Kim The president: Jin Tae, Kim	39-5, Yusan-dong, Yangsan-shi, Kyeong-nam, Korea	Seal products, fluororesin products, production, import and export, real estate business [bank] Busan	Joint venture form (Nippon Valqua Co., Ltd. 17%)

(continued)

Table 8.5 (continued)

Year	Company name	Year when it advanced	Capital	The number of employees	Representative	The head office and factory	Activities	Characteristic
	Jeil & Rex Co., Ltd.	October, 1969	550 million won	230	The chairman: Seong Keun, Kim The president: Jin Tae, Kim	39-5, Yusan-dong, Yangsan-shi, Kyeong-nam, Korea	Asbestos yarn, asbestos cloth, asbestos packing, all rubber products, production, import and export [bank] Busan	Joint venture form (Jeil Chemical Co., Ltd. 70%)
2000	Jeil Chemical Co., Ltd.	February, 1977	550 million won	220	Jin Tae, Kim	39-5, Yusan-dong, Yangsan-shi, Kyeong-nam, Korea	Seal products, fluororesin products, production, import and export [bank] Busan	Joint venture form (Nippon Valqua Co., Ltd. 17%)
	Jeil Chemical Co., Ltd.	February, 1977	550 million won	220	Jin Tae, Kim	39-5, Yusan-dong, Yangsan-shi, Kyeong-nam, Korea	Asbestos yarn, asbestos cloth, asbestos packing, all rubber products, production, import and export [bank] Busan	Joint venture form (Jeil Chemical Co., Ltd. 70%)
2002	Jeil E&S Co., Ltd.	February, 1977	3 billion won	210	Jin Tae, Kim	39-5, Yusan-dong, Yangsan-shi, Kyeong-nam, Korea	Seal products, fluororesin products, production, import and export [bank] Busan	Technical cooperation: Nippon Valqua classified in category of chemistry

Source: Toyo Keizai Nippo-sha, *Toyo Business Directory (Korea Business Directory)*, various issues

That it was engaged in the manufacture of textile products using crocidolite may be cited as a factor explaining the higher incidence of illness from exposure to asbestos dust in the workers from the Jeil Asbestos factory. This is further supported by documentation relating to Japan Asbestos. According to Hidero Onodera, the book *Shutting the Heat Out: the 80-year History of Japan Asbestos* contains remarks on foreign factory operations in 1973, noting that two plants had been set up in South Korea: one in Busan and another in Inchon. It notes that the Busan plant produced about 15 metric tons of crocidolite yarn per month, about half of which was destined for Japan, while the remaining half went to other foreign countries. In addition, it explains that the Inchon plant was engaged in the production of rock wool (Onodera 1973:9).The Jeil Asbestos plant and the Korea Rock Wool plant listed in Table 8.4 correspond to the two plants referred to in the Onodera book.

According to testimony given about the first 10 years of Jeil Chemical operations by former workers who had been engaged as mechanics, when the firm went into operation in 1969 it only had a line for the production of chrysotile textile products. When the Jeil Asbestos plant was founded in 1971, a new line for the production of crocidolite textiles was added. Japanese asbestos companies were involved in the operations when Jeil Asbestos was set up, and provided manufacturing technologies and production machinery in addition to investment capital. Almost all the production machinery had been manufactured in Japan; it had been dismantled for importation and reassembled with the assistance of Japanese engineers. The products manufactured with chrysotile were destined for the domestic market, while products utilizing crocidolite were destined for markets overseas (primarily Japan).[6] The testimony that goods made with crocidolite were meant for overseas markets was consistent with statements made in the Onodera book.

The 1973 Onodera book states that the destinations for crocidolite yarn products were divided almost equally between Japan and other foreign markets (Onodera 1973:9). Presumably this refers to arrangements which were in effect just after Jeil Asbestos went into operation. In addition, entries in Table 8.5 relating to the scope of the business from 1982 to 1994 show that the company listed Japan and West Germany, or Germany, as its principal countries for business. For comparison, in the *One Hundred Year's History of NICHIAS*, it is noted that Jeil Asbestos was established in Busan in 1971, produced asbestos textile products, and exported its merchandise primarily to the USA (NICHIAS Corporation 1996:12). Although the company's primary export markets may have changed year by year, it seems safe to say that at a minimum, the foreign destinations for its products were countries where regulations relating to asbestos had already been implemented, or where the public already had some awareness of the hazards of asbestos.

Returning again to the data in Table 8.5, entries for 1996 and later are once again for a pair of firms (Jeil & Rex and Jeil Chemical). The entry for 2002 concerns only Jeil E&S, the current name of the company, and it is no longer described as an asbestos company in the description of its business content. This presumably reflected an effort to end its status as a member of the asbestos industry. Jeil Chemical, the company that appeared in 1996 and in later editions of the *Directory of Korea and Japan*, established a joint-venture arrangement and technology tie-up with Nippon Valqua Industries, Ltd. (hereafter "Nippon Valqua"). Nippon Valqua was one of Japan's top asbestos firms. In its corporate time-line, Jeil Chemical also indicates that it had entered into technology tie-ups with Nippon Valqua in 1990.[7] In the *Directory of Korea and Japan*, its listed business is the manufacture of sealing products and fluorine resin products. Those details suggest that the company had implemented technologies developed by Japanese companies for the manufacture of nonasbestos products and asbestos substitutes, and that it had shifted its operations to the manufacture of nonasbestos products.

As detailed above, Jeil Chemical was established as an overseas manufacturing base for a Japanese asbestos firm, thrived as a manufacturer of asbestos products, expanded its operations through tie-ups with foreign firms as one of the best-known asbestos textile manufacturers in South Korea, and then shifted to the manufacture of nonasbestos products after domestic regulations had been strengthened. However, in 1990 it built an asbestos textile mill in Indonesia (currently operating locally as Jeil PARJAR), thus continuing to run asbestos product manufacturing operations abroad.

8.3.3 Dong Yang Asbestos Industry

In terms of operating scale, the Dong Yang Asbestos Industry is considered to have been the second largest asbestos factory in Busan after Jeil Chemical. It was located slightly west of Busan city center, in an industrial zone with few residential zones nearby. The Dong Yang Asbestos Industry was yet another joint venture set up with the participation of a Japanese company.

The Japanese company behind this joint venture was Kanazawa Asbestos Industries Co., Ltd. (hereafter "Kanazawa Asbestos"), an asbestos textile mill located in the Sennan district of Osaka Prefecture that had begun operation in 1965. It has since changed its name, and is now known as Tosho Kogyo.

Table 8.6 provides a sampling of listings for the Dong Yang Asbestos Industry as entered in the *Directory of Korea and Japan*. From the 1976 edition (the year of the first entry) to the 1998 edition, the Dong Yang Asbestos Industry was listed as a joint venture in which Kanazawa Asbestos held a 66.7% equity stake. Until the 1988 edition, the entries listed Kanazawa Asbestos as an overseas branch. Given that its products were listed as entirely for export by all entries until the 1996 edition, the Dong Yang Asbestos Industry could be considered to have been a Busan-based branch factory operation of Kanazawa Asbestos. Also worth noting is the fact that it was listed as an overseas facility of an asbestos firm based in Osaka's Sennan district.

At one time, Sennan was the single largest center of asbestos textile production in Japan, and home to countless numbers of small-scale factory operations. In every annual edition of the Osaka factory register in which it was listed, Kanazawa Asbestos was always described as a company in the 10–19 workers category, and was considered to be one of the mid-tier operations in the Sennan district.[8]

Sennan has traditionally had a large population of residents of Korean ancestry or nationality. Following World War II, there were numerous instances of Korean nationals

Table 8.6 Activities of the Dong Yang Asbestos Industry and Kanazawa Asbestos Industries

Year	Company name	Capital(won)	Number of employees	Head office and factory	Activities	Characteristics
1976	Dong Yang Asbestos Industry. Co., Ltd.	150 000 000	100	532-1, Deogpo-Dong, Busanjin-Gu, Busan, Korea	Manufacture of asbestos yarn, asbestos fabric, asbestos tape, asbestos products with gold wire, and other asbestos products	Joint venture form (Kanazawa Asbestos Industries 66.7%). Its products were listed as entirely for export
1977	Dong Yang Asbestos Industry. Co., Ltd.	150 000 000	100	532-1, Deogpo-Dong, Busanjin-Gu, Busan, Korea	Manufacture of asbestos yarn, asbestos fabric, asbestos tape, asbestos products with gold wire, and other asbestos products	Joint venture form (Kanazawa Asbestos Industries 66.7%). Its products were listed as entirely for export
1979	Dong Yang Asbestos Industry. Co., Ltd.	150 000 000	100	373-9, Deogpo-Dong, Puk-gu, Busan, Korea	Manufacture of asbestos yarn, asbestos fabric, asbestos tape, asbestos products with gold wire, and other asbestos products	Joint venture form (Kanazawa Asbestos Industries 66.7%). Its products were listed as entirely for export
1980	Dong Yang Asbestos Industry. Co., Ltd.	150 000 000	100	373-9, Deogpo-Dong, Puk-gu, Busan, Korea	Manufacture of asbestos yarn, asbestos fabric, asbestos tape, asbestos products with gold wire, and other asbestos products	Joint venture form (Kanazawa Asbestos Industries 66.7%). Its products were listed as entirely for export
1982	Dong Yang Asbestos Ind. Co.	1 000 000 000	100	373-9, Deogpo-Dong, Puk-gu, Busan, Korea	Manufacture of asbestos yarn, asbestos fabric, asbestos tape, asbestos products with gold wire, and other asbestos products.Main trade countries for export: Japan, Europe, USA, and others	Joint venture form (Kanazawa Asbestos Industries 66.7%). Its products were listed as entirely for export

(continued)

Table 8.6 (continued)

Year	Company name	Capital(won)	Number of employees	Head office and factory	Activities	Characteristics
1984	Dong Yang Asbestos Ind. Co.	145 000 000	100	373-9, Tokpo-dong, Bug-Gu, Busan, Korea	Manufacture of asbestos yarn, asbestos fabric, asbestos tape, asbestos products with gold wire, and other asbestos products. Main trade countries for export: Japan, Europe, USA, and others	Joint venture form (Kanazawa Asbestos Industries 66.7%). Its products were listed as entirely for export
1987	Dong Yang Asbestos Ind. Co.	145 000 000	70	373-9, Tokpo-dong, Puk-gu, Busan, Korea	Manufacture of asbestos yarn, asbestos fabric, asbestos tape, asbestos products with gold wire, and other asbestos products. Main trade countries for export: Japan, Europe, USA, and others	Joint venture form (Kanazawa Asbestos Industries 66.7%). Its products were listed as entirely for export
1988	Dong Yang Asbestos Ind. Co.	145 000 000	70	373-9, Tokpo-dong, Puk-gu, Busan, Korea	Manufacture of asbestos yarn, asbestos fabric, asbestos tape, asbestos products with gold wire, and other asbestos products. Main trade countries for export: Japan, Europe, USA, and others	Joint venture form (Kanazawa Asbestos Industries 66.7%). Its products were listed as entirely for export
1990	Dong Yang Asbestos Ind. Co.	150 000 000	64	373-9, Tokpo-dong, Puk-gu, Busan, Korea	Manufacture of asbestos yarn, asbestos fabric, asbestos tape, asbestos products with gold wire, and other asbestos products. Main trade countries for export: Japan, Europe, USA, and others	Joint venture form (Kanazawa Asbestos Industries 66.7%). Its products were listed as entirely for export
1994	Dong Yang Asbestos Ind. Co.	150 000 000	40	373-9, Tokpo-dong, Puk-gu, Busan, Korea	Manufacture of asbestos yarn, asbestos fabric, asbestos tape, asbestos products with gold wire, and other asbestos products. Main trade countries for export: Japan and USA	Joint venture form (Kanazawa Asbestos Industries 66.7%). Its products were listed as entirely for export

1996	Dong Yang Asbestos Ind. Co.	145 000 000	20	373-9, Tokpo-dong, Puk-gu, Busan, Korea	Manufacture of asbestos yarn, asbestos fabric, asbestos tape, asbestos products with gold wire, and other asbestos products	Joint venture form (Kanazawa Asbestos Industries 66.7%). Its products were listed as entirely for export
1998	Dong Yang Asbestos Ind. Co.	140 000 000	30	373-9, Tokpo-dong, Puk-gu, Busan, Korea	Manufacture of asbestos yarn, asbestos fabric, asbestos tape, asbestos products with gold wire, and other asbestos products	Joint venture form (Kanazawa Asbestos Industries 66.7%). Its products were listed as entirely for export
2000	Dong Yang Asbestos Ind. Co.	140 000 000	25	373-9, Tokpo-dong, Puk-gu, Busan, Korea	Manufacture of asbestos yarn, asbestos fabric, asbestos tape, and other asbestos products	
2002	Dong Yang Asbestos Ind. Co.	140 000 000	10	373-9, Tokpo-dong, Puk-gu, Busan, Korea	Manufacture of asbestos yarn, asbestos fabric, asbestos tape, and other asbestos products	

Source: Toyo Keizai Nippo-sha, *Toyo Business Directory (Korea Business Directory)*, various issues in Japan purchasing supplies of used textile machinery and starting up their own asbestos factories. By some accounts, after Japan tightened its regulations on asbestos, some Korean factory owners returned home to the Korean peninsula and restarted their operations, or set up branch factories in Korea with assistance from relatives.[9] The Dong Yang Asbestos Industry, a joint-venture with Kanazawa Asbestos, arguably demonstrated that elements of the Sennan asbestos industry had actually been relocated to South Korea. However, aside from the Dong Yang Asbestos Industry example, no reliable documentation has been obtained thus far which accurately details the number of companies and workers that relocated from the Sennan district each year. Furthermore, because so many asbestos companies based in Sennan were traditionally small-scale operations, it was understandably difficult to leave any documented record of their business operations after having relocated to South Korea.

Judging at least from the entries in the "highlights" section of each table 6 entry from 1976 to 1996, the Dong Yang Asbestos Industry exported all of its merchandise. Entries for the years 1982–1994 list Japan, Europe, the USA, and "others" as the primary countries for the company's export business. These details support the conclusion that most of its output was destined for export markets.

As of September 2007, the Dong Yang Asbestos Industry was in the process of retooling its factory's processing lines for the manufacture of nonasbestos products, and accordingly transitioning its business away from the asbestos industry.

8.3.4 Union Asbestos

Unlike the other companies, Union Asbestos was a venture set up entirely with foreign capital with its operations in the Masan Free-Trade Zone. These points would suggest that it was set up purely to take advantage of the available tax incentives, cheap labor, and regulatory gaps (i.e., the "double standard"). A listing of companies licensed to set up in the Masan Free-Trade Zone as of September 30, 1970, shows the entry "Union Asbestos; Teiki Nemoto; investment scale: US$150000; six product lines including asbestos yarn; wholly owned subsidiary (branch factory)" (Kang 2002:46).

Table 8.7 lists entries for Union Asbestos in several annual editions of the *Directory of Korea and Japan*, demonstrating that it was in operation until the date of the 1982 edition. Its representative director was an individual named Teiki Nemoto. Although "President: Watanabe Asbestos" is listed as the company's representative, judging from the data in the "highlights" column, Union Asbestos was conceivably a local South Korean subsidiary of Watanabe Asbestos Co., Ltd.

The Masan Free-Trade Zone had been created for the purpose of building foreign reserves by encouraging export industries to set up local operations. In return for huge breaks on income taxes, corporation taxes, property taxes, customs duties on imported goods, and other taxes, companies that set up in the free-trade zone were obligated to export all their manufactured goods to markets outside South Korea. It may be concluded that Union Asbestos, as one such firm operating in this free-trade zone, exported all its manufactured output.

Table 8.7 Activities of Union Asbestos

Year	Company name	Year when advanced	Capital	Number of employees	Representative	Head office and factory	Activities	Characteristics
1973	Union Asbestos Co., Ltd.	May, 1971	60 million won	Not described	Chairman: Nemoto Teiki President: Watanabe Ishiwata	975, Yangdeog-Dong, Masan-Si Gyeong-Nam, Korea	Asbestos products and accessories for asbestos products	Foreign affiliated company (Nemoto Teiki)
1974	Union Asbestos Co.	May, 1971	0.2 million dollars	90	Chairman: Nemoto TeikiPresident: Watanabe Ishiwata	975, Yangdeog-Dong, Masan-Si Gyeong-Nam, Korea	Asbestos products and accessories for asbestos products	Foreign affiliated company (Nemoto Teiki, Watanabe-Sekimen Co., Ltd.) 100%
1976	Union Asbestos Co., Ltd.	May, 1971	80 million won	90	Chairman: Nemoto Teiki President: Watanabe Ishiwata	975, Yangdeog-Dong, Masan-Si Gyeong-Nam, Korea	Asbestos products and accessories for asbestos products	Foreign affiliated company (Nemoto Teiki, Watanabe-Sekimen Co., Ltd.) 100%
1977	Union Asbestos Co., Ltd.	May, 1971	96 million won	74	President: Teiki Nemoto	975, Yangdeog-Dong, Masan-Si Gyeong-Nam, Korea	Asbestos products and accessory for asbestos products	Foreign affiliated company (Nemoto Teiki, Watanabe-Sekimen Co., Ltd.) 100%
1979	Union Asbestos Co., Ltd.	May, 1971	96 million won	74	Representative director: Notatsu Saido	975, Yangdeog-Dong, Masan-Si Gyeong-Nam, Korea	Asbestos products and accessories for asbestos products	Foreign affiliated company (Nemoto Teiki, Watanabe-Sekimen Co., Ltd.) 100%
1980	Union Asbestos Co., Ltd.	May, 1971	96 million won	74	Representative director: Notatsu Saido	975, Yangdeog-Dong, Masan-Si Gyeong-Nam, Korea	Asbestos products and accessories for asbestos products	Foreign affiliated company (Nemoto Teiki, Watanabe-Sekimen Co., Ltd.) 100%
1982	Union Asbestos Co., Ltd.	May, 1971	96 million won	74	Representative director: Notatsu Saido	975, Yangdeog-Dong, Masan-Si Gyeong-Nam, Korea	Asbestos products and accessories for asbestos products	Foreign affiliated company (Nemoto Teiki, Watanabe-Sekimen Co., Ltd.) 100%

Source: Toyo Keizai Nippo-sha, *Toyo Business Directory (Korea Business Directory),* various issues

It was not possible to determine whether this company is still in existence. It is likely that it has already gone out of business or withdrawn its operations from South Korea.

8.4 Asbestos Disasters in Japan and South Korea and the "Exportation of Pollution"

8.4.1 The Exportation of Pollution and the Social Costs

This section explores the relocation of Japan's asbestos industry to South Korea from the perspective of exporting pollution, and paying special attention to the subject of the preceding section, namely, the trend among Japanese asbestos firms to set up local operations in South Korea.

First, we review two basic tenets of Teranishi's "exportation of pollution" thesis. The exportation of pollution exploits regulatory gaps (the so-called "double standard" delineated by differences in the regulations applied by two nation states), and the incentives that are the result of that double standard. To companies, these are the benefits of reduced labor costs and spending on measures to prevent workplace accidents and pollution, and to the host nations they are the benefits of economic development, etc. (Teranishi 1992:65–66). Under the influence of systemic factors such as regulatory disparities and national policies on economic growth, the costs of certain forms of corporate conduct may be shifted onto the shoulders of third parties, or societies in other nations, as "unpaid costs." The "exportation of pollution" may be perceived as a factor that generates such social costs.

By its nature, the manufacture of asbestos products involves processes that have a tendency to release dust. Indeed, asbestos has frequently been cited as a classic case of an industry where regulations affecting the workplace environment were easily applied, and the exportation of polluting processes was readily pursued (Ishi 1988:230–232). The implication is that dangerous manufacturing processes could be transplanted abroad, ultimately exposing the host country and its society at large to asbestos pollution.

Insights and conclusions about the exportation of pollution may be derived from a confluence of factors, including the existence of regulatory gaps, South Korean policies on economic growth that sought to attract foreign investment, the thriving state of the South Korean asbestos industry, and the timing of Japanese corporate efforts to set up operations abroad. Here, the actual extent to which asbestos-related pollution and health problems were exported from Japan to South Korea will be explored, with special attention to the case studies of three firms discussed in the preceding section, and in terms of three stages. These are the stage prior to the launch of local factory operations in South Korea, the stage during which those factory operations were under way, and the stage comprising the withdrawal of those operations.

8.4.2 The Exportation of Asbestos Pollution to South Korea

8.4.2.1 The Stage Prior to the Launch of Local Factory Operations

The first stage was prior to the move by Japanese asbestos firms into South Korea. At this stage, the focus was on assorted preconditions in Japan and South Korea that would facilitate that move. All three companies reviewed earlier set up their operations in South Korea in the first half of the 1970s. Hence, the policies and regulatory structures in place in both countries at that time deserve closer attention. In South Korea, for example, policies promoting economic development were reaching their prime, and efforts were fully under way to lure in foreign investment with the goal of cultivating the country's export-oriented industries by harnessing the manufacturing technologies of the advanced industrial economies and utilizing foreign capital. Under those conditions, South Korea viewed Japan as a pioneer in the achievement of rapid economic growth powered by industrialization, a geographically close neighbor, and the most immediate target for its foreign investment incentives. The industrial park development policies which South Korea pursued in the interests of economic growth were as described earlier.

It is known that while South Korea was in the process of cultivating an environment which would be attractive to foreign investment, Japanese interests were working to persuade South Korea to allow Japanese firms to set up local operations on South Korean soil. Lobbying efforts were also initiated by the Japan–Korea Cooperation Committee (JKCC), a grouping of Japanese and South Korean legislators and business entrepreneurs that had formed in 1969. On April 22, 1970, a meeting of the JKCC was presented with a proposal for long-term Japan–Korea economic cooperation (the "Yatsugi proposal"). Under that proposal, the two countries would be designated an economic cooperation zone, with a framework which allowed Japanese firms to set up business in South Korea and pursue joint-venture operations there (Kang 2002:40–42; Working Group for Recording Historical Japan–South Korea Relations 1976:155–157).

In contrast to the economic development policies then being pursued by South Korea, around 1970 Japan faced a different situation. Owing to increasingly severe levels of industrial pollution, new legal frameworks and regulations were being implemented to deal with various forms of environmental pollution. On the topic of asbestos alone, provisions in the Specified Chemicals Ordinance of 1971 stipulated that asbestos was a substance that should be strictly controlled, and mandated that businesses engaged in operations involving the handling or processing of asbestos must install dust-collection equipment and ensure that employees used face masks and other protective gear as appropriate. In other words, as long as a company was operating an asbestos product factory within Japan, it had to implement countermeasures aimed at preventing workplace disasters which were attributable to asbestos dust. The circumstances in Japan and South Korea during the first half of the 1970s had the following structure. Whereas tighter regulations were inflating the cost of production in Japan, South Korea had no significant regulations to speak of, and offered an environment that welcomed penetration by foreign firms. Consequently, South Korea experienced an influx by many manufacturing firms from various industries, including the asbestos industry, which were engaged in operations that were known to cause pollution. In fact, from 1966 to July 31, 1973, a sum total of 493 Japanese firms (459 if counting from 1970 alone) set up operations in South Korea (Working Group for Recording Historical Japan–South Korea Relations 1976:160). Needless to say, that total included several asbestos companies.

8.4.2.2 The Operational Stage

The second stage was when asbestos companies from Japan set up and ran factory operations in South Korea under the preconditions for both countries that had been agreed in the first stage. The benefits for Japanese firms that set up operations in South Korea deserve consideration. Although Japan in the 1970s had a domestic demand for asbestos products, domestic manufacturing operations faced rising costs with the addition of expenditure on measures against dust pollution. In South Korea around that time, the average worker's wage was lower, and regulations for the prevention of dust pollution did not exist. Accordingly, moving their facilities from Japan to South Korea would enable manufacturers to hold their labor costs down, and cut corners in their expenditure on dust prevention measures.

Furthermore, in keeping with one element of South Korean policy on economic development – the goal of cultivating export-oriented industries – the products manufactured by the three asbestos start-ups in South Korea discussed above would be destined for export to other countries, and Japan in particular, rather than satisfying demand in the South Korean domestic market. In effect, for Japanese asbestos companies, this arrangement served not only to reduce their operating costs, but also to enable them to split off their hazardous manufacturing processes and relocate them to South Korea. The processes involved in the manufacture of asbestos products can easily expose workers to airborne asbestos dust, and thus involves a certain level of risk. Relocating those processes that can be sources of pollution to foreign countries is considered to be consistent with the defining features characterizing the exportation of pollution.

Here, we need to review the asbestos disaster experienced by residents living around the Jeil Chemical facility in Busan in particular. The incidence of asbestos-induced illness was more concentrated in the vicinity of this facility because Jeil Chemical was utilizing crocidolite, one of the more strongly toxic forms of asbestos. Products made with crocidolite were destined for export (primarily to Japan). The 1960 report by Wagner et al. encouraged spreading international awareness of the hazards of crocidolite (Wagner et al. 1960). Knowledge of those hazards is considered to have reached Japan no later than the 1970s. On May 22, 1976, the Director of the Labor Standards Bureau of the Ministry of Labor issued a notice concerning the promotion of measures to prevent health problems due to exposure to asbestos dust. Above all, in view of the strong toxicity of crocidolite, the notice specifically urged the priority promotion of substitute materials. Prior to this notice, labor standards inspection offices around Japan had apparently been stepping up their efforts to distribute administrative guidance for facilities that handled crocidolite. In fact, the amount of crocidolite used as a raw material for asbestos products fell sharply during the first half of the 1970s. From 1968 to 1972, yearly editions of the *Yearbook of Building Material Statistics* listed the volume of crocidolite consumption at above 1 000 metric tons. By 1973, the listed volume had fallen to 163 tons, and by 1974 to only 6 tons.[10] As these statistics demonstrate, in the early 1970s the utilization of crocidolite in particular became increasingly difficult in Japan due to the material's high toxicity. Jeil Chemical constituted an example of the transfer to South Korea of a manufacturing process involving crocidolite, a material that had become too problematic for continued use in Japan. As one outcome of this transfer, Jeil Chemical caused an asbestos disaster in Busan. This was a classic example of the exportation of a manufacturing process that would become a source of pollution.

8.4.2.3 The Withdrawal Stage

The third stage comprised the withdrawal of Japanese asbestos companies from South Korea. The points for consideration here include the corporate reasons for withdrawal, the decline of the asbestos industry in both countries, and the ban on asbestos itself. All of the Japanese asbestos firms that set up operations in South Korea to manufacture asbestos products had already left that country. Further, all South Korean asbestos firms had either changed to other lines of business or relocated their asbestos product manufacturing operations to other countries. Although the three Japanese companies that had set up operations in South Korea eventually withdrew at different times. Their principal reasons for withdrawing were that the changing regulatory and economic conditions in South Korea undermined their ability to reap the benefits that motivated them to start those operations in the first place, and also that there was no longer any necessity to source products from overseas manufacturing operations.

First, consider the changes affecting conditions in South Korea. When Japanese asbestos companies set up their local operations in the 1970s, South Korea did not have any notable regulations on asbestos. However, in 1984, it began conducting frequent measurements of airborne asbestos concentrations at asbestos facilities, and in 1988, under the provisions of the Industrial Safety and Health Act, it established a benchmark concentration ceiling of 2 f/cc for workplace settings. Then, drawing on regulatory trends in Japan and the West, it took the step of implementing asbestos-specific regulations, and in the late 1980s is believed to have begun demanding that factories should adopt dust countermeasures, thus echoing developments seen in Japan in the early 1970s. Beginning around 1995, asbestos industry operations in the country entered a steep decline.[11] After becoming aware of the drive for an all-out ban on asbestos in Japan, South Korea began preparing for its own total ban on the material. In terms of their bans on the manufacture, importation, and use of asbestos products, Japan and South Korea closely resemble one another, and in both countries today asbestos is essentially no longer a viable industry.

Next we examine some specific factors behind the withdrawal of the three Japanese-affiliated factories that operated in South Korea. Japan Asbestos was replaced by Rex, a German company, as the partner in the Jeil Asbestos joint-venture with Jeil Chemical, and this was conceivably an outcome of negotiations by all the companies involved. However, after being replaced by Rex in that arrangement, in the early 1980s Japan Asbestos embarked on a number of changes. Specifically, it changed its name to NICHIAS in 1981, and began transitioning to nonasbestos lines of business (starting product development work on asbestos substitutes and shutting down its asbestos product manufacturing operations).

Kanazawa Asbestos is known to have maintained its joint-venture arrangement until around 1998, thus becoming the last of the three Japanese enterprises to withdraw from South Korea. That step is thought to have been purely an outcome of having downsized its manufacturing operations in response to dwindling demand in Japan. In fact, the consumption of asbestos products in Japan followed an uninterrupted downward trend throughout the 1990s.[12]

The reasons for the exit of Union Asbestos from South Korea are unclear. It was subject to local criticism over problems relating to the working environment at its facility and the hours its workers were on duty, and is assumed to have pulled out in the 1980s without ever achieving the operational performance it initially anticipated. Criticisms of its operations were noted in a report entitled "The Masan Free-Trade Zone Field Survey" in 1974. They included the following. "Union Asbestos, 92 employees. Asbestos. Runs two shifts per day: a day shift of 10 h and a night shift of 12 h. Owing to the unusual nature of the operations (exposure to excessively heavy levels of asbestos dust), work is limited to 6-h shifts, but in reality, workers are forced to do an additional 5 h of overtime" (Korea Justice and Peace Committee, and Catholic Priests' Association for Justice 1975). In addition, "Workers wear Japanese-made dust masks, but the risk of exposure remains owing to the exceptionally heavy levels of dust in the workplace. The noise in the workplace is so severe that it is typically impossible to hear coworkers speaking. The die mold machinery operates at temperatures of 300°C, at times causing workers respiratory distress." (Korea Justice and Peace Committee, and Catholic Priests' Association for Justice 1975:44). Foreign-affiliated companies that set up in the free-trade zone were criticized in general for imposing harsh working conditions and exploiting low wage scales, and Union Asbestos was one of the firms subject to such complaints.

Judging from the foregoing portrayal, the Japanese firms that set up in South Korea to take advantage of regulatory and wage differentials eventually saw the differentials diminished and the benefits canceled out, and as the asbestos industry in both countries transitioned into its final period of decline (with the need for its manufacturing operations eliminated), the three Japanese-affiliated manufacturing operations withdrew.

8.5 The International Proliferation of Asbestos Disasters

This chapter has examined the case studies of three Japanese asbestos companies that set up manufacturing operations in South Korea. All shared certain similarities in terms of the systemic conditions facilitating their operations, the factors that motivated them to move into South Korea, and the trends surrounding their withdrawal. On the whole, their decisions were motivated by the conclusion that overseas operations would be able to enjoy reduced operating and labor costs as long as the benefits of asbestos continued to win societal acceptance in Japan. They withdrew those operations once their basis for establishing them had been lost, i.e., after the benefits of asbestos had been rejected by society, and the expected savings in operating costs, etc., were no longer available. Regulatory disparities and double standards are generally cited as definitive systemic factors that enable the phenomenon of exporting pollution to take place. However, if we pursue the reasons further, it is probably more realistic to point to bilateral differences in social awareness.

The historical pattern of mass asbestos consumption began with enthusiastic levels of utilization in the West. That pattern spread to Japan, and from there to South Korea and other Asian countries. South Korea followed in Japan's footsteps by implementing an all-out asbestos ban of its own. Conversely, levels of asbestos consumption continue to trace an upward trend in other Asian countries that are still moving through various stages of economic growth. Today, China, India, and Thailand are the world's top three consumers of asbestos. Indonesia, Vietnam, and other Asian economies are also heavy consumers. It is as if disparities in social awareness regarding the hazards of asbestos have become the prime determinant behind the international proliferation of asbestos disasters. In a nutshell, overcoming this factor will call for efforts to promote a shared societal awareness at the international level. Prior to pursuing and achieving that objective, however, it will be necessary to identify public policies on asbestos (in particular, those that put priority on awareness of, and measures against, asbestos as a hazardous toxin) at the national level that also function as influential determinants. Unless disparities in social awareness are resolved, other countries will be doomed to repeat the mistakes of relocating asbestos product manufacturing operations, and causing asbestos disasters like those observed in Japan and South Korea.

Notes

1. Based on a report by the South Korean Institute on Workplace Environmental Health, compiled from Geological Research Institute documentation.
2. Based on interviews conducted on September 5, 2007, by Park Sang-Gue, affiliated journalist to Busan MBC Broadcasting, and Prof. Dong mug Kang of Busan University Medical School.
3. The local interview-based survey above, and the Kang report delivered at the International Asbestos Conference in Celebration of the 20th Anniversary of BANJAN, held November 23–24, 2007, in Yokohama.

4. *Toyo Business Directory (Korea Business Directory),* Tokyo: Toyo Keizai Nippo-sha, various issues.
5. *Japanese Multinationals Facts and Figures.* Toyo Keizai Inc., various issues.
6. Based on September 6, 2007, interviews with former Jeil Chemical workers.
7. Timeline section on Jeil E & S home page, http://www.jeilens.co.kr/Eng/company/history.htm. Accessed November 11, 2007.
8. Osaka Prefectural Government, *Factory Directory in Osaka Prefecture Handbook of Factories in Osaka Prefecture,* various issues.
9. Based on an interview-based survey of former asbestos factory owners conducted on March 7, 2007, at the Makino Citizens Center in Sennan, Osaka.
10. Research and Statistics Department, Minister's Secretariat, Ministry of International Trade and Industry *ed., Yearbook of Building Materials and Statistics,* various issues, Tokyo: Investigation Commission of the Ministry of International Trade and Industry. Crocidolite is listed here as "blue."
11. Although totaling around 90 000 metric tons in 1995, annual imports of raw asbestos had fallen to approximately 30 000 tons by 1998, and continued a downward trend thereafter. Ministry of Employment and Labor of Korea (2007). "Measures for Asbestos Control."
12. At its peak, annual asbestos consumption reached 300 000 tons in Japan, but later declined to about 100 000 tons by the second half of the 1990s (Morinaga 2006:21).

References

Akiyama N et al (1992) As environmental problem and policy in Asia NIES – the debt of intensive industrialization and urbanization. In: Fujisaki S (ed) Environmental problems on a developing country (in Japanese). Institute of Developing Economies, Tokyo

Choi J, Paek D, Paik N (1998) The production, the use, the number of workers and exposure level of asbestos in Korea. Korean Ind Hyg Assoc J 8(2):242–253 (in Korean)

Ishi H (1988) Global environmental report (in Japanese). Iwanami Shoten, Tokyo

Kang S (2002) Japanese Economic Cooperation with South Korea: Korean and Japanese economic cooperation in and around the Masan Free Zone. J Study Mod Soc Cult (in Japanese) 23:37–55

Korea Justice and Peace Committee, and Catholic Priests' Association for Justice (1975) Investigation of the Masan Free Export Zone. SEKAI (in Japanese) 354:23–55

Kurumatani N, Kumagai S (2007) Epidemiology of mesothelioma related with asbestos exposure in the neighborhood. The Medical Frontline (in Japanese) 62(1):36–38

Miki S (1973) Judicial conditions of industrial development strategy and the attracting of business in Korea. J Jpn Inst Int Bus Law (in Japanese) 1(11):16–21

Morinaga K (ed) (2006) In asbestos pollution and its health effect (in Japanese), 2nd edn. Nihon-Hyouron-sha, Tokyo

NICHIAS Corporation (1996) One hundred year's history of NICHIAS (in Japanese). NICHIAS Corporation, Tokyo

Onodera H (1973) Shutting the heat out: the 80-year history of Nippon Asbestos Co. Ltd. (in Japanese). Japan Industry Shimbun, Tokyo

Paek D (2001) Asbestos problems in Korea: its history and present situation. Inf Jpn Occup Saf Health Resour Cent (in Japanese) 271:39–41

Park J, Yoon C, Paik D (1995) A study on exposure among asbestos textile workers and estimation of their historical exposures. Korean Ind Hyg Assoc J 5(1):16–39 (in Korean)

Sonokawa K (1944) East Asian co-prosperity sphere asbestos mining field survey (in Japanese). Japanese Agency of Industrial Science and Technology, Tokyo

Teranishi S (1992) Political economics of environmental problems on the earth (in Japanese). Toyo Keizai Inc., Tokyo

The Government-General of Korea, Industrial Expansion Bureau, Mining Deparmnet (ed) (1933) The Korean Asbestos Mining Industry (in Japanese). Korean Mining Association, Seoul

Wagner J, Sleggs C, Marchand P (1960) Diffuse pleural mesothelioma and asbestos exposure in the North Western Cape Province. Br J Ind Med 17:260–271

Working Group for Recording Historical Japan–South Korea Relations (ed) (1976) Documents: historical Japan–South Korea relation I (in Japanese). Contemporary History Press, Tokyo

Chapter 9
Inaction on Asbestos Disasters and Delayed Countermeasures*

Norio Obata

9.1 Introduction

What might have changed had there been no Kubota Shock? Local residents viewed asbestos problems as workplace accidents and rarely considered them to have a pollution-related dimension. However, it appears that the issue abruptly took on that dimension in 1987 with mounting public concern over the removal of spray-on asbestos materials from school facilities, and reports of asbestos dumping at the Yokosuka naval base. However, even in these cases, the issue was treated as a construction-related problem with sprayed asbestos applications, and it was deemed unlikely to have occurred at all had workers followed proper control procedures.

Then in June 2005 came the news that many residents living in the neighborhood of a Kubota Corporation asbestos facility had been exposed to asbestos and were suffering from asbestos-related illnesses. This was the so-called Kubota Shock. The national government displayed a surprisingly quick response, moving within the relatively short period from September to December 2005 to coordinate information on asbestos victims, instruct its various agencies to gather relevant data and documentation, and roll out a comprehensive set of countermeasures. However, persistent activist efforts by Ban Asbestos Network Japan (BANJAN) and the Asbestos Center can be credited for prodding the government into action. On March 27, 2006, the government enacted a new law aimed at providing relief for asbestos victims (the "Act on Asbestos Health

---

*Translated and revised from the Japanese original, Obata, N (2008). Inaction on Asbestos Disasters and Delayed Countermeasures. *Policy Science* (in Japanese), Supplementary Volume, 13–27, with permission of the Policy Science Association of Ritsumeikan University.

N. Obata (✉)
College of Policy Science, Ritsumeikan University, 56-1 Tojiin Kitamachi, Kita-ku,
Kyoto 603-8577, Japan
e-mail: obatan@sps.ritsumei.ac.jp

K. Miyamoto et al. (eds.), *Asbestos Disaster: Lessons from Japan's Experience*,
DOI 10.1007/978-4-431-53915-5_9, © Springer 2011

Damage Relief"). As a preliminary step, in December 2005 the government had announced a comprehensive package of countermeasures to deal with the asbestos issue. As many as 3 000 different categories of asbestos products were in use, which reflected the life cycle of asbestos. Several government agencies launched studies to identify the risks and the extent of exposure and harm from these products. A huge volume of documentation and data was gathered which identified its use on school buildings, the asbestos content in different types of products, the status of discarded scrap materials, concentrations of asbestos in the air and inside factories, and the methods of control applied. Drawing from these extensive sources of information, I intend to demonstrate the causal factors that influenced delays in the preparation of countermeasures against asbestos contamination.

9.2 A Diversity of Causes and the Realities of Asbestos Contamination and Harm

It has been known for many years that inhaling asbestos is harmful to human health. According to Morinaga (2005), the first cases of pulmonary asbestosis were identified in the UK in 1906, in the US in 1918, and Japan in 1929. The first cases of asbestos-induced lung cancer in the UK and US were recorded in 1935, and in Japan in 1960. The first cases of mesothelioma in the UK were reported in 1935, in the US in 1960, and in Japan in 1973. It has been reported that the causal link between asbestos and asbestosis was identified in the 1930s. In effect, the causes of illnesses attributable to asbestos exposure had been discussed and debated for some 70–80 years.

The people affected by asbestos-induced illnesses represent a wide variety of occupations: shipbuilding, the construction industry, electric power stations and substations, the manufacture of construction materials, asbestos mining, seaport and dockyard operations, asbestos factories, the insulation industry, electric train manufacturing, railroad maintenance and repair, automobile maintenance and repair, maritime crew positions (engine rooms), breweries (beverage filtration equipment utilized asbestos materials), glass manufacture (asbestos was used for insulation), dental technicians, plumbing-related occupations (asbestos was at one time used in piping), boiler engineers, maintenance and repair-related occupations, electrical equipment manufacturing (asbestos was used as insulating material around power units), chemical engineering, automotive industry-related facilities (asbestos was added and utilized in research and manufacturing operations), and pump-related operations (asbestos was utilized in gasket materials). Although the harm from asbestos is caused by inhalation, people within 10 m of an occupational activity involving the use or handling of asbestos are at risk of secondary exposure even if they do not directly handle asbestos materials themselves (BANJAN and Asbestos Center 2004).

Moreover, by some accounts, current applications for asbestos number close to 3000, including spray-on asbestos materials, asbestos slate, asbestos wire mesh, asbestos pipes, and asbestos materials in filters, dryers, baby powder, clothes irons, automobiles, modular kitchens, electric water heaters, electric bicycles, and so forth.

Records of the number of patients designated for coverage under the workers accident compensation system provide a clear window into the current scale of suffering caused by asbestos-induced illnesses in Japan. Data from various government agencies and corporate records on the number of affected workers give an understanding of the extent of the harm suffered by families and residents of areas neighboring industrial sites of asbestos-related accidents.

No accurate estimates of previous cases of asbestos exposure and harm have been made. According to the World Health Organization's revised International Classification of Diseases (ICD), demographic tables did not begin incorporating useful statistics on asbestos-induced mesothelioma deaths until 1995. Prior to that time, one has to rely on estimates. Although it should be possible to develop a clear picture using statistics which are available now, it is unclear whether those statistics accurately reflect data published by various corporate sources. Identifying the actual number of asbestos victims should be the first task. Asbestos-induced lung cancer accounts for up to twice as many patients as those with malignant mesothelioma. To date, most patients have not been interviewed to determine their histories of asbestos inhalation, and almost no testing for asbestos particles or fibers has been performed. As a consequence, in many cases the diagnosis has pinned the blame on tobacco, and has overlooked the asbestos connection completely. Procedures for the diagnostic evaluation and treatment of conventional lung cancer and asbestos-induced lung cancer are essentially the same. What are needed are interviews and improved testing methods to identify potential relationships with asbestos. Field surveys that take these approaches into full consideration are also needed (Natori 2006).

The Japan Association of Mesothelioma and Asbestos-Related Disease Victims and Their Families conducted a questionnaire survey of its members in 2004. According to the findings, 10% of patients who responded and 17% of family members who responded had not heard the term "mesothelioma" as the name of a disease when they or their loved ones first presented symptoms, and 10% of patients and 33% of family members had never even heard the term "asbestos." It is understandable that a fair percentage of individuals would not be clearly aware of any relationship between mesothelioma and the daily occupational utilization of asbestos. As of 2004, virtually every questionnaire survey of patients eligible for workers accident compensation has identified levels of awareness similar to the survey findings cited above. This state of affairs plausibly reflects the inadequacy of efforts by government and the corporate sector to familiarize employees and industrial consumers with the hazards of asbestos. Suffice it to say that many patients who are unaware of the realities of asbestos may end up failing to apply for relief before the time limit on their eligibility expires.

9.3 Government Inaction on the Asbestos Problem

9.3.1 The Government Response to Date

On its handling of the asbestos problem, the government has concluded that it was not guilty of inaction. Specifically, based on the findings of several reviews, all government agencies concerned were deemed to have acted in accordance with the scientific knowledge then available to them in every case, and the charge of government inaction was thus rejected (Tables 9.1 and 9.2). Nonetheless, it has conceded that it did not have a full awareness of precautionary approaches at the time (that is, an awareness that even in the absence of complete scientific certainty, countermeasures must not be delayed should a situation pose a risk of serious injury or harm), and that inter-agency levels of coordination were not entirely adequate. Hence, although it expressed regrets, the government concluded it was not guilty of inaction.[1] Within the context of the asbestos issue and related contamination and harm, no government employees were guilty of inaction. However, in the government's view, there was a need to reflect on its handling of the matter. For one thing, it lacked a full awareness of precautionary approaches, inter-agency coordination proved inadequate, and the end-of-line policy perspectives of the former Environment Agency were still strongly influential.[2]

Table 9.1 History of the asbestos problem overseas

1930s	– Causal relationship confirmed between pulmonary asbestosis and asbestos
1950s	– Asbestos confirmed to cause lung cancer
1956	– Sprayed asbestos applications become commonplace
1960s	– Causal relationship confirmed between mesothelioma and asbestos
1960	– Victims exposed to asbestos while playing on a garbage dump
1965	– Symptoms appear in family members of asbestos workers; homes located 0.5 miles from the factory
1972	– Asbestos declared carcinogenic by IARC
	– Asbestos declared to be the cause of occupational cancers by an ILO expert committee
1973	– Spray-on applications of asbestos fireproofing and insulation materials banned in the US
1983	– Iceland bans the use of asbestos
1986	– US lowers the ceiling for workplace asbestos concentrations to 200 particles per liter
1989	– WHO recommends bans on the use of crocidolite and amosite
	– US Environmental Protection Agency phases in restrictions on the manufacture and importation of asbestos
1993	– EU bans the use of crocidolite and amosite
	– Germany bans the use of asbestos with certain exceptions
	– US bans the use of some asbestos products
1996	– France implements a partial ban on the use of asbestos
1999	– With certain exceptions, the EU calls on all member countries to completely ban asbestos by 2005
	– UK bans the use of asbestos

Source: Data from various documents

Table 9.2 Actions taken in Japan

1972	– Study report on the biological effects of asbestos – In testimony before a Lower House special committee, the then Director of the Welfare Ministry's Public Health Bureau declared a "need to consider health examinations for local residents"
1975	– Sprayed applications of asbestos banned in principle (sprays with an asbestos content of 5% or more)
1976	– Established a tolerable asbestos concentration ceiling of 2 000 particles per liter of air in workplace settings
1986	– Issued a manual for the control of asbestos emissions
1987	– Public uproar concerning the removal of asbestos from school buildings. The Ministry of Education, Science, and Culture had discovered sprayed asbestos in 1 337 structures at public primary, middle, and high schools nation-wide – Japan Asbestos Association bans the use of crocidolite
1989	– Book of guidelines for countermeasures against environmental asbestos prepared – Provisions of the Air Pollution Control Law mandate an on-site asbestos concentration ceiling of 10 particles per liter of air
1991	– Japan Asbestos Association sets a voluntary asbestos concentration ceiling of 1 000 particles per liter of air
1995	– Revised enforcement regulations for the Industrial Safety and Health Act ban the production, transfer, supply, and use of crocidolite and amosite – The definition of asbestos-containing materials revised from an asbestos content of more than 5% to more than 1% – Spray-on applications with an asbestos content of less than 5% also banned
2004	– Japan bans the use of asbestos
2005	– The benchmark concentration ceiling for asbestos in workplace settings is limited to 150 particles per liter of air – Regulations for the prevention of asbestos health damage enacted

Source: Data from various documents

It is not easy to comprehend the wording of the government's statements. Although it claims to have responded in line with the scientific knowledge then available, was that really the case? Table 9.1 summarizes the key developments in the history of the asbestos problem overseas.[3]

Judging from these two chronological lists, although Japan took steps to manage the utilization of asbestos and implement precautionary measures and bans, it was slower than other countries in taking such actions. Can it claim to have taken action as necessary even if that action was delayed? The fact is that these countermeasures were delayed out of consideration for the asbestos industry.

Article 17 of Japan's Constitution contains the following provision: "Every person may sue for redress as provided by law from the State or a public entity if he has suffered damage through an illegal act of any public official." Pursuant to this provision, Japan has established a State Redress Law, holding the state accountable for damage compensation. Liability for damage also applies in cases of inaction. Inaction is a legal term that essentially means a deliberate failure to take action. A person is guilty of the unlawful act of inaction, or nonfeasance, in the event that he or she is responsible for preventing the occurrence of a fact that should be deemed a crime, but, despite his or her competence, has allowed that fact to occur through a failure to take precautionary action. Under criminal law, nonfeasance and contrivance are both considered to be a type of "act." However, compared to specific acts that are treated as crimes, with cases of inaction there is the risk that ambiguity will cloud the boundaries of what is considered prosecutable. Although the question is less problematic in cases of bona fide nonfeasance where an obligation has been clearly prescribed, in cases of non-bona fide nonfeasance, the obligation has not been clearly stipulated, and for that reason, the principles of legality may be compromised unless the obligations in these cases are clearly set forth (Hashimoto 2006).

Because the concept of inaction is itself somewhat ambiguous, one recourse would naturally be in litigation. In May 2006, the first class-action lawsuit seeking damages for asbestos-related injuries was filed against the national government. A group of eight citizens, including former employees of an asbestos-related factory in the Sennan district in Osaka Prefecture as well as local residents who lived near that facility, filed a lawsuit in the Osaka District Court seeking consolation compensation from the government of 30 million yen each for failing to take action despite knowledge of the hazards of asbestos.[4] According to the filed complaint, the Sennan district had been the subject of an epidemiological survey in the second half of the 1930s by a team of physicians affiliated to the former Ministry of Home Affairs, and had also been surveyed and studied in the postwar era. However, the national government had ignored the Sennan asbestos issue despite the possibility of implementing appropriate regulations under the provisions of the Labor Standards Law and Air Pollution Control Law.

Even the government is at risk of being held legally accountable for an act of nonfeasance. However, unlike a private company, it is not in danger of being put out of business, and any monetary compensation it may be compelled to pay would come out of public tax revenues. That being the case, acts of nonfeasance or inaction present more of a risk to residents in terms of asbestos exposure and an increase in the number of victims with asbestos-related diseases than they do to the government in terms of being targeted by lawsuits, and also pose a risk for citizens in the form of unnecessary tax expenditure. Taken in this context, measures to prevent indirect exposure to asbestos, ban its manufacture and sale, and implement steps in administrative guidance had the effect of curbing risks on two levels: the risk of negative health effects on the one hand, and the risk of increased fiscal outlays on the other (Hagiwara 2005).

9.3.2 A Failed Legislative Drive

The Japanese government had numerous chances to implement a ban on the manufacture and use of asbestos products. One such opportunity came on December 3, 1992, with the submission by a group of legislators led by the Socialist Party of Japan to an extraordinary session of the Diet of a legislative bill[5] for the regulation of asbestos. Although submitted, the measure was killed without any deliberation. As it stated in its Article 1 provisions, the proposed legislation would have the purpose of helping to protect the health of citizens and preserve the residential environment by applying regulations as necessary from a public health perspective to the manufacture of products that utilized asbestos (hereafter "asbestos products"). On the subject of imposing manufacturing restrictions, Article 2 stipulated that asbestos products, other than those compliant with certain defined criteria, must not be manufactured, imported, sold, accepted, or provided by anyone. Article 3 on recovery directives, Article 6 on field inspections, and Article 8 on financial accommodation stipulated that the national government should endeavor to provide financial accommodation as needed should local municipal governments implement necessary projects for the removal of asbestos materials from facilities for public use, or in the public interest, or otherwise for the prevention of asbestos-induced illnesses. In addition, Article 11 stipulated that a Council for Prevention Measures on Asbestos Health Hazards should be created within the Ministry of Health and Welfare, thus calling for the establishment of a comprehensive investigative agency.

Had this bill been approved, investigations into the causes, and compensation for victims, would conceivably have been substantively different. However, it reportedly failed because of opposition from the Liberal Democratic Party and the influence of a massive campaign by the Japan Asbestos Association aimed at assuring the public that in the years ahead, industry workers would not suffer asbestos-related diseases, and the environment would not be compromised by the health hazards of asbestos, despite the fact that the dangers of this material had already been demonstrated. Cozy ties between the Japan Asbestos Association and MITI bureaucrats (at that time) have also been cited as a factor. Granted these points, it should be clear that the Japan Asbestos Association and the government share a burden of responsibility.

9.4 The Uncoordinated Response of Government Agencies

9.4.1 Efforts by the Government and Its Agencies

An extensive volume of literature, including reports and other data sources, has been compiled on the harm caused by asbestos. The mountains of documentation prepared by, or submitted to, the government after the Kubota Shock drive home this point. However, the survey findings and other documentation prepared by several agencies reveal severe inconsistencies. In particular, not enough data have been produced even on the number of patient deaths, and there are other signs that bureaucracy is not aware of the true scale of the problem.

The asbestos problem stretches across many occupational fields and affects practically every aspect of their endeavors. It has compelled various government agencies to become even more involved in those areas under their jurisdiction: the Ministry of the Environment (MOE) with issues about air pollution; the Ministry of Health, Labour, and Welfare (MHLW) with workplace safety and industrial waste disposal issues; the Ministry of Economy, Trade, and Industry (METI) with the promotion of alternative materials and the affairs of industrial supervision;

the Ministry of Land, Infrastructure, Transport, and Tourism (MLIT) with the construction industry (previously the purview of the former Construction Ministry) and automobiles (previously managed by the former Ministry of Transportation); the Ministry of Education, Culture, Sports, Science, and Technology (MEXT) with the issue of asbestos in schools; the Defense Ministry with the management of military bases and the naval fleet; and the Fire and Disaster Management Agency (FDMA) with issues relating to insulation materials. Inadequate levels of communication between these agencies have already been cited in reports, conjuring analogies of an orchestra without its conductor. Why were they unable to set up a task force with a comprehensive mandate?

Given the diverse array of issues – in the manufacturing sector (under the jurisdiction of METI), workplace accidents and human health (MHLW), community affairs (Ministry of Internal Affairs and Communications, or "MIC"), and the environment (MOE) – Cabinet-level meetings on the asbestos problem were primarily focused on exploring interim countermeasures coordinated by all sectors of the government. Apparently the idea was to explore all the problems in a closely coordinated way. Conversely, though, it meant that all would share the blame, and that the meetings would not insist of finding whether anyone should bear a significantly larger share of that blame. It is not known what powers of authority the members participating in these Cabinet-level meetings on asbestos actually have.

A range of interim countermeasures to deal with asbestos-related problems were declared[6]: steps to contain the spread of the health-related problems would include exhaustive measures to prevent the dispersal of asbestos dust during the demolition or dismantling of buildings (MHLW, MLIT, MOE); implementation of an all-out ban on the manufacture and use of asbestos at an early date (MHLW, METI); and countermeasures against asbestos exposure in schools and other structures (MEXT, FDMA, the Defense Agency). The drive to alleviate public anxiety over the asbestos issue would comprise active steps for the disclosure of public information (MIC, MHLW, MLIT, MOE), the creation of health counseling centers for workers, retirees, and their family members, as well as residents living near asbestos facilities (MHLW, MOE), and the preparation and release of FAQ sheets to alleviate citizen's anxieties and doubts (MEXT, MHLW, MLIT, MOE). In addition, to aid previous victims of asbestos, the government planned to launch a drive aimed at fully familiarizing the public with the workers accident compensation insurance system and related mechanisms (MHLW, MLIT, FMDA), and to review previous government measures (documentation from seven government agencies attached). Furthermore, in order to strengthen its factual understanding of the issues, the government will: implement field surveys concerning the use of sprayed asbestos applications, and release its findings at an early date (MLIT, MIC, MEXT, MHLW); carry out on-site inspections of business establishments (MHLW); instigate field surveys of airborne asbestos concentrations in neighborhoods surrounding asbestos product manufacturing facilities (MOE); carry out studies of the risk of asbestos-induced mesothelioma and cancer (MHLW, MEXT); and gather information about asbestos regulations and victim compensation trends in other leading industrial nations (METI, MHLW, MOE).

To complement these undertakings by the national government, many prefectural governments around the country have set up websites on asbestos-related topics. In terms of information disclosure, efforts by the public sector have been extensive.

Questions remain as to whether this ostensibly far-reaching set of government initiatives has actually resulted in the implementation of asbestos surveys and measures on a commensurate scale. Considering that initiatives are usually implemented within a vertically structured government apparatus, inter-agency communications and information sharing are a must. However, there are no tangible signs that members of the bureaucracy have engaged in the discussions and debates that might have removed the solid barriers isolating them from one another. Individual agencies have been in the practice of revising the content of some initiatives at their own discretion. For example, in 1987, the MLIT unilaterally adopted a policy of not using fluffy dust-prone asbestos or asbestos slate in buildings ordered by the government.

9.4.2 A Less-than-Sweeping Precautionary Approach

The Ministry of the Environment has pointed out that its forerunner (the Environment Agency) shared with other members of the bureaucracy a perception that its mission was limited to end-of-line environmental measures, and that asbestos was considered to be essentially a cause of workplace accidents inside factory facilities. As a consequence, it lacked a complete understanding of the asbestos problem. Had asbestos been considered to be a substance that could cause environmental contamination, active countermeasures should have been taken whether the problem was limited to the factory workplace or a line of products. This backdrop supports the conclusion that the Environment Agency did not engage in sufficient levels of information-sharing or collaboration with other agencies concerned.[7] Moreover, MOE notes that precautionary approaches would also have been difficult to implement because expert study committees assembled by the Environment Agency had declared that the risks posed to the general public by asbestos were low or small.

Although the 1992 Rio Declaration on Environment and Development does not have legally binding force, the precautionary approach described in Principle 15 is relatively detailed, and was the approach most widely accepted by the international community at this time. Principle 15 includes the following statement:

"Where there are threats of serious or irreversible damage, lack of full scientific certainty shall not be used as a reason for postponing cost-effective measures to prevent environmental degradation."[8]

Prior to the Rio Declaration, the precautionary principle had been integrated into routine levels of decision-making on environmental and public health issues in the US. The first significant achievement derived from that trend was the Wingspread Statement[9] of January 1998. This statement discusses three elements: (1) the risk of harm; (2) scientific uncertainty; and (3) precautionary action. "When an activity raises threats of harm to human health or the environment, precautionary measures should be taken even if some cause and effect relationships are not fully established scientifically.... The process of applying the Precautionary Principle must be open, informed, and democratic, and must include potentially affected parties. It must also involve an examination of the full range of alternatives, including no action."

Until the Kubota announcement of 2005, the huge volumes of government agency documentation cited earlier contained nothing to indicate that the ideas underpinning the precautionary approach had been assimilated. In effect, nothing had been achieved at all. To be sure, reports from some of the investigative studies did note that the government's precautionary approach had been inadequate. Apparently there had been some soul-searching on the matter. If the precautionary approach is deficient, then steps should be taken to integrate the precautionary principle into new legal mechanisms, among other measures.

9.5 Inadequate Asbestos Management Rules and Standards

9.5.1 From Occupational Illnesses (Workplace Disasters) to Industrial Pollution

The preface to the guide for the Ordinance on Prevention of Hazards Due to Specified Chemical Substances (announced in 1971 by the director of the Industrial Safety and Health Department of the Labor Standards Bureau) states that strengthened regulations on hazardous substances will not only help to protect the health of workers, but also contribute to the prevention of pollution. These remarks in the explanatory text for a public ordinance were pointing out that emissions of hazardous substances into the air could be toxic to workers and cause environmental damage.[10]

Paragraph 5 of the government notice, "On the Promotion of Preventive Measures against the Health Hazards of Asbestos Dust" (May 22, 1976, Circular Notice by the Head of the Employment Standards Administration 408), contains the following passage (in Japanese): "Work clothing contaminated by asbestos can be a secondary source of asbestos dust. It has been noted that there is a risk of asbestos contamination not only to workers currently engaged in asbestos-related operations, but also to family members from the work clothing that workers wear or carry back to their homes. For this reason, workers involved in such operations should wear special work clothing solely for the purpose, and keep work clothing contaminated with asbestos away from other clothing. Advise them to remove asbestos from work clothing by washing in a way that prevents the release of asbestos dust, and avoid removing this work clothing from workplace premises. Furthermore, encourage workers to routinely wash their hands and faces and gargle as necessary after they have finished their work duties."

However, regarding the promotion of preventive measures, the same document acknowledges that the guidance may not reach everyone for whom it is intended, noting that "complete compliance with these recommendations will conceivably be difficult because the applicable industrial sector is broad in scope and many of the companies are small and medium-scale establishments."

This notification contained explicit warnings regarding the risk of pollution. However, did the government assume that it would be effective enough for the purpose of alerting local residents? The themes of workplace disasters and public health rarely intersect.

9.5.2 *The Necessity of Systematic Standards Reflecting an Awareness of the Asbestos Lifecycle*

Government bureaucracy resembles the highly segmented and specialized fields of modern medicine. What are needed are comprehensive decisions, not merely decisions that deal with a single aspect of a given problem. Although a variety of standards on asbestos exist, they have to be studied in a systematic way. Decisions to revise benchmark values are not something that can be left to the discretion of a single agency.

Current standards for asbestos include benchmarks for the assessment of workplace environments, worksite borderline benchmarks set under the Air Pollution Control Law, Waste Disposal and Public Cleansing Law standards for waste products under special management, the Building Standards Law, and so forth.

Lifetime risk is a benchmark rate routinely used to assess asbestos and other carcinogenic substances. The lifetime risk rate measures the probability of contracting cancer after exposure to a carcinogenic substance over the lifetime of the average person (hypothetically 70 years).

Based on estimates by the US Environmental Protection Agency (EPA), 0.004 particles per liter of air is the maximum asbestos concentration required to keep the lifetime risk to within no more than a one-in-a-million increased chance of developing cancer. The corresponding concentration for a lifetime risk rate equivalent to a one-in-a hundred thousand increased chance would be 0.04 particles per liter of air, and for a one-in-ten-thousand increased chance, 0.4 particles per liter. The guideline benchmark used by the World Health Organization (WHO) is one person per 100 000 population (Hirose 2005). In Japan, the allowed concentration for asbestos in workplace settings was set at a ceiling of 2 000 particles per liter of air in 1976. In 1991, the Japan Asbestos Association set a voluntary benchmark ceiling concentration of 1 000 particles per liter, and in 2005 the benchmark for the assessment of workplace environments was revised downward to a ceiling concentration of 150 particles per liter. In other words, from 1976 to 2004, the accepted benchmark ceiling concentration was 2 000 particles per liter. As long as concentrations were kept within the limits set by law, they did not constitute legal violations even if they went above the benchmarks in government guidelines for the industry. It thus follows that administrative guidelines of this nature lack any effectiveness. Among revisions made to the Air Pollution Control Law in 1989, the Environment Agency set a worksite borderline ceiling concentration of 10 particles per liter. Although the MOE has its own views concerning risk, there was no clear stance behind the 1989 revisions. Or maybe they had not given the matter serious thought. Although Japan's MOE appears to be striving to position itself as an institution comparable to the EPA in the US, its numerical targets remain ambiguous.

Estimating levels of environmental risk provides important perspectives. Although discussions within WHO and the US government center around clear-cut benchmarks, the position of Japan's government remains unclear. It is as if the prevailing philosophy is to let sleeping dogs lie. When legislators sought to pass a law regulating asbestos products in 1992, their efforts met resistance from worker unions in the asbestos industry and ultimately ended in failure. Moreover, questions about tolerable limits for environmental risk have been quelled without serious debate.

Considering the life cycle of asbestos as a material or product involving the processes of manufacture, installation, utilization, and maintenance, a set of several benchmark concentrations is needed. Preparing appropriate standards and guidelines for each of these stages calls for the creation of an organization with a comprehensive remit.

9.6 The Responsibilities of Large Companies

We now examine some of the public disclosures made by the Kubota Corporation and the NICHIAS Corporation – two companies that manufactured and sold asbestos products – and determine if they actually feel any responsibility or remorse.[11]

First we look at the 2006 and 2007 CSR reports released by Kubota. The 2006 CSR report is a 62-page document comprising a statement of commitment by top management, basic perspectives on CSR management, an economic report, a social report, an environmental report, and a section with independent comments. Only one page in the entire document is devoted to the subject of asbestos. In the comments section, the company president is lauded for being candid about asbestos-related health issues in his admission that they constitute "a serious problem that pains me deeply," and his commitment to "never conceal anything, no matter what it be." However, no mention of the MHLW is to be found in the CSR report. Although asbestos is mentioned by the ninth economic report in the 2006 version, in the 2007 CSR report it is treated as one of the top items in the economic reports section. That discussion covers the establishment of provisions for relief payments to patients or families of patients with asbestos-related diseases living in the vicinity of the defunct Kanzaki plant, medical assistance for patients with asbestos-related diseases (Hyogo College of Medicine, Osaka Medical Center for Cancer and

Cardiovascular Diseases), and the status of Kubota employees (and retirees) with asbestos-related diseases.

The section with independent comments contains remarks suggesting that Kubota has been working in good faith to address the asbestos issue and other problems in the public eye, and also has information that casts the company in a negative light. The independent comments in the 2006 edition of the CSR report contain no direct references to the asbestos issue.

In NICHIAS' 2005 environmental report, the president in his prefatory message states that his company would release asbestos-related information (deaths from asbestos-related causes, number of patients undergoing treatment, lists of past asbestos products, etc.) in the interests of disclosure, and then offers his apologies and condolences. However, something seems strange. There is a misunderstanding about the facts relating to the victims of asbestos-induced diseases, as is apparent in the following comments by the president (in Japanese): "Many workers joined the company prior to 1971, before asbestos was recognized to be a carcinogen, were exposed to high concentrations of asbestos in their work without adequate precautions, and have either died of asbestos-related diseases or are undergoing treatment." The asbestos-related article is at the beginning of the environmental report and expresses a sense of commitment. However, most of the other articles discuss undertakings or accomplishments involving nonasbestos products.

In the 2006 report, the asbestos article has been placed near the end, and contains details on nonasbestos products together with an undertaking to provide relief under the terms of the Act on Health Damage Relief. Unsurprisingly, the amount of content on the transition to nonasbestos products stands out.

In its fiscal 2006 environmental report, NICHIAS has a section that discusses the termination of asbestos joint sheet product sales. It explains that effective from December 31, 2006, the company halted sales (shipments) of joint sheet even for the limited applications approved by the MHLW because, with the international trend toward restrictions on asbestos, procurements had become increasingly difficult. Because the MHLW has established limitations on the applications for asbestos joint sheet, it is possible to read this section and gain the impression that the company halted sales as a matter of compliance with the MHLW. Why, then, did NICHIAS not advance its schedule for the termination of sales in the light of the health damage asbestos was known to be causing? Was its position that no action was necessary unless some MHLW decision mandated it? The underlying assumption appears to be that all responsibility in these matters rests with the government and none with NICHIAS. Elsewhere in the report, NICHIAS states that it participated in a subcommittee of the Study Group on Asbestos Alternatives set up by the MOE and MHLW in preparation for the all-out ban on products containing asbestos, and agreed to cooperate with efforts to promote and implement that ban. However, it is unclear how it has actually made its decisions and acted to date. If a company has been involved in a scandal, it must investigate thoroughly to determine the facts.

According to NICHIAS' 100-year timeline, its sales revenue multiplied 4.3-fold in the 7 years from 1955 to 1962, 3.6-fold in the 9 years from 1963 to 1972, and twofold in the 12 years from 1973 to 1985 (NICHIAS Corporation 1996).

The company reaped profits on this ballooning growth in sales while asserting that asbestos substitutes did not exist. Why was it unable to begin investing some of those profits in the development of substitutes at a time when the contamination and negative health effects caused by asbestos were already known?

The foregoing section has briefly reviewed the CSR reports and environmental reports of these two companies. Although the reports leave the impression that both firms have a sense of responsibility and regrets about their past, neither has demonstrated a commitment to the coherent management of hazardous materials.

9.7 The Act on Asbestos Health Damage Relief: Establishment and Selected Problems of Funding Obligations[12]

The "New Law" for the provision of relief to victims of asbestos-related diseases was promulgated on February 10, 2006, and brought into force in April 2007. Since the problems with the content of the New Act were pointed out in Chap. 2, we focus here on the funding aspects.

The investigative commission on the funding obligations of business establishments for the relief of asbestos victims (August 2008) has calculated that 9.05 billion yen will be required each year from FY 2007 through FY 2010 as funding to offset payments of condolence money and medical expenses for the relief of victims of asbestos-related diseases. The commission estimates that the national government and local municipalities will be responsible for 1.67 billion yen of that total, in the form of administrative fees, with the remaining 7.38 billion yen to be an obligation of the private business community.[13] As this suggests, the idea is to make the entire business community fund the corporate burden by collecting 7.38 billion yen through increases in the workers accident compensation insurance premiums that some 2.6 million business establishments nation-wide are required to pay. Further, the four companies with the heaviest responsibility in terms of their use of asbestos will be instructed to pay a combined sum of close to 340 million yen. This funding obligation for these four firms, including Kubota, NICHIAS, and two other "special enterprises" (their corporate names have not been disclosed), will be no more than 1.8%. In effect, a massive burden will be placed on the shoulders of businesses that already pay workers accident compensation insurance premiums, including many companies that have nothing to do with the asbestos industry. Having the entire industrial community bear the burden of funding for the relief of a national-scale asbestos disaster represents an institutional approach without precedent.

Obviously the basic idea underpinning this method of calculating funding obligations is to seek fast-track relief for asbestos victims, thus bypassing the provisions of the Law Concerning Civil and State Liability for Damage Compensation, and the process of demonstrating causal relationships to establish liability on a case-by-case basis, or collecting damage compensation from the parties actually liable. In other words, the law is designed to provide relief for victims, not pinpoint the causal agents and place blame.

Yet another issue with the law is that it assigns special enterprise status to no more than four companies. This number is far too small considering that 84 companies were members of the Japan Asbestos Association (as of the start of FY 2000) and 31 were the subjects of "Statistics on Victims of, and Deaths from, Asbestos-Related Diseases," a report released on July 8, 2005, just prior to the all-out ban on asbestos use. Furthermore, the funding burden for these four firms is considered to be too small. As one of these so-called special enterprises, NICHIAS Corp. in the year ending March 31, 2006, posted 5.4 billion yen in net profit, paid shareholder dividends of approximately 1.2 billion yen, and had a year-end earned surplus of 24.0 billion yen (data from NICHIAS' financial report. That same business year, Kubota registered 81.0 billion yen in net profit, paid out shareholder dividends totaling approximately 13.0 billion yen, and had a year-end earned surplus of 522.4 billion yen (data from Kubota's business report). The four special enterprises are expected to contribute a combined total of 338 million yen to the relief fund each fiscal year. That is equivalent to no more than 2.4% of the 14.2 billion yen combined total in shareholder dividends paid out by NICHIAS and Kubota together in 2006. The industrial business community has also sought a convincing rationale for these funding burdens.

The strategy of placing uniform obligations on all business establishments nation-wide is yet another huge problem. Trends in market share for single-family home roofing materials show that asbestos-containing veneer slate shingles – a product for residential applications that was manufactured up to 2004 – expanded their share of the market from 21.7% in 1987 to 32.1% in 1995, for a gain of approximately 50%. By contrast, the share of the market commanded by clay shingle products tumbled in the same period from 41.6 to 27.2%. It is the height of absurdity to insist that companies in the clay shingle manufacturing and spraying industries, which saw their market share and business performance hurt by competition from rival asbestos shingle products, should assume a funding burden for asbestos victim relief under the same terms and conditions as companies that were profiting from the manufacture and sale of asbestos products without following adequate safety precautions.

9.8 Closing Remarks

The national government is clearly guilty of inaction on the asbestos issue. It is projected that the scale of asbestos harm and contamination will continue to grow in the years ahead. The Act on Health Damage Relief has been established and is now in force. As noted in the review of selected provisions, the New Law has quite a few problems. This is a law that should be revised with a preamble that clarifies the national government's liability and recognizes the asbestos crisis to be the stock pollution disaster that it is.

Notes

1. Cabinet Ministerial Conference on the Asbestos Crisis (2005). "Review of Past Government's Handling of Asbestos Issues (Supplement)."
2. Minutes of the Meeting of the Committee on the Environment, House of Councillors, 164th Diet Session, No. 2, p. 4, 2006.
3. Compiled from available documentation.
4. Nihon Keizai Shimbun Newspaper, May 26, 2006.
5. Socialist Party of Japan (1992). "Asbestos-Containing Product Control Act".
6. Cabinet Ministerial Conference on the Asbestos Crisis (2005). "Interim Measures against the Asbestos Crisis."
7. Ministry of the Environment (2005). "A Review of Past Handling of Asbestos Issues by the Ministry of the Environment".
8. United Nations Conference on the Environment and Development (UNCED) (1992). Rio Declaration on the Environment and Development.
9. Joel et al. (1998). The Precautionary Principle in Action: A Handbook. The Science and Environmental Health Network. http://www.sehn.org/. Accessed May 5, 2010.
10. Yomiuri Shimbun Newspaper, July 22, 2005.
11. Kubota Corp. http://www.kubota.co.jp/. Accessed May 5, 2010. NICHIAS Corporation. http://www.nichias.co.jp/. Accessed May 5, 2010.
12. Kita (2006).
13. Meeting on Employer's Burden Relating to Asbestos Health Damage Relief, the Ministry of the Environment (2006). "Basic Standpoint on Employee's Burden Relating to Asbestos Health Damage Relief".

References

BANJAN, Asbestos Center (eds) (2004) Toward non-asbestos society: the challenge posed by killer dust surrounding us (in Japanese). Kamogawa Press, Kyoto

Hagiwara I (2005) The role of local governments, administration accused for omission on asbestos issue. Glob Environ (in Japanese) 36

Hashimoto Y (2006) Pluralistic structure of Liability Law (in Japanese). Yuhikaku, Tokyo

Hirose H (2005) The issue of government's inaction on asbestos. Wkly Econ (in Japanese) 3779

Joel T et al (1998) The precautionary principle in action, a handbook. The Science and Environmental Health Network. Accessed 5 May 2010

Kita M (2006) Contradictions in policy of having all businesses contribute to the asbestos victim relief fund (in Japanese). October 11, 2006. http://www.news.janjan.jp/living/0610/0610092515/1.php. Accessed 5 May 2010

Morinaga K (ed) (2005) Occupational asbestos exposure and asbestos-related diseases: fundamental knowledge and workers accident compensation (in Japanese). Sanshin Tosho, Tokyo

Natori Y (2006) Current situation and issue of victims (in Japanese). Asbestos Center. http://www.asbestoscenter.jp/about/propositions/victims20060115.pdf. Accessed 5 May 2010

NICHIAS Corporation (1996) One hundred year's history of NICHIAS (in Japanese). NICHIAS, Tokyo

Chapter 10
Process Tracing of Asbestos Politics in Japan: Focus on Fiscal Years 2005 and 2006*

Michiya Mori

10.1 Introduction

In June 29, 2005, the Mainichi Shimbun reported Kubota Corporation's announcement that scores of its workers had contracted asbestos-related diseases. It has been estimated that 79 Kubota personnel had died from asbestos-linked causes during the period from 1978 through 2004. This disclosure had prompted Kubota to quickly begin providing consolation payments to workers and residents living near its decommissioned Kanzaki factory, and also put pressure on other asbestos industry firms, as well as government officials, to pursue remedial action. Previous studies conducted in Japan and abroad had already pointed out that many of the diseases caused by asbestos manifest themselves only after a long period of dormancy, extending anywhere from 20 to 50 years. Furthermore, in 1987 the utilization of asbestos materials in Japanese public school structures received heavy coverage in the news media and gained status as a serious social issue. This was the so-called "school panic." However, the latest series of events, collectively described as the Kubota Shock, have evolved into a political issue and have drawn more public attention and concern than ever.[1] Further, as exemplified by a pending case of litigation launched on May 26, 2006, and seeking damage compensation from the national government, debate centering around the theme of "government inaction," namely in terms of the government's failure for many years to implement countermeasures against asbestos, has become even more intense than the criticism aimed at private corporations in the asbestos industry.[2]

*Translated and revised from the Japanese original, Mori, M. (2008). Process Tracing of Asbestos Politics in Japan: Focus on Fiscal Years 2005 and 2006. *Policy Science* (in Japanese), Supplementary Volume, 203–225, with permission of the Policy Science Association of Ritsumeikan University.

M. Mori (✉)
Graduate School of Public Policy, Ritsumeikan University, 1 Nishinokyo-Suzaku-cho, Nakagyo-ku, Kyoto 604-8520, Japan
e-mail: mori-m@pp.ritsumei.ac.jp

K. Miyamoto et al. (eds.), *Asbestos Disaster: Lessons from Japan's Experience*, DOI 10.1007/978-4-431-53915-5_10, © Springer 2011

However, the fact that the relevant actors have tentatively been credited for their fast response, coupled with the structural preoccupation of the "government inaction" controversy with events predating the Kubota Shock, has conversely nurtured a pattern of misguided perceptions about the political impact of the asbestos issue in the aftermath of the Kubota Shock, and the impact at the national government level in particular.[3] In other words, questions remain over how key actors at the national government level implemented asbestos policies in the wake of the Kubota Shock, and the outcomes that their policies have had for society as a whole.

In this chapter, I strive to answer these questions by detailing interactions between the ruling party, core government executives, ministries and government offices, the corporate sector, and local governments (Sects. 10.3–10.7), while examining assorted trends in asbestos-related politics with a focus on the years 2005 and 2006 (Sect. 10.2). The principal implications deriving from that exercise are that the ministries and agencies of the national government, through their interactions with other actors, limited the options for an asbestos policy, effectively leaving no alternative other than the formulation of new laws, and were also behind the curve in the implementation stages for the new legislation. Finally (Sect. 10.8), I take these conclusions into consideration, and identify issues for further study.

10.2 Asbestos and the Political Context

This section reviews information in two contexts that will serve as a basis for the arguments presented in the following sections: (1) developments prior to and following the asbestos crisis that unfolded in 2005 and 2006; and (2) the conditions for the government administration then in power, policy priorities, and other elements of the political context or "environment" prevailing at that time.

In considering point (1), I pay attention to the frequency of occurrence of the terms "asbestos" and "*ishiwata*" in Asahi Shimbun Newspaper articles and national parliamentary deliberations. Taken together with content to be discussed later, the data presented in Table 10.1 reveal three features. First, although neither term was mentioned much at all prior to the Kubota Shock, their respective frequencies of use rose sharply after that event. Second, following coverage of asbestos issues by the news media, deliberations within the Diet moved forward, prompting renewed media coverage. Third, the use of the terms "asbestos" and "*ishiwata*" in newspapers remained at a high frequency even after media coverage of the Kubota Shock had subsided.[4] One point worth noting here is that within the context of the "government inaction" arguments, whether subconsciously or without careful consideration, the term "Kubota Shock" might seem to have been emblematic of the asbestos crisis during this period. However, Table 10.1 suggests that the term is not equivalent to the asbestos crisis.

In pursuing a study of the political process, three potential scenarios have to be taken into account in relation to context (2) above, as indicated by discussions concerned with agenda-setting[5]: (a) when social trends (e.g., in the newspaper and television media) precede political trends; (b) vice-versa, i.e., when political trends precede social trends; and (c) when these trends are not connected. For example, the so-called "school panic" and the ensuing lull thereafter would serve as evidence for scenario (c). To add to this case example, if the political process developed in the sequence described by scenario (a), we would need to consider what factors or conditions fostered the formulation of asbestos countermeasures at the government level and independently of the social factors, namely, a rapid surge in the volume of information. That necessity would lead to a consideration of context (2).

I now provide a brief review of the political trends under way during the period in question. The Kubota Shock came in June 2005, coinciding with the final days of the second Junichiro Koizumi Cabinet. At that time, domestic politics were focused on the battle over proposed legislation for the privatization of the postal service. Koizumi had circumvented the "procedure" of the Liberal Democratic Party (LDP) General Council and submitted the legislative bill to the House of Representatives for deliberation. Following the bill's rejection on August 8, Koizumi chose to confront his resistance forces by dissolving the House of Representatives outright. On September 11, the LDP went on to a landslide victory in a general election which was marked by selective targeting of resistant candidates for defeat, and on October 14, the new Diet followed up with the passage of a bill deemed to be a flagship of legislative reform. As pointed out by many studies,[6] Koizumi demonstrated the clout he had earned through efforts at electoral reform which he had pursued since the 1990s, and then carried through with the implementation of policies he himself had backed in a variety of fields, including proposed legislative bills for administrative reform and an overhaul of the health-care system. However, the reverse side of the coin, as highlighted in an analysis of this process by Ryuunoshin Kamikawa, was that policies that Koizumi either did not support or had no interest in were arguably at risk of being further delayed or neglected entirely.[7]

To elaborate on this point, although not limited to this particular period, it needs to be kept in mind that while several public policies were being dealt with or implemented in parallel, some were considered to be politically contentious whereas others were not. Although I have not prepared any systematic comparisons with other policy domains, one point I want to make in this context is that the asbestos issue was not only of concern to the general public, but it had also continued to draw the attention of the national Diet on a level equivalent to one of the larger issues of the day, namely, government policies pertaining to privatization of the postal service system (Table 10.1). Granted that assumption and restating the investigative focus of this paper, how were these two threads of concern inter-related, and more specifically, how were asbestos policies adopted during the time of the third Koizumi Cabinet, and through what political process did they evolve thereafter? Sections 10.3–10.7 examine this process with reliance on the facts as presented in Asahi Shimbun newspaper articles. Each section explores changes in the courses of action taken by different groups of actors at the national government level.

Table 10.1 "Asbestos" and "*Ishiwata.*" Frequency of use in Asahi Shimbun Newspaper Articles and National Diet Deliberations, FY 2005–2006

Year/month	Asahi Shimbun (no. of news articles)	National Diet (no. of plenary sessions and committee meetings)
April 2005	3[196]	4[19]
May	4[111]	0[25]
June	9[145]	6[42]
July	467[266]	18[40]
August	456[1 898]	7[11]
September	547[1 496]	5[11]
October	402[297]	32[60]
November	365[123]	1[3]
December	281[110]	0[0]
January 2006	72[68]	17[7]
February	150[49]	22[26]
March	131[37]	32[31]
April	69[57]	8[21]
May	63[37]	10[20]
June	70[69]	12[16]
July	51[68]	0[0]
August	48[43]	0[2]
September	72[98]	0[2]
October	54[71]	10[14]
November	55[103]	3[10]
December	32[73]	6[12]
January 2007	27[44]	0[3]
February	39[33]	3[6]
March	45[28]	5[16]
Total	3 512[5 520]	201[397]

Note: Data in brackets are the frequencies of use of the term "*yuusei min'eika*" ("postal service privatization")
Sources: Asahi Shimbun database (Kikuzo II Visual for Libraries) and the full-text database system for the Minutes of the Diet

10.3 The Repercussions of the Kubota Shock: July 2005

This section focuses on the initial response by the relevant actors immediately following the Kubota Shock up to late July. Prompted by reports of asbestos-linked deaths among Kubota employees, business visitors to the Kubota factory, and local residents, on June 30, 2005, the City of Amagasaki announced that it would launch a retroactive investigation.[8] Moreover, the Ministry of the Environment (MOE) and Hyogo Prefectural officials began questioning Kubota personnel on the matter. The following day – July 1 – the Ministry of Economy, Trade, and Industry (METI) sprang into action, demanding production records and information about asbestos victims from seven industrial groups. On July 5, the NICHIAS Corp., the largest company in the asbestos industry, announced that asbestos deaths had occurred at its facilities in Ibaraki, Kanagawa, Shizuoka, Gifu, and Nara Prefectures, prompting officials in Nara, the prefecture with the most deaths, to demand an investigation the following day. Amid revelations that families of Kubota employees had also been victimized by asbestos contamination, on July 7 the Ministry of Health, Labour, and Welfare (MHLW) decided to launch on-site investigations at some 300 businesses with reported asbestos victims, and demanded that current and former employees undergo health examinations at each employer's expense. On July 8, news began to spread of asbestos victims in Ohita, Fukushima, and other prefectures; additional disclosures were made by Sumitomo Heavy Industries on July 11, and IHI Corp., Mitsui Engineering & Shipbuilding Co., Ltd., and Mitsubishi Heavy Industries, Ltd. on July 13. Responding to the revelations of this widening crisis, the Minister of Health, Labor, and Welfare, Hidehisa Otsuji, held a press conference after a meeting with other cabinet ministers and reported that his ministry would set up counseling centers and step up research on treatment methods for mesothelioma. The MHLW thus solidified its position that the Industrial Safety and Health Act should be revised with attention to industry trends, and that a full-scale ban on the utilization of asbestos should be implemented by 2008.

However, providing victims with clear-cut relief and pursuing legal regulations for asbestos faced various obstacles in the form of existing legal frameworks and the nature of asbestos as a substance that remains dormant in the human body for long periods before the symptoms of exposure become manifest. First, with regard to relief for victims, the MHLW's administrative vice-minister suggested that under the current workers accident insurance compensation framework, providing relief compensation to family members would probably be impossible. In addition, the Minister of the Environment, Yuriko Koike, distanced herself from the view that asbestos contamination warranted treatment as an issue in environmental pollution, noting that it would be necessary to carefully weigh whether that idea was consistent with the spirit of laws that treat air pollution as a phenomenon of a significantly broad scale (July 10, 2005). In response, Hyogo Prefectural Government, for example, urged that the MOE give asbestos victims the same coverage under the provisions of the Act on Pollution-Related Health Damage Compensation as was provided to victims of Minamata disease and *itai-itai* disease. On this matter, Cabinet Secretary Hiroyuki Hosoda expressed support for the Hyogo Prefectural Government's request, remarking that it would have been better had a ban been implemented at an earlier date, and that with the huge amounts of asbestos that had accumulated in the environment thus far, the issue appeared in his view to be a stock pollution disaster (July 12, 2005). Immediately afterward, however, discrepancies began to appear in the positions taken by core executives and relevant government ministries and agencies. Responding to repeated requests by the Hyogo Prefectural Government that the government should extend assistance on grounds that the asbestos crisis was nation-wide in its reach, and not limited to Hyogo alone, the MOE replied that the adverse health effects of asbestos on local residents were somewhat limited in their scope relative to accepted models of environmental pollution, and instructed the relevant ministries and agencies to act in a coordinated manner after identifying the actual extent of asbestos contamination (July 13, 2005). Incidentally, Hyogo Prefectural officials were also quick to act on this response. At the National Governors' Association of Japan on July 14, they proposed, and gained, approval for an emergency request seeking rapid action by the national government (July 14, 2005),[9] and on July 15, they set up an asbestos countermeasures promotion council that would be chaired by the vice-governor himself.

Ironically, on the regulatory front, on July 1 the government enacted the Ordinance on the Prevention of Asbestos Hazards, which had been established in February 2005 to replace the existing ordinance on the prevention of hazards due to specific chemical substances, and thus reinforced efforts for safety management (for further information, see the Japan Industrial Safety and Health Association 2005). However, deep-rooted doubts about the viability of this initiative were voiced, partly as a reflection of the growing awareness of the aforementioned cases of asbestos contamination and victims.

Nevertheless, it was urgent that actors at the national government level should take action. For example, on July 15, METI disclosed that a total of 374 workers in a wide range of industrial fields, including the shipbuilding, automotive, and construction sectors but not including subcontracting businesses, had died from asbestos-related causes and another 88 workers were struggling with asbestos-induced illnesses. Commenting on these data, spokespersons for METI's Housing, Industry, Ceramics, and Construction Materials Division noted that the extent of contamination and harm had been more serious than anticipated, and conceded that investigations into the matter should have been launched earlier (July 15, 2005). With ironic timing, the same day a plenary session of the House of Representatives unanimously approved the Convention Concerning Safety in the use of asbestos that the ILO had adopted in 1986. This was something that the administration had been calling for since October 2004, after concluding that conditions then existed for a ban in principle on the manufacture and use of products containing asbestos. Given that perspective, it seems apparent that the hazards of asbestos were already well recognized at the government level even before the Kubota Shock, and that certain steps to control its use had been taken. Still, the political schedule for this process had not yet been arranged.

Nonetheless, Tetsuzo Fuyushiba, party secretary for NEW KOMEITO, the party that had entered into a ruling coalition with the LDP, noted in remarks at Kubota's Hanshin office that legislative measures would have to be taken should many residents (near Kubota's factory) be found to be suffering from asbestos-linked illnesses. As this remark illustrated, politicians were growing increasingly alarmed about the issue (July 14, 2005), and it was against that backdrop that the Diet at last began deliberating the matter in earnest on July 20. During intensive deliberations of the asbestos issue by the Committee in the House of Representatives on Health, Labor, and Welfare, LDP representative Shinji Inoue joined opposition party members in blaming the government for negligence. This criticism constituted a major development in the light of the relations then existing between the ruling and opposition parties. However, in defending the record on regulation-based management, the director of the Labor Standards Bureau of the MHLW stressed that the government had done everything it could have done. In addition, the director of the Manufacturing Industries Bureau of METI remarked that while total regulation was an approach that had been adopted only in selected areas of Europe, the internationally accepted view was that asbestos could still be utilized if it were properly managed, and on that understanding, members of the industry had announced their opposition to regulations and the government had acceded to their demands (July 20, 2005).

However, attention then focused on the admissions by high-level bureaucrats that the government's response to the problem had been inadequate. In other words, while recognizing that the problem was one involving factory "workplace environments," the government failed to fully acknowledge the "air pollution" dimensions of the problem that affected local residents. On that point, the Vice-Minister of the MHLW, Hiroyoshi Nishi (NEW KOMEITO), commented that the government had not properly followed up on the matter despite its awareness of the facts, and that in his view, this constituted a decisive failure. In effect, the implication was that the government had reached an impasse in assigning responsibility because the asbestos problem spanned three administrative dimensions: labor administration, environmental administration, and health and welfare administration. Still, the administrative vice-minister of the MHLW questioned whether there had actually been any failure at all (July 21, 2005).[10] In the course of these Diet deliberations, it became apparent that members of the ruling party, the core executive branch, and the administrative bureaucracy had differing perceptions of the problem.

On the other hand, how did private industry handle the problem once it had been brought to light? With tensions mounting amid news of the sheer scale of the disaster, the spotlight turned to Kubota as it took a step on the night of July 12 to extend aid to local residents living around its defunct Kanzaki factory. Until that time, it had gone no further than offering to refer citizens with health-related complaints to hospitals or aid organizations. However, in view of the strong demand from local residents for health examinations, Kubota stated that it would investigate ways of providing assistance, including helping to offset their medical bills, and then decided to move forward and sponsor health examinations at its own expense. The NICHIAS Corp., for its part, implemented free examinations as well. Although NICHIAS contended that it viewed its own cases of asbestos contamination to be internal or workplace disasters, and that it was not ready to disclose information on such matters (July 17, 2005), its actions can be interpreted as an effort to follow Kubota's lead. The Tatsuta Industry Co., a NICHIAS subsidiary, held a briefing session at which it issued a formal apology. In the eyes of the public, Kubota's response was taken as an expression of good faith, and as such contributed to the identification of additional asbestos victims.

July 20 witnessed an outpouring of news on the following developments. First, there were reports that NICHIAS might have even more worker deaths from asbestos. The YANMAR Co., Ltd. disclosed that some of its former workers had died from asbestos-linked diseases and that others were undergoing treatment. Amagasaki's mayor declared that her city had an obligation to protect the health and well-being of its citizens, was actively considering a program of health examinations, and was searching for ways to effectively implement that idea. Further, the MHLW announced preliminary estimates that the number of individuals in FY 2004 with asbestos-linked illnesses who had been certified for coverage under the workers accident compensation insurance framework had increased 1.5-fold from the year before (July 21, 2005).

Prompted by this mounting public reaction, the Diet finally decided to take up the issue. The relevant government ministries and agencies in particular also stepped up their efforts in response, as detailed below. At a press conference called on the morning of July 21, Chief Cabinet Secretary Hosoda announced that the government had many reasons to regret the position it had taken on the matter to date. He added that because the various ministries and agencies concerned had overlapping jurisdictions, he wanted to see them coordinate their efforts without any lapses or oversights. Hosoda's remarks may have had an influence. For instance, the MHLW and METI later demanded that the Petroleum Association of Japan, the Japan Iron and Steel Foundation, and 18 other related industrial organizations should gradually ban the use of asbestos, even for applications that had been exempted due to a lack of readily available substitutes. The industry groups agreed to this (July 21, 2005).

By July 22, efforts were afoot to strengthen ministry and agency coordination on the understanding that measures to ban the use of asbestos should be implemented ahead of schedule. Specifically, a decision was made to promote managerial-level bureaucrats to bureau director status, and instructed them to gain a full understanding of the extent of the harm asbestos that had caused to human health, as well as to work out and announce countermeasures within the month. This prompted the MHLW and MOE to inaugurate their own task force teams – the former to collaborate with experts in exploring the need to give local residents health examinations, as well as the scope of these and other countermeasures, and the latter to put the administrative vice-minister in the top position and to coordinate with all relevant departments on pollution damage compensation, air pollution, and the processing of scrap asbestos materials.

However, this is not to imply that progress on these countermeasures always went smoothly. For example, on July 25 it was disclosed that construction material manufacturers had been selling off their asbestos product inventory, which was outside the scope of the October 2004 revisions to the Industrial Safety and Health Act enforcement regulations that mandated a ban on the manufacture, sale, and importation of asbestos. To deal with this situation, the MHLW's Chemical Hazards Control Division, acting through its channels of administrative guidance, decided to impose an immediate halt to such sales. Although the prevailing view within the MOE was that the MHLW had jurisdiction over workplace accidents involving asbestos, at a regular press conference held the same day, the administrative vice-minister for the MOE expressed the view that (compensating residents near asbestos factories) was a role that the MOE was prepared to share, thus indicating that his ministry was ready to assume a more active role. However, environment minister Koike, had taken the position that actual countermeasures would be contingent upon the findings of field surveys aimed at investigating the extent of harm caused by asbestos, and the imperative of conclusively identifying the link between asbestos and mesothelioma. Regarding her ministry's decision to hammer out countermeasures by the end of September, Koike sparked additional controversy with the comment that it would be difficult, if not impossible, to apply the provisions of the Law Concerning Pollution-related Health Damage Compensation and Other Measures (July 30, 2005). In other words, still reeling from the "first wave" of the Kubota Shock, key players at the national government level were only just beginning to consolidate their views on relief for asbestos victims and the need for new legal measures.

10.4 National Government Policies on Asbestos, Part 1: August 2005

This section organizes all available information for the period from late July through late August 2005. On July 22, it was reported that the conference of ministry and agency personnel that had only recently been conducted at the bureau director level would now be lifted to the cabinet level. With that news, it was evident that government-level efforts had entered an entirely new phase. Headed by the Chief Cabinet Secretary, Hosoda, the first cabinet ministerial conference on the asbestos crisis convened on July 29. One issue at this meeting was a decision to put off until the end of September its conclusions regarding relief measures for victims of asbestos who had already died but had not received any workers accident compensation, family members who were not eligible for accident compensation benefits, and local residents (but it did release an outline of its proposals at the end of August). However, the conference did issue a written statement entitled "Interim Measures Against the Asbestos Crisis" (hereafter "Interim Measures"),[11] which was concerned with the following topics: preventing further asbestos contamination and harm, steps to alleviate public anxiety, measures to assist previous asbestos victims, and a review of the government's past handling of asbestos issues.

Although Chief Cabinet Secretary Hosoda declared that the government would faithfully consolidate its actions in line with that document (July 30, 2005), the steps taken by the various ministries and agencies did not convey to the public that a systematic approach had actually been devised.[12] Consider some of the actions being implemented by elements of the bureaucracy at the time. The MHLW, for example, indicated that it intended to streamline workers accident compensation insurance benefit application procedures for affected workers who were unable to prove their work histories because their employers had gone out of business, and also that it would notify active workers of the names of approximately 230 workplaces that had certified asbestos patients on their payrolls receiving workers accident compensation insurance benefits. In addition, the MOE announced that among other actions, it would take steps to rebuild its capacity to track the disposal of scrap asbestos, a task it had abandoned in a measure to reduce office workloads after its restructuring in 2000. Without question, efforts of this nature suggested that relevant government ministries and agencies had begun implementing countermeasures aimed at regulating asbestos. However, from the general public's vantage point, albeit possibly due to a lack of professional or technical understanding, the situation still deserved the cynicism reserved for the traditional evils of a vertically structured bureaucracy.[13]

By this time, many politicians had become seriously involved. Within the LDP, Genichiro Sata, the Chairman of the Joint Policy Division on Asbestos Countermeasures, declared that the issue demanded a political decision, not more bureaucratic management, and added that while future questions would revolve around whether action should comprise new legislation, budget measures, or government decrees, he personally preferred to move forward with a sound legislative proposal (July 30, 2005). In another move on July 29, the ruling coalition launched an asbestos countermeasures project team headed by the House of Councillors member Tadashi Taura (LDP) and including leading members from various policy divisions on health and labor, the environment, and the economy and industry, and assigned it the mission of submitting recommendations to the national government before the end of August.

In the end, however, the independent responses by different players in the national government – namely, the cabinet (core executives), ministries and agencies (administrative bureaucracy), and the ruling party (legislators) – stood out. To reiterate yet again, this approach to asbestos policy-making failed to send the public a clear and coherent message. In contrast, among efforts by the opposition, the Democratic Party of Japan (DPJ) distinguished itself by submitting to the Committee in the House of Representatives on Health and Labor a proposed revision to the Industrial Safety and Health Act aimed at providing relief for families of asbestos victims beyond the 5-year post-death time limit for the submission of workers accident compensation insurance applications.

I now summarize events up to late August 2005, when the government submitted the text of its revised "Interim Measures" and its "Review of Past Government Action on Asbestos" (hereafter "Review of Past Action"). As already noted, the legislative and administrative branches had already become actively engaged. However, it is worth noting that by August, the spotlight of public attention had focused on efforts in the private sector and at the local government level. Given that these developments can be construed as having had a stimulating effect on the ensuing response of actors at the national government level, I would like to discuss them here in some detail.

First, in early August, one of the experts leading the MHLW research team on mesothelioma criticized the government's record. In the expert's view, no other area in Japan had experienced as much harm from asbestos exposure than the neighborhood around Kubota's defunct Kanzaki factory, and it would not be surprising if the government was labeled negligent for its failure to even make plans for an in-depth epidemiological survey of local residents (not limited solely to death cases) (August 8, 2005). Acting on Amagasaki Health Center projections of an increase in mesothelioma patients, in mid-August the City of Amagasaki decided to begin offering free health examinations to all citizens who wanted them, starting on August 19. On the first day, 29 citizens received examinations. In an emergency move, the city boosted its operating budget for this program to 10 million yen, and by the evening of August 24, a total of 253 citizens had reportedly applied for the examinations. In addition, on August 22, Amagasaki's City Office decided to move forward with a study to measure the concentrations of asbestos particles in the air. Around that time, Amagasaki's mayor described the situation as a case of environmental pollution, predicted that the number of victims would rise, and disclosed plans to call on the national government to come up with new measures to assist asbestos patients (August 25, 2005). The mayor's comments proved to be on the mark. Approximately 1 month later, by September 20, a total of 404 citizens had undergone health examinations, and 124 were determined to be in need of more thorough tests.

Amid these developments, Kubota decided to provide condolence and consolation payments to an additional seven residents living near its defunct Kanzaki factory, bringing the total number of beneficiaries to 24 citizens in all. Although some insiders reportedly conceded that the situation had already developed into a full-scale environmental disaster, Kubota's public relations office countered with the explanation that the causal relationships were not yet fully known, but that the company was not striving to avoid accountability for past problems (August 25, 2005). On top of this, NICHIAS announced that it would offer damage compensation coverage even to delivery personnel and other business visitors to its asbestos facilities, and that it would launch a program of free health examinations for local residents. It was further reported that the Sumitomo Osaka Cement Co., Ltd. (the company that assumed the operations of the former construction materials firm Kansai Slate Co.) would make consolation payments to home-care patients who had lived near its factory facilities in the past. The public took notice of the relative substance and speed of the measures being taken by these members of the private business community compared with efforts at the national government level.

The City of Amagasaki was not the only municipality to become involved. I can cite other examples of communities and prefectures across the nation that began taking action. On August 17, the governor of Tottori Prefectural Government stated that under current laws, a comprehensive response to the asbestos problem was not possible, and it was unclear who had responsibility for asbestos management and the handling of asbestos-containing materials from demolished structures. He added that he planned to submit a proposal to the regularly scheduled prefectural assembly in September in the interests of curbing the airborne dispersal of asbestos dust (August 18, 2005). Fukui Prefectural Government also began taking steps to do something. Its governor issued a directive calling for the establishment of an ordinance urging fast-track, inter-agency studies without waiting for the national government to act (August 23, 2005). These examples compare with the leadership shown by local governments in putting together measures against pollution in the 1960s and 1970s,[14] and could be one of the factors that fueled public scrutiny of the response at the national government level.

10.5 National Government Policies on Asbestos, Part 2: September 2005

This section covers events from late August to late September, 2005. On August 26, the government issued a revised version of its "Interim Measures." That document is noteworthy for the relief framework it presents for the handling of past asbestos-related damage and victims. However, in terms of facilitating a better understanding of the actions taken by key actors, the first half of this section will comprise an overview of the written "Review of Past Action." That document is a massive work divided into a summary and a review of actions by seven related ministries and agencies. This chapter is concerned not with an appraisal of specific parts of that document, but with an investigation of its political impact. Accordingly, the discussion herein focuses on the differences in the way government documents from different ministries and agencies have been worded, the implications of recriminatory exchanges among members of the bureaucracy,[15] and the significance of ruling party comments on the asbestos issue ahead of general elections called following the dissolution of parliament over the loss of the vote on postal privatization.

The summary section in the Review of Past Action concludes that relevant ministries and agencies did not adequately coordinate their actions in the past and have room to improve. It further states that a study throughout September 2005 will be devoted to the creation of a framework that facilitates more thorough levels of inter-office coordination on problems of this kind in the years ahead. With regard to the question of whether Japan has trailed behind other nations in its regulation of asbestos, the review states that further study on that matter will be needed. However, the Asahi Shimbun makes some interesting points regarding the background to the statements on inter-office coordination, as follows (August 27, 2005, italicized excerpts translated from Japanese):

> *As an example of poor inter-agency coordination, the report cites a government conference on measures against asbestos that convened under the aegis of the former Environment Agency in 1990. Although it is disclosed that the conference only met once more 3 years later, and existed in name only, ministry and agency stances on that assessment vary.*
>
> *The Ministry of Health, Labour, and Welfare expressed its view that a gathering of manager-class personnel from eight ministries and agencies would likely have been able to discuss issues on a serious level, and achieve effective coordination on countermeasures if they met several times. When queried about the reasons why the conference did not continue to meet, the MHLW spokesperson replied that he did not know because the conference was sponsored by the Ministry of the Environment. Reacting to these comments, the MOE insisted past events were being used to portray it as an organization suffering from inertia. These responses suggest the two ministries are trying to pin the blame on each other.*

In other words, while leaving in the comment that further study will be needed to determine whether Japan has trailed behind other nations in its regulation of asbestos, the summary section of the review (according to the Asahi article) suggests that in reality, inter-agency coordination could not be described in general terms.

Furthermore, among government insiders, there was a deep-rooted belief that it was still too early to say whether government policies had aggravated the scope of asbestos contamination and harm, because in their view many of the victims currently showing symptoms of asbestos-linked illnesses were individuals who had been exposed to asbestos prior to 1975, when the regulations first went into effect (August 26, 2005). If that were indeed the case, why would the government release an official document at this time that could be interpreted as an admission of policy failures? The Asahi Shimbun offers the following explanation (August 27, 2005, italicized excerpts translated from Japanese):

> *Although we are not prepared to address the issue of administrative accountability, we have no choice but to actively deal with the spread of asbestos contamination and harm. The decision by the government to introduce new legislation for the relief of local residents exposed to asbestos was a product of the "mutually influenced" (according to prime minister's residence insiders) intentions of cabinet ministers and a ruling party highly sensitive to public opinion in the lead-up to the general election.*
>
> *Originally, the topic of interest on this day was the review of past actions on asbestos. The government stance (according to one high-level official) is that the asbestos problem is in no way comparable to the issue of HIV-contaminated blood products. That is the argument used to persuade the public that there have been no problems with the government's administrative handling of the issue. However, news of the Diet's dissolution changed the atmosphere completely. Fear that public opinion could not be contained spread throughout the government and ruling party. Hence, the initial date of the end of September for a decision on victim relief was moved up.*
>
> *The New Act is great news for asbestos victims. In view of the high publicity accorded this theme, it seems clear the government and ruling party were motivated to treat the law as a high-profile statement of policy.*

Although the government conceivably had a range of policy options for action at its disposal, on August 25 it decided to take an approach involving the establishment of a new law. That was the day prior to the release of its *Review of Past Action*. Accordingly, the manifesto for the September 11 general election incorporated a pledge to the effect that new legislation would be established to properly address the issues involving workers who had died without access to workers accident compensation, their surviving family members, and residents in asbestos-exposed neighborhoods (LDP), and that sweeping measures, including new legislation, would be implemented (a priority policy of the ruling coalition) (August 26, 2005). However, despite all the emphasis on speed, Chief Cabinet Secretary Hosoda disclosed that formulating the specifics of the new legislation was a task yet to be started. As epitomized by that comment, the government still foresaw a range of problems, including identifying the causal links between asbestos and the adverse health consequences attributed to it after a long period of dormancy within the bodies of victims, budgeting resources to cover the cost of measures against asbestos (assuming that it was in fact shown to be a causal agent of disease), and exploring the structure of legislative measures of relief for workers as well as local residents harmed by workplace accidents (MOE) (August 27, 2005).

However, there was a reason why the ruling party was able to take action of this kind while the administrative bureaucracy had no choice but to follow suit. Considering the political context in effect at that time, it seems plausible that these actions were influenced in particular by Prime Minister Koizumi, who demonstrated an awareness of flaws in the government response, and showed support for quick action with his remarks that the government had reason to feel regret, could not deny asbestos was causing harm, and had failed to identify the risks (August 27, 2005).[16] In connection with these new actions, it is worth noting that the LDP, the DPJ, NEW KOMEITO, the Japanese Communist Party, and the Social Democratic Party clearly supported the need for the New Act in response to a public inquiry issued ahead of the general election to eight political parties by the Ban Asbestos Network Japan (BANJAN), an alliance of labor unions and citizens groups that have been working to alert the public to the dangers of asbestos.

Incidentally, the Asahi Shimbun conducted a questionnaire survey of its own on the asbestos issue around this time. One of the questions asked who bears responsibility for the increase in asbestos victims. In reply, 65% of the respondents assigned blame to the national government, 18% to corporations and business groups, 5% to legislators, 5% to labor unions, 4% to the news media, 1% to people with a poor awareness of the problem, and 2% to other entities (September 10, 2005). Considerations about the wording of the question itself or the choices of response aside, these results appear to suggest that the public was aiming its criticism toward politicians (the political parties) and private companies, and even more so toward the national government as symbolized by the bureaucracy and bureaucratic elite. Echoing this finding, on September 9, four labor unions in the construction industry urged the MHLW to implement its all-out ban on the manufacture and utilization of asbestos products earlier than the scheduled 2008 fiscal year.

To avoid such action, relevant government ministries and agencies deemed it necessary to impress upon the public that progress on the development of asbestos substitutes was under way. Although the quest for alternatives to asbestos had been an issue needing to be addressed for some years, companies in the field stressed one after another that from a safety standpoint, it was still too early to rely on the use of alternative materials. By some accounts, this position was also supported by an entrenched belief that asbestos itself is safe as long as it is properly managed (September 26, 2005). However, in the light of the situation as described above, the MHLW emphatically urged that substitutes should be found and used.

The few examples cited do provide a glimpse into some of the criticisms leveled against the policies of the national government. However, we now look at some specific examples of public reaction at that time. On August 27, the day a group of asbestos victims was organized, the findings of an epidemiological survey were released showing that the incidence of mesothelioma deaths within a radius of up to 500 m from Kubota's defunct Kanzaki factory was 9.5 times higher than the national average. Although Kubota spokespersons insisted that the causal links between the Kubota factory and mesothelioma were unclear given the limited scope of the survey, they did not discount the view that the data were enough to suspect Kubota as the underlying cause of mesothelioma cases even though the results were "inconclusive" (August 28, 2005).[17] In time, more information became available and indicated an increase in patients with asbestos-linked illnesses along with rising public anxiety over that trend. For example, in September there were reports estimating that the number of workers accident compensation applications filed by former national railway employees has been rising under the influence of the Kubota Shock. In addition, a joint conference of MHLW and MOE officials heard a report that approximately 1 000 citizens had applied to receive health examinations independently sponsored by the NICHIAS Corp. Furthermore, it was reported that NICHIAS was prepared to provide consolation payments. With regard to Kubota, a survey by the City of Amagasaki identified 50 mesothelioma deaths in the period from fiscal years 2002 to 2004, a figure which was five times the national average. On September 21, it was disclosed that Kubota had made consolation and condolence payments to an additional eight patients. However, on the subject of corporate accountability, it is assumed that these steps did little if anything to close the rift in perceptions dividing victims' families and the companies involved (September 22, 2005).

In connection with perceptions surrounding the issue of accountability, the MHLW and MOE led efforts to take action on the findings of the Review of Past Action in view of the need to achieve a uniform solution with a new law covering the surviving family members of asbestos victims and local residents who were otherwise ineligible for workers accident compensation insurance benefits once the 5-year time limit for applications submitted after a patient's death had expired. On September 22, the government unveiled a draft of the Act on Asbestos Health Damage Relief (the New Act) that it had prepared with the aid of the two aforementioned ministries, and though the details remained unclear, announced plans to submit the proposed legislation for deliberation by the regular session of the National Diet in 2006, and to begin accepting applications that autumn.[18]

On September 29, the draft of the New Act began to take shape in a cabinet meeting that focused on the revised Interim Measures and the Master Framework for Asbestos Health Damage Relief. However, the reaction of the ruling party's asbestos countermeasures team was still one of disbelief (September 30, 2005). That was because the draft did not contain any clear provisions on the amount of benefits to be paid, the scale of funding measures for that purpose, the size of the cost burden to be borne by the national government and private companies, and the scope of the funding burden for the corporate sector. As will be shown in Sect. 10.6, these ambiguities developed into major sources of controversy.[19]

I now touch briefly on the relationships with another document, the Review of Past Action (supplement), that was released on the same day (September 29). One purpose of that document was to express the government's regrets for a poor awareness of preventive approaches, and for the lack of inter-office coordination among the various ministries and agencies concerned. On the other hand, it emphasized that administrative guidance resulted in a phase-out of asbestos and did not cause any delays in government action, and concluded that the government was not guilty of inaction (September 30, 2005). It is, of course, important that the government vowed to implement "seamless" measures of relief, and disclosed plans to set up a new liaison conference on hazardous chemicals. Nonetheless, if one takes into account the criticisms aimed at the government's stance later, it would appear that the public did not fully understand that various government-level "asbestos measures" were components of a single, integrated policy package.

One point that seems clear from this is that while the relevant government ministries and agencies were taking public criticism on the firing line and hammering out a series of policy measures, the ruling party sought to improve its public image by articulating a policy stance that set it apart from the bureaucracy.[20] Another point is that the public was not concerned about this apparent difference, and simply took a critical view of the new policies as a whole. In other words, public awareness and approval of the various actors involved in the formulation of asbestos policy was both mixed and complex. This situation was characterized, for example, by an anxiety betrayed by the MOE's administrative vice-minister, who had to act while bearing the provisions of the New Act in mind; he said that the hard part was yet to come (September 30, 2005).

10.6 The Decision-making Process Over the Act on Asbestos Health Damage Relief: October 2005 to January 2006

This section describes the lead taken by local governments from September 2005 amid the drive to pass new legislation, as well as the national government efforts in apparent response from October 2005 to late January 2006. On September 14, Fukui became the first prefectural government to submit a proposal (approved October 11) for an ordinance aimed at regulating asbestos. It was disclosed that the Tottori Prefectural Government would submit a similar legislative proposal on September 15. Although they varied in their considerations of the imposition of penalties, if any, the Kyoto Prefectural Government sought to submit a proposal on September 21 and the Osaka Prefectural Government on September 28. Tokushima Prefectural Government and others soon followed with proposals of their own. On September 30, the Naruto City Council approved an opinion paper calling for substantive policy improvements, and on October 4, the Wakayama City Council unanimously approved an opinion paper addressed to Prime Minister Koizumi and other top government officials. Furthermore, Toyama, Osaka, Mie, and Gifu Prefectural Governments announced plans to offer low-interest financing to cover the cost of removing asbestos materials. Moreover, Hyogo Prefectural Government urged the MHLW and MOE to establish a 10-point support framework for the removal of asbestos.

Developments at this time did not stop with the efforts at the local government level. For example, 13 mayors of municipalities in Kyoto Prefecture stepped up their calls on the national government to take more action, and submitted a proposal for asbestos countermeasures to the Kinki Association of City Mayors This proposal was approved on October 28. Hokushinetsu Association of City Mayors followed up with a decision to pursue similar actions of its own. Local governments had the option of pushing ordinances through their own legislative assemblies in policy domains where the national government was unable to move quickly, if at all. With such efforts, they demonstrated that they were maintaining their administrative legitimacy within their own communities. As such, their actions were clearly patterned on the efforts in August that were discussed at the end of Sect. 10.4.

During this time, what were actors at the national government level doing? On October 7, one MHLW research team that had been working since August reported that there were 953 mesothelioma deaths in 2004. That number constituted a roughly twofold increase over the corresponding annual total a decade earlier. On October 13, the MOE came out with an estimate that some 15 000 people would die of asbestos-linked mesothelioma or lung cancer in the 5-year period to 2010. Given such concrete numbers, addressing this issue developed a new sense of urgency. To that end, on October 25 the DPJ – the largest opposition party – tendered its draft of a Law for the Promotion of Comprehensive Asbestos Countermeasures for deliberation by the House of Representatives, thus pre-empting the ruling coalition on the legislative front.

To counter that move, what legislative proposals would the government, the ruling party, and the administrative bureaucracy pursue in preparation for the next regular Diet session in 2006? What follows is a review of the debate centering on three facets of the government response: namely, the relationships between workers accident compensation and relief measures under the New Act, and more specifically, the scale of special condolence payments to the surviving families of deceased asbestos patients prior to implementation of the New Act, the entities responsible for funding these payments, and their share of that financial burden.

On October 16, the Minister of Health, Labor, and Welfare, Hidehisa Otsuji, met asbestos patients and their families, listened to their concerns and expectations, and as a first step, declared that the government would cover the travel expenses for the long-distance outpatient care of mesothelioma patients approved to receive workers accident compensation. This measure was taken because, given the scarcity of qualified medical specialists, the traditional condition that outpatient treatment should be sought at health-care facilities within 4 km from the patient's home was simply unrealistic. On the following day, Otsuji expressed the opinion that given the pain and suffering that patients endure during diagnostic examinations, in his view it seemed appropriate to consider asbestos to be the cause of mesothelioma. In making that point, he then declared that he would also like to review the cases of patients who had been denied workers accident compensation eligibility because asbestos could not be conclusively demonstrated to be the cause of their illnesses, and then proposed that the criteria for eligibility be relaxed (October 18, 2005). This set the stage for the MOE and MHLW to establish an expert committee with the task of exploring diagnostic criteria for the identification of patients who were eligible for relief.

On October 21, the MOE released an estimate that a cumulative 9 993 people had died of mesothelioma in Japan since 1970. Although the New Act was designed to provide relief to asbestos victims who were not eligible for workers accident compensation, it did provide for a lump-sum payment of 2.4 million yen to surviving families of such victims (a special condolence benefit), with an estimated 9 500 citizens deemed to be eligible for that payment. After these estimates had been released, the question turned to one of exploring how the burden would be shared by members of the corporate community. To deal with this issue, the countermeasures project team, headed by Genichiro Sata, had settled on the idea of assigning funding responsibilities which were proportional to the number of employees with mesothelioma or lung cancer who had been approved for workers accident compensation. However, it was determined that this approach would assign about 90% of the funding burden to companies in the manufacturing and construction sectors. It would appear that this echoed the position taken by the MOE and MHLW: namely, that based on the polluter-pays principle, companies in related industries also would be asked to share the funding burden for the relief of asbestos victims (October 22, 2005).

However, the task of working out these details did not go very smoothly. First of all, in calculating the special condolence benefit for surviving families, a "balance" had to be struck with the policies the government had followed on Minamata disease and other pollution disasters. Although the sum of 2.4 million yen was given renewed consideration with a view to lifting the payment to 2.6 million yen, NEW KOMEITO Diet member Masahiro Tabata was not satisfied with that amount, and additional requests for increases in the payment failed to obtain a consensus. NEW KOMEITO reportedly placed importance on the fact that a lump-sum payment of 3 million yen had already been established for surviving families under the workers accident compensation framework.

Members of the industrial community objected to the share of the funding burden they would be expected to bear. Whereas the executive director of the Japan Asbestos Association argued, in hearings by the LDP joint subcommittee, that the funding obligations under the New Act should be shared with the government by importers, product manufacturers, and corporate users of asbestos, representatives for the manufacturing sector stressed that asbestos used by chemical plants was not at risk of being released into the air. Other objections were also raised. For example, in a routine press briefing held on November 15, the chairman of the Shipbuilders' Association of Japan insisted that asbestos-linked illnesses and deaths among asbestos workers' families and residents around asbestos factories had nothing to do with industrial accidents. He suggested that the share of the funding obligation for compensation paid to asbestos victims should be based not on the number of certified cases of mesothelioma or lung cancer industry-by-industry, but on the actual amounts of asbestos each industry was determined to have used, and declared that he would work to persuade the government to adopt that approach (November 16, 2005).

In the meantime, a group of patients and their families visited the Cabinet Office and urged reconsideration of the 2-year time limit for compensation provided to patients still undergoing treatment. They were requesting an alternative approach to the 2-year limit on relief under the existing arrangement for patients who received subsidies for medical treatment in addition to workers accident compensation payments equivalent to 80% of their average salary while on medical leave. Speaking on behalf of asbestos victims, their supporters pointed out that patients with asbestos-linked illnesses could not receive any compensation under the established framework, and called for an end to the time limit on the grounds that such illnesses went through a long-term period of dormancy before symptoms even became manifest (November 10, 2005). In a related move, the government decided to make the New Act offset the medical expenses which patients paid while alive by adding an equivalent amount to the compensation benefits provided to their surviving families. The MHLW stated that it would take into consideration medical costs and other reasonable expenses that facilitated compliance with other existing frameworks (November 12, 2005). Not long thereafter, the government put together a supplementary budget backed by increased tax revenues, and incorporated appropriations for countermeasures against the adverse health effects of asbestos. (Ultimately, a sum of 180.5 billion yen had been earmarked for this purpose as of December 20, 2005.)

On November 16, a study group set up by the MOE and MHLW decided at its first meeting in principle to extend relief eligibility to all mesothelioma patients. In addition, it determined that the total in lump-sum compensation payments and reimbursements for medical costs paid to eligible patients from FY 1970 through FY 2010 would reach 70 billion yen. Furthermore, the group decided to set the public funding burden for the national government and all local governments at a four-to-one ratio for the approximately 40 billion yen in total compensation paid up to FY 2006, and to call on private corporations with an involvement in the asbestos crisis to cover the entire funding burden from FY 2007 forward. Subsequent deliberations were concerned almost exclusively with sharing of the funding burden. As indicated above, several benchmarks were proposed for the determination of a given industry's obligations in this context, including the number of workers receiving accident compensation benefits, or the volumes of asbestos utilized. It was around this time that a two-step proposal was tendered, calling for Kubota, NICHIAS, and other companies in the asbestos industry to assume a heavier share of the funding burden, with the rest to be spread more thinly and broadly among the construction, automotive, shipbuilding, and other industries (November 17, 2005).

More in-depth discussions of these topics emerged on November 25. To obtain relief funding for the period from FY 2007 through FY 2010, the government decided "in the interests of fairness" to collect a sum of approximately 10 billion yen per annum from the 2.6 million companies enrolled in the workers accident compensation insurance program. It was reasonable to assume that this would be problematic given that even companies in the financial and information technology sectors, among others, would be expected to share the overall funding burden. Nonetheless, the government disclosed that it would set logical benchmarks for consistency, and engage in efforts to win over members of the industrial community by FY 2007 (November 25, 2005). On November 28, MOE officials visited the offices of Nippon Keidanren (Japan Business Federation) and reportedly explained that their objective was to establish a viable framework as soon as possible, and not to pursue the issue of corporate responsibility. Keidanren officials displayed a wait-and-see attitude, stating in response that they had no comment on the matter at that time (November 30, 2005).

However, executives for a certain IT company commented around the same time that it was only natural that industries with no connection to asbestos would want to be considered for exemption from the funding obligations. Finally, about a week later, the chairman of Nippon Keidanren stated their position that first and foremost, the national government itself should be responsible for funding the compensation program. While expressing understanding for the idea of making companies that produced or utilized asbestos products assume a share of the funding burden, he considered it unacceptable to make the entire business community bear those funding obligations (December 6, 2005).

Assuming a share of the fiscal burden was something that met with mixed reviews, not only among members of the corporate community. In a routine press briefing, the MOE's administrative vice-minister stated that METI and the Ministry of Land, Infrastructure, Transport, and Tourism (MLIT) were leading negotiations with the business sector. Executives within other government ministries and agencies were reportedly stunned by that remark and the outdated assumption that companies would willingly pay up if bureaucrats so demanded. Furthermore, not all local governments were ready to accept a funding burden either. The National Governors' Association maintained its insistence that the national government should foot the entire bill (November 30, 2005).

Although the national government had to put up with all this controversy, on November 29 it finalized its draft version of the New Act, which is compared with the workers accident compensation framework in Table 10.2. (Note, however, that changes were made in the final stages, as described later and illustrated in Fig. 10.1.) However, one victim of asbestos exposure felt compelled to state that the proposed New Act fell below expectations, for although it alleviated the burden for families of workers who had never received any relief, as well as for affected residents living near asbestos factories, it did not reflect the voices of the victims who had sought compensation which was equivalent to that provided through the workers accident compensation framework (November 30, 2005). In contrast with this, in informal talks with a group of asbestos victims and their families on October 26, Environment Minister Koike reportedly remarked that although asbestos disasters were different from workplace accidents in some ways, she was interested in improvising workable solutions, and at a press briefing held afterward, declared that she had listened to the group and felt committed to accelerating the legislative process (November 27, 2005). On a related note, in deference to NEW KOMEITO, on December 6 the ruling party agreed in principle to raise the proposed special condolence payment for surviving families by 200000 yen to a total of 2.8 million yen, which, together with a special funeral payment of 200000 yen, would add up to 3 million yen.[21]

Incidentally, although the momentum of the aforementioned actors could have had an impact, Kubota, while maintaining that it did not know whether there were any causal links between asbestos and certain diseases, conceded in late December that it had a moral obligation, and announced that it would begin providing local residents with relief compensation in lieu of the condolence and consolation payments it had paid since June (December 24, 2005). On that occasion, Kubota's president issued a formal apology, stating that his company was not absolutely certain that asbestos dust had not been released from the premises of the Kanzaki factory site. He went on to declare that effective April 2006, when the New Act was expected to be approved by the next ordinary session of the Diet, Kubota would

Table 10.2 Comparison of workers accident compensation and relief under the Act on Asbestos Health Damage Relief

In cases where the patient is eligible for workers accident compensation	
Medical costs	No out-of-pocket expenses
Outpatient expenses	Mesothelioma patients compensated
Work leave compensation	80% of daily wage
At time of death	Surviving family pension (starting from the average wage for 153 days for one dependent), special lump-sum payment of 3 million yen, and education assistance for children (12 000 yen/month per elementary school student)
Funeral assistance	315 000 yen plus the average wage for 30 days
• *Special additional compensation program of Kubota*	
Additional compensation to employees eligible for workers accident compensation	15 million yen
Compensation during sick leave to those otherwise ineligible	20% of daily the wage
Benefit to the surviving family upon employee's death	17 million yen benefit
In cases of relief under the Act on Asbestos Health Damage Relief	
• *Worker deceased more than 5 years and who was unable to apply for surviving family compensation under the workers accident compensation insurance framework*	
Surviving family benefit	Surviving family pension of approximately 2.4 million yen
Note: Not eligible for sick leave compensation or funeral assistance owing to the 2-year time limit	
• *Affected local residents and families of affected workers*	
Medical costs	No out-of-pocket expenses
Patients receiving care	Approximately 100 000 yen allowance for care
Funeral assistance	Approximately 200 000 yen
Surviving family of deceased patient	Special condolence payment of 2.6 million yen plus funeral assistance

Source: Asahi Shimbun Newspaper, November 30, 2005

begin paying local residents compensation on a scale which was equivalent to that provided to the company's own employees (December 26, 2005).[22]

In an interesting coincidence, the following day, December 27, saw the Koizumi Cabinet approve a set of comprehensive measures on the asbestos issue that had three objectives: to provide seamless relief for asbestos victims, to prevent further asbestos exposure and harm, and to quell public anxiety over the crisis. From this point, all actors stood by and waited for the New Act to be submitted to the next ordinary Diet session. Early in the New Year, on January 17, 2006, the LDP approved the government version of the proposed legislation. It was then passed by the House of Representatives on January 31, by the House of Councillors on February 3, and promulgated on February 10 (Fig. 10.1).[23]

10.7 Relief for Victims Enters a New Stage: February 2006 to March 2007

This section covers trends through the first year following enactment of the New Act by focusing on victim applications for relief, the response by the corporate sector, and action taken at the national government level.

Beginning around March 20, 2006, prior to the implementation of the New Act and the acceptance of applications for assistance, efforts to provide relief to victims were in a confused state. On February 9, the MHLW and MOE had projected that somewhere between 2 000 and 3 000 applicants would file for assistance in the first year under the New Act. However, because local health centers were not yet prepared to start accepting and processing applications, the initial application drive was launched with the participation of local environmental centers, labor standards inspection offices, and the Environmental Restoration and Conservation Agency of Japan (ERCA) only. At this time, debate was still under way on the eligibility criteria for workers accident insurance. The turmoil on this day was further stirred up by notices that the MHLW sent out to Labor Bureau directors around the nation, stating that the eligibility criteria would be relaxed.[24] Some personnel in the field felt that it would be ridiculous to expect that their offices would be able to begin accepting and processing applications within less than 2 months since the passage of the New Act. There were many reasons for this. First, on February 21–22, the ERCA implemented recruitment examinations for an additional 50 new employees to augment its existing work force of 110. Applications from affected local residents were to be handled by the MOE, and applications from workers by the MHLW. On top of that, local governments had to contract out their application-related operations to local health centers (February 26, 2006).

Amid these developments, the family of one mesothelioma patient expressed anxiety over not knowing whether their loved one would be approved, and questioned why application procedures for people who were ill were not more streamlined. Noting the likelihood that patients with certain medical conditions

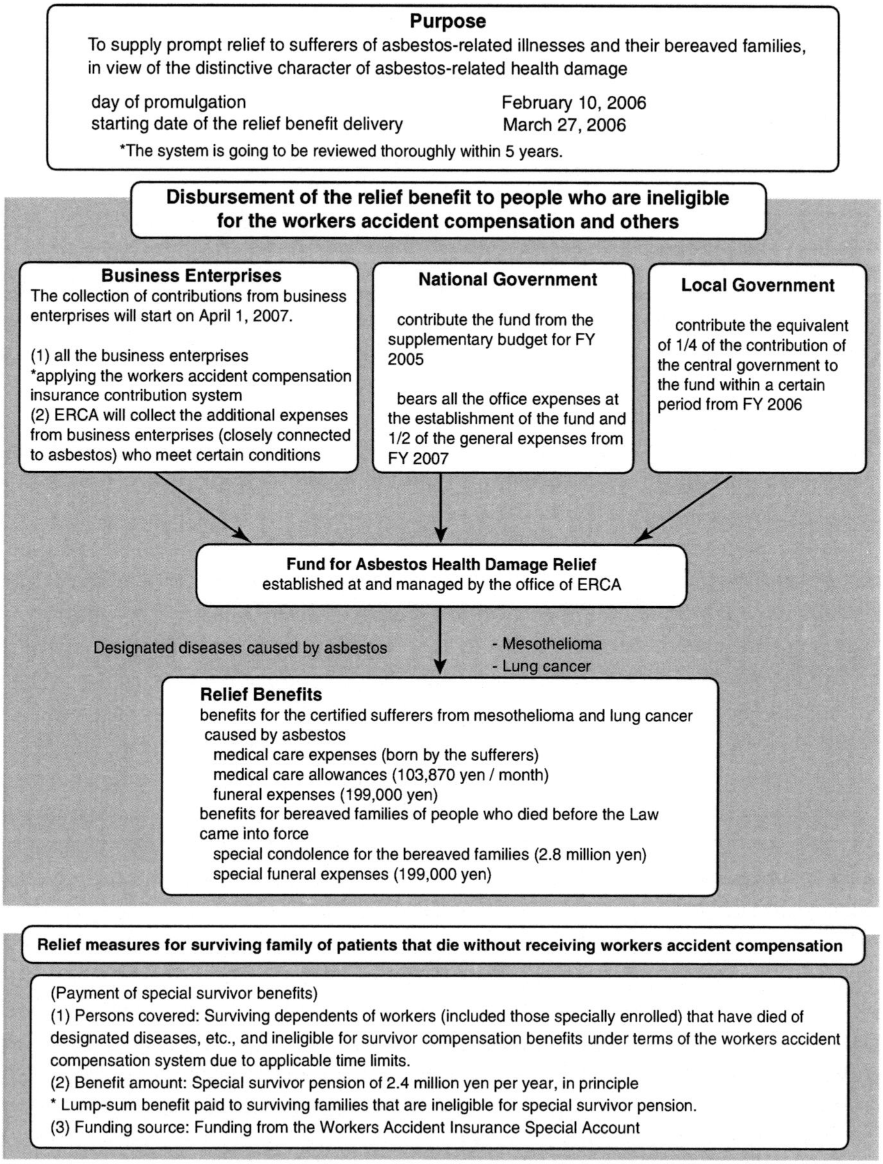

Fig. 10.1 Overview of the Act on Asbestos Health Damage Relief. *Sources*: Ministry of the Environment and Ministry of Health, Labour, and Welfare, "Enforcement of the Act on Asbestos Health Damage Relief," *Environment*, March 2006 Issue. Gyosei, 3

were not eligible for compensation to offset their medical costs, one supporter pledged that he would not give up his campaign until the national government acknowledged that delays in regulating asbestos had created more victims, and until all victims were receiving relief. Supporters of asbestos victims have continued their quiet push for answers, pointing out that because asbestos was so prevalent in all facets of everyday life, few people were aware of its reality as a health hazard (March 9, 2006).

On March 7, the Cabinet approved the ordinance that would bring the New Act into force on March 27. The starting date for the acceptance of applications – March 20 – was just around the corner. By MOE and MHLW counts, as of April 6, close to 2 000 applications for benefits had been received from affected local residents and dependents of workers, and of the 1 100 applications submitted by worker families, 782 were from surviving dependents of deceased workers. In effect, ministry projections cited earlier on the number of applications were met in the short space of just over 2 weeks. Among decisions reported on the applications for benefits, in May the Amagasaki Labor Standards Inspection Office approved special survivor condolence payments for the surviving families of 14 workers who had lost eligibility for assistance under the workers accident compensation framework owing to the effective time limit for submission of applications. In addition, the ERCA approved applications from 64 citizens for similar reasons.

However, many of the applications submitted involved patients who were difficult to certify as victims of asbestos-linked illnesses. In response to a request from ERCA in June, a subcommittee of the Asbestos Health Damage Commission, an advisory body to the Environment Minister, issued its first decision, certifying 27 applicants as *bona fide* asbestos victims. Reviewing the data for approximately 3 months, the MOE and MHLW reported that of 4 000 applications received, 320 were approved, and of those, about 70% had been submitted by affected local residents living near asbestos facilities. However, in some cases the applicants were mesothelioma patients with only short life expectancies after their symptoms became manifest, while some others were patients who had died while waiting for their applications to be approved. In some communities, diagnostic testing standards had still not improved, many applications were routinely passed on to the aforementioned subcommittee for final judgment, and the percentage of patients receiving benefits failed to rise substantially. From application desks to the inbox of the subcommittee itself, the entire process remained a struggle. Later, in March 2007, a prefectural labor bureau reversed a decision by a labor standards inspection office not to provide benefits to a surviving family who had been deemed ineligible owing to a time limit. This reversal came after the family filed a claim demanding an administrative review. As such cases illustrated, the application approval process itself could not be described as steady or smooth-flowing.[25]

We now examine the response by private firms, focusing on Kubota in particular. First, according to a researcher involved in an epidemiological survey of Kubota's defunct Kanzaki factory site, the findings supported a causal link between asbestos and mesothelioma symptoms within the factory and surrounding neighborhood, and demonstrated that the impact was spread over a broader area than was initially anticipated. The researcher added that anyone working at, or living near, the Kanzaki site for a given length of time during the period that the factory used asbestos was considered to be at risk, and stated that steps should be taken to identify new patients and devise necessary measures of relief (April 12, 2006) (for related literature, see Kurumatani and Kumagai 2006). On April 16, amid news of a drive to obtain a resolution through a Japanese-style class-action lawsuit seeking damages for victims of asbestos exposure in the Sennan district in Osaka Prefecture, Kubota, having met a group of patients four times for negotiations in March, proposed to pay relief compensation of 46 million yen per patient,[26] an increase from the maximum of 32 million yen offered initially. A senior managing director for Kubota noted that the company wanted to alleviate the suffering and psychological pain experienced by residents living near the Kanzaki factory as quickly as possible, but at the same time he expressed relief that his company had averted litigation. An article in the Asahi Shimbun provides details of the net profit posted by Kubota in its consolidated financial statement, and its analysis concludes that the financial results gave the company the latitude to handle the relief compensation burden (April 18, 2006). Given that details were then coming out about the scale and scope of corporate contributions to the government relief fund for asbestos victims, as discussed in the previous section, Kubota's decision to offer relief compensation seems to have been well timed. Although the MOE expressed puzzlement at the Kubota move, noting that it seemed unusual for a company to make large relief compensation payments while denying any causal relationships (April 18, 2006),[27] judging from the editorials[28] issued by the leading newspapers on April 19, Kubota was in general highly applauded for its actions, and that was also the assessment voiced by most members of the business community.

Two points are worth noting here. First, a positive correlation can be observed between company scale and the size of relief compensation packages. Second, opinions varied on the correct geographic scope of compensation provided to "local residents." In the context of point one, a sense of unfairness is assumed to have arisen among families of workers who were formerly employed by small companies or companies that went out of business. For example, compensation amounts per beneficiary reached 30 million yen at most from NICHIAS and 20 million yen at most from its subsidiary Tatsuta Kogyo – less than half the maximum compensation payout from Kubota.[29] Also, bear in mind that while the companies involved in this crisis were at last beginning to take serious steps toward providing relief for the victims, their problems had not gone away.[30] For example, if a NICHIAS-sponsored health examination turned up anomalies, the patients were declared in need of further observation. The Tatsuta Industry Co. was also cited as a clear cause of asbestos-related exposure and harm.[31] Furthermore, in the case of NICHIAS in particular, former employees formed a labor union for collective bargaining purposes, and later reiterated their demands for compensation.

In addition to these more obvious cases, there were also news articles that sought to identify and encourage relief measures for latent asbestos victims. In terms of identifying latent victims, one of the more-intriguing topics relating to point two above was Kubota's decision on August 24 to define "local residents" as residents living within no more than a 1.5-km radius from the Kanzaki factory. Although this was not as large as the 4-km radius recommended by one researcher, it did result in relief compensation for 13 additional residents. Other high profile actions were taken by Kubota thereafter. It was disclosed, for example, that it provided 500 million yen to cover the cost of diagnostic tests for the treatment of mesothelioma, and paid relief compensation even to former employees of partnering subcontractors.

Of course, the action strategies chosen by the companies involved were not the only factor that influenced the latent or hidden nature of asbestos-linked health damage. According to one labor consultant on social insurance, the extent to which public death statistics were retained or released (November 29, 2006), the speed of development of new drugs and approvals, and other factors of a professional or administrative nature also had an impact. While the approval process for the drug Pemetrexed (Alimta) was swift, on the subject of the increase in deaths from mesothelioma, the mayor of Yokohama stated that it would difficult to identify the number of such deaths city-wide (February 27, 2007). This could be indirect evidence that the long period of dormancy before the onset of asbestos-induced symptoms can have the effect of delaying the formulation of administrative measures.[32] Beyond focusing on each and every failure to identify latent or hidden problems, it is important to note that the collection and organization of these data became possible through an interest in observing how the latent problem is (or can be) visualized.

In conclusion, there is a point which I wish to reaffirm in the context of the time frame covered by this section. Namely, in contrast to the flexible and sustained response by companies in the private sector, and except for a lawsuit claiming reparation from the state, national government actors appeared not only to have failed in following up with action on the asbestos crisis, but also seem to have let it slip away from their awareness as a key issue. In addition, as a consequence of this, the "seamless" relief declared by the Cabinet conference on the asbestos crisis was apparently evident only within the context of the aforementioned administrative processes, i.e., at the so-called street-level bureaucrats. One implication to be drawn from this is that criticism aimed directly at the administrative bureaucracy – compelled as it was to take action – may have been on the wane.

10.8 Concluding Remarks

Granted the almost complete absence of efforts to organize information on developments following the Kubota Shock chronologically, in this chapter I sought to shed light on the asbestos policies adopted in FY 2005–2006 by actors at the national government level, and chronologically retrace events thereafter with the Asahi Shimbun articles as my primary information resource. Although much of the paper focuses on exploratory and speculative themes, I believe it affirms the following four points in relation to the administrative bureaucracy formed by the national government ministries and agencies. These points are specifically within the political context shaped by the intensity of the public spotlight of attention, the existence or influence of the Koizumi administration when policies were being carried forward, and the bold response demonstrated by private companies and local governments. First, the bureaucracy at virtually every stage of the political process remained in the firing line despite its efforts to avoid public criticism for flaws in its policies. Second, the bureaucracy was a target of blame not only for the opposition parties, but also for the ruling coalition (Sects. 10.3–10.5). Third, inter-agency bickering, particularly during the decision-making stages involved in the formulation of the New Act, created a heightened atmosphere of distrust throughout the bureaucracy at large and consequently undermined the autonomy of action within the policy arena (Sect. 10.6). Fourth, these factors had a consistent and sustained impact on the policy implementation process (Sect. 10.7).[33]

Nonetheless, these interpretations are tentative. First and foremost, although the time frame for analysis herein covers an overwhelming volume of information on the political history of the asbestos issue, the conclusions presented in this essay remain rough-edged at best. Further, this chapter, in association with my companion paper,[34] places heavy emphasis on the continuity of documentation spanning 60 years of postwar history. Hence, for verification purposes, it relied exclusively on articles in the Asahi Shimbun that were consistent with that requirement. However, the inherent bias of that approach had to be borne firmly in mind. Granted that this was a trial endeavor, readers may need to cross-reference and complement the content with information from other papers and publications. In particular, sections of text covering the period from FY 2007 onward contain less information, albeit not as scarce as that available on the period prior to the Kubota Shock (see Table 10.1 and Note 4.) Consequently, it will conceivably be crucial to accumulate additional information through an approach that combines a rich diversity of documented resources with interviews covering the same time periods.

However, not all the remaining problems have to do with documentation or information. Recalling the long period of dormancy that intervenes before symptoms of asbestos-induced illnesses become manifest, it will be essential to focus our attention on the period ahead, when the regulations implemented at the national government level from the 1970s forward begin to demonstrate benefits. In tracing the political processes surrounding asbestos, it will be necessary to direct our thoughts to those subjects that cannot be captured as snapshots by the approaches applied in case studies.[35]

Notes

1. Table 10.1 lists the number of newspaper articles in an Asahi Shimbun database on the so-called school panic and Kubota Shock over a 2-year period that contained the words "asbestos" and/or "*ishiwata*," the original Japanese term for asbestos. There were 119 such articles on the school panic (FY 1987–1988) and 3512 articles on the Kubota Shock (FY 2005–2006). The number of articles containing these terms averaged 74 per year throughout the entire postwar period (FY 1945–2006), three per year from 1945 (the year World War II ended) to 1986 (the year prior to the school panic), and 51 per year in the period from 1989 to 2004 (the year prior to the Kubota Shock). Although these statistics are nothing more than examples, they do support the conjecture that among postwar asbestos-related problems, both of these events received significant news coverage, and the latter in particular had an enormous impact.

2. For my discussion of these points with consideration of the conclusions of earlier studies, including the literature listed in Note 4, see Mori, M. (2008). That discussion may be summarized as follows (*Translated from Japanese*). "Regardless of whether or not the government's position is labeled as 'inaction,' at a minimum the justifications for governmental policies on this issue are demonstrably flawed. Additionally, I contend that the flaws were in the administrative and core executive branches of government, not the political party (the ruling party). Specifically, it is shown that the key to exploring current asbestos-related issues resides in an understanding of the absence of Liberal Democratic Party (LDP) involvement (a 'political vacuum') in policy affairs relating to the labor environment and air quality, and behind-the-scenes LDP maneuvering ('hidden politics') in the realm of policy on economic issues. Further, as a theoretical basis for this interpretation, it is argued that features of the asbestos problem and inter-related linkages within the policy arena intentionally or unintentionally limit the policy preferences, strategy formulations, and behavior of key actors at the government level, and substantially prescribe the outcomes." (pp. 45–46) In this chapter, I discuss the features of Japan's politico-historical "shock within a shock" as an extension of the aforementioned concerns (see Note 1).
3. Numerous publications have reported in detail on the activities of asbestos victims and their supporters. However, documentation produced to explain trends at the government level is unusually scarce. Although it only covers the period up to late November 2005, Awano (2006) provides useful insights, with a focus on political developments, while also taking into account the activities of victims and support groups. While Awano pursues several themes, his discussion complements the chronological arrangement of subject matter in this chapter. The following work is instructive for its presentation of asbestos-related issues within the context of an analysis of government efforts to re-regulate society (see Nottage 2006).
4. This trend remained consistent even if the periods prior to and after those covered by Table 10.1 are factored in: for example, the first 3 months of 2005 (3, 4, and 9 instances of usage, respectively) and the months of April through September 2007 (with 39, 49, 42, 29, 23, and 29 instances, respectively).
5. For a convenient reference, see Ryu (1988).
6. Although diverse in their presentational tone, see Takenaka (2006), Iio (2006), Uchiyama (2007), Yamaguchi (2007), Otake (2006), Machidori (2005), and Machidori (2006).
7. Kamikawa (2007). In addition, Kamikawa points out that Koizumi's power base shifted from a Cabinet consensus to a top-down approach after the general election.
8. For more recent details on the Kubota Shock, see Kataoka (2006).
9. The following was referenced for press documentation on "National governors' Association," http://www.nga.gr.jp/upload/pdf/2005_7_x68.PDF. Accessed 30 October 2007.
10. In other Diet debates, during intensive deliberations by the Special Subcommittee for the Promotion of Science and Technology, attention was focused on points made by committee member Hidekatsu Yoshii (the Japanese Communist Party). Specifically, Yoshii described the government position as analogous to the defensive response given by the director of the MHLW's Public Health Bureau in 1972 when asked to explain the harm caused by the exposure of local residents to asbestos dust. On that occasion, the director explained that the risk of actual harm had been circumvented, but that if the risk persisted, health examinations for citizens should be considered (Asahi Shimbun Newspaper, July 21, 2005).
11. Further information on this document and other national government-level material covered therein were examined on the official website of the prime minister's residence, under the heading "Cabinet-level conference on the asbestos problem," http://www.kantei.go.jp/jp/singi/asbestos/index.html, Accessed October 30, 2007.
12. That is not to say there was nothing worthy of notice. Reference materials for the "Interim Measures" demonstrated that government action in the past chiefly comprised administrative guidance in the form of notices and advisories. Of the 73 notices and advisories on asbestos which the government had issued up to March 2005, 23 were determined to have been issued in the period from 1987 through 1988 alone, when the so-called school panic was at its height. On this point, one retired MHLW veteran commented that perhaps the bureaucracy thought it

could control the industry with advisories alone. By some accounts, though, the industry generally ignored them in practice (Asahi Shimbun Newspaper, July 30, 2005). The following episode serves as evidence of just how "lax" the policy of advisory-based administrative guidance could be. In a meeting of the Committee in the House of Representatives on Health, Labor, and Welfare that convened immediately after a related meeting of Cabinet ministers, the Director of the Labor Standards Bureau had to apologize for his failure to examine an advisory that had been issued in a junior manager's name for the purpose of relaxing the application procedure for workers accident compensation insurance benefits. This episode apparently left patients, their families, and their supporters with the clear impression of a lack of transparency when it came to matters of administrative accountability.

13. Awano (2006) is a representative example in the literature on this topic.
14. For example, consider Kitayama (1995).
15. See Asahi Shimbun Newspaper (August 27, 2005) for one synopsis of the points made in ministry and agency documents.
16. At the time, Prime Minister Koizumi was preoccupied with putting his utmost into the drive to privatize the postal service, and apparently this was the only occasion when he had anything to say about the asbestos issue. Indeed, as pointed out by Nottage, Koizumi may not have had any time or latitude to obsess about this matter (Nottage, 2006: 252). However, that is not to say he was simply unconcerned because, as indicated earlier, asbestos countermeasures were one of the seven priorities listed in his party's manifesto for the September 11 general election. Indeed, considering the context, the fact that Koizumi commented on asbestos at all may in itself be taken as something quite significant. Here, the point is to affirm that Koizumi was neither indifferent nor actively opposed to asbestos countermeasures.
17. A follow-up epidemiological survey of this district identified mesothelioma death rates of 18.1 times the national average for women, 9.8 times the national average for men, and 11.7 times the national average for both genders combined, thus highlighting the dangerous situation in the community surrounding the Kubota factory (Asahi Shimbun Newspaper, November 24, 2005).
18. However, spot inspections of private businesses by the MHLW reportedly uncovered a significant number of regulatory violations, thus placing the viability of these plans into doubt.
19. The course of this debate also mattered in terms of the impact it could have on the victim relief efforts pursued by individual companies. In general, whereas victims of asbestos contamination attributable to Kubota, NICHIAS, or other comparably large-scale firms could expect to receive generous amounts of compensation, that was not the case for victims of asbestos contamination caused by small-scale businesses. Indeed, considering that even NICHIAS – then the largest firm in the asbestos industry – had declared that the burden of victim compensation was too large for any single company(Asahi Shimbun Newspaper, September 30, 2005), one can readily imagine how smaller firms felt about the issue. Determining who would bear the compensation funding burden, how much of the burden they would share, and the channels through which their contributions would be made were all realistic questions that demanded solutions.
20. Kent Weaver (1986). For example, if the ruling party had been exposed to criticism in the questionnaire survey cited earlier, the party's behavior could have been perceived as not claiming credit, but avoiding blame.
21. Within the context of requests and expectations toward the proposed legislation, it is worth noting that Hannan City in Osaka Prefecture, which historically had a high concentration of asbestos-related industries, had asked the MOE, MHLW, and MLIT to designate it a district with relief priority.
22. These steps were not limited to Kubota. NICHIAS reportedly sustained the trend by having 486 workers at its Gifu facility undergo health examinations, thus setting the stage for the industry to benefit from a show of action. Later on, NICHIAS subsidiary Tatsuta Kogyou would begin making its first payments of condolence money.
23. Air Pollution Control Law, Building Standards Law, Waste Disposal and Cleaning Law, and Local Finance Law were all amended and revised when the New Act was established. Although these developments were also important, they were outside the scope of this chapter.

24. According to the Asahi Shimbun Newspaper, compared with trends in FY 2004, the number of applications for workers accident insurance ballooned eightfold, with an increase of over two-fold in those actually approved (April 5, 2006). By MHLW estimates, the number of applications rose around 9-fold, with those which were approved being almost 4 times the corresponding level in FY 2004 (May 31, 2006). Furthermore, it is assumed that there was also a sharp surge in the volume of work involved in processing these applications. A spokesperson for the Compensation Division in the Workers Compensation Department of the MHLW noted that because the period to the onset of symptoms in most cases of mesothelioma was so long, patients and their health-care providers rarely made any associations between this disease and the patient's occupational history. Although the Division made an effort to familiarize patients and health-care facilities with asbestos-linked illnesses and the workers accident.
25. See the Asahi Shimbun Newspaper (March 5, 2007) for similar cases concerned with the approval of patients with asbestos-linked lung cancer.
26. Although cases of litigation have not been adequately covered in this chapter, they can be cited for their importance to political and public administration-related research on the power of the courts. (I have included some notes on the paper listed above.) Some court-related developments are briefly listed here. In April 2006, the Osaka pneumoconiosis and asbestos legal team filed the first class-action lawsuit, and as briefly mentioned in the introduction section of this chapter, sought damage compensation from the national government. Around this time, a series of pneumoconiosis-related lawsuits concluded in victory for the plaintiffs, and there were impressions this would have an impact on the outcome of asbestos-related litigation as well. Although one case is currently pending in the Osaka District Court, future developments will be closely followed.
27. However, in the Asahi article, the reporter adds that MOE officials considered Kubota to be a cause of many workplace accidents, and surmised that its decision on compensation reflected an awareness that local residents shared the same view. The reporter concludes that Kubota had in reality conceded that causal links exist (Asahi Shimbun Newspaper April 18, 2007).
28. The headings of various editorials published by leading papers on April 19, 2006, read as follows; "Kubota finds its resolve, but ...." (Asahi Shimbun Newspaper); "Kubota quickly decides to pay relief compensation" (Yomiuri Shimbun Newspaper); "After asbestos compensation, what next for Kubota?" (Nihon Keizai Shimbun Newspaper); "'Kubota-style' asbestos relief compensation serves as a model" (Mainichi Shimbun Newspaper).
29. Afterward, Kansai Electric Power Co., Inc. apparently paid surviving families compensation sums of close to 60 million yen. In addition, Kawasaki Heavy Industries, Ltd. reportedly did away with age discrimination in the payment of condolence money.
30. For further details on this factor, see the Asahi Shimbun Newspaper feature series, "The Kubota Shock one year later" (June 27–29, 2006).
31. There are many other related examples, including those involving such firms as Mitsubishi Materials Corp., Sumitomo Heavy Industries, Ltd., and Bridgestone Corp. Another example, albeit different, is a reported increase in the number of former employees of the Nagasaki Shipyard and Machinery Works of Mitsubishi Heavy Industries, who have become victims of asbestos-linked illnesses and have received asbestos health management handbooks.
32. Let me make a point relating to local government. It is well known that decisions made at the national government level are not always implemented at the local level. One example would be the subsidies budgeted by the MLIT for the prevention of asbestos dust dispersal from private structures. In some cases, prefectural frameworks for the provision of these subsidies have not been set up. As of November 2006, 11 prefectures had completed their subsidy frameworks, 11 had no plans to develop them, and 18 more were either in the framework development or study stages. According to an analysis by the Asahi Shimbun Newspaper (February 6, 2007), the main reason cited for this state of affairs is that the prefectures have been asked to provide half of the subsidy funding, and certain quarters are still discontented and feel that the national government should assume the entire burden of funding because its regulatory policies have failed. Taken together with reports that the National Governors' Association has been divided over the burden of funding for medical costs and condolence

money payments under the New Act, these examples may be interpreted as independent manifestations of resistance.

33. One document supporting the interpretations in this paper from a risk-awareness research-based approach is Nakayachi (2005). Although Nakayachi's analysis is focused on Kubota's response to the asbestos crisis, it contains insights that apply to an analysis of the political and administrative dimensions from the perspective of "restoring public trust."
34. The previously cited Mori paper chiefly explored trends in the National Diet debate. See Note 2.
35. For a paper dealing with the significance of treating emerging problems as case studies, see Otake (1996), 69–70. Also, as concisely pointed out by Falleti, the elaboration of process-tracing methods remains a challenge. See Falleti (2006).

References

Awano M (2006) Asbestos disaster (in Japanese). Shueisha, Tokyo

Falleti T (2006) Theory-guided process-tracing: something old, something new. Newsletter of the Organized Section in Comparative Politics of the American Political Science Association 17(1):9–14

Iio J (2006) Japan's structure of government (in Japanese). Chuokoron Shinsha, Tokyo

Japan Industrial Safety and Health Association (2005) Ordinance on prevention of asbestos hazards (in Japanese). Japan Industrial Safety and Health Association, Tokyo

Kamikawa R (2007) Transformation of the policy process after the election of 2005 (in Japanese). Jpn J Elect Stud 22:54–68

Kataoka A (2006) The backdrop to the asbestos disaster at Kubota's defunct Kanzaki factory in Amagasaki City: facing a reality far worse than imagined. Res Environ Disruption (in Japanese) 35(3):49–54

Kent Weaver R (1986) The politics of blame avoidance. J Public Policy 6:369–398

Kitayama T (1995) Local policy initiatives. In: Muramatsu M, Kume I, Iqbal F (eds) Local government development in post-war Japan. Oxford University Press, Tokyo, pp 222–241

Kurumatani N, Kumagai S (2006) A significant clustering of mesotheliomas among residents around an old asbestos type production plant: epidemiologic aspect and bitter lesson. J Occup Health (in Japanese) 48:143–145

Machidori S (2005) Political reforms and their impact on the long administration by Prime Minister Koizumi. Chuokoron (in Japanese) 120(4):176–186

Machidori S (2006) The strong Prime Minister' becomes common. Chuokoron (in Japanese) 121(4):174–184

Mori M (2008) Strategic environmental regulations? A case of asbestos politics in postwar Japan. Kagawa Law Rev (in Japanese) 8(1):43–129

Nakayachi K (2005) Why was Kubota forgiven? The asbestos crisis as reflected in corporate responses and environmental risk psychology. Risk Manag Bus (in Japanese) 20(11):20–23

Nottage L (2006) The ABCs of product safety re-regulation in Japan: asbestos, buildings, consumer electrical goods and Schindler's lifts. Griffith Law Rev 15(2):242–286

Otake H (1996) Political and economic power in contemporary Japan, expanded edition (in Japanese). San-ichi Shobo, Tokyo

Otake H (2006) A study of populism: Koizumi Jun-ichiro (in Japanese). Toyo Keizai, Tokyo

Ryu K (1988) The concept of 'pre-decision' in the policy making process. Kyoto Law Rev (in Japanese) 123(4):48–71; 124(1):91–125

Takenaka H (2006) The Prime Minister's rule (in Japanese). Chuokoron Shinsha, Tokyo

Uchiyama Y (2007) Koizumi administration (in Japanese). Chuokoron Shinsha, Tokyo

Yamaguchi J (2007) The Cabinet system (in Japanese). University of Tokyo Press, Tokyo

Chapter 11
Future Challenges for Asbestos Policy in the Construction Industry*

Kazuhiko Ishihara

11.1 Introduction

In September 2006, restrictions were brought into effect on virtually every type of asbestos product for industrial use. In effect, with this all-out ban on asbestos products, all questions about the use of asbestos in new buildings and structures were essentially averted. Future asbestos-related challenges for the construction industry may be boiled down to problems in two key areas: measures to deal with the "stock" of asbestos that has accumulated in existing buildings, and measures to deal with asbestos-related diseases among workers in this industry. Following a review of current conditions and trends, this chapter endeavors to identify measures that have been taken thus far to deal with asbestos accumulations in buildings and asbestos-related diseases among workers, and shed light on the issues that should be addressed in the years to come.

11.2 Asbestos Measures in the Construction Industry: Current Conditions and Challenges

With the exception of a temporary grace period for six product categories, effective September 1, 2006, revisions to the ordinance for enforcement of the Industrial Safety and Health Law imposed an all-out ban on the manufacture of products with

*Translated and revised from the Japanese original, Ishihara, K. (2009). Asbestos Damage and Remedial Measures in the Construction Industry. *Research on Environmental Disruption* (in Japanese), 38(4), 55–60, with permission of the Editing Committee of *Research on Environmental Disruption*.

K. Ishihara (✉)
College of Policy Science, Ritsumeikan University, 56-1 Tojiin Kitamachi, Kita-ku, Kyoto 603-8577, Japan
e-mail: ishihara@sps.ritsumei.ac.jp

K. Miyamoto et al. (eds.), *Asbestos Disaster: Lessons from Japan's Experience*,
DOI 10.1007/978-4-431-53915-5_11, © Springer 2011

an asbestos content of at least 0.1%. Looking back on the history of the asbestos problem in Japan, the all-out ban could only be described as a step which was long overdue. Nonetheless, it still counted as an achievement in that it cut off any new supplies of products that could only make the asbestos crisis worse than it had already become.

Measures against asbestos used in existing buildings may be cited as one of the key asbestos-related challenges facing the construction industry in the years ahead. Buildings accounted for 80–90% of the asbestos used in Japan. Most of that was in the form of sprayed asbestos coatings for fireproofing purposes in steel-reinforced structures, or molded asbestos panels used as an interior ceiling and wall material. Sprayed asbestos applications are prone to the airborne release of asbestos fibers, particularly after they have degraded with age, and as such, present a high exposure risk. Sprayed coatings therefore demand regular maintenance. By contrast, molded asbestos panels are not that likely to release asbestos dust under conditions of normal use. However, during the structural demolition or scrap material processing stages, even these products are likely to release asbestos dust into the air. Asbestos materials in buildings require proper upkeep on a routine basis, as well as appropriate handling during structural demolition and other handling stages. Measures to deal with the asbestos that has accumulated in buildings (hereafter "stock asbestos") face two major hurdles: the costs associated with maintenance and disposal, and the task of ensuring total implementation.

Measures to deal with asbestos-related diseases among workers constitute the other key challenge for the construction industry. The Kubota Shock focused the public spotlight on these diseases among workers in factories involved in the production of asbestos-related products, as well as the problem of environmental asbestos exposure among residents living around such factories. Construction workers are at risk of exposure to asbestos dust at construction sites, and the incidence of asbestos-linked disease has been rising sharply in recent years. However, the medical equipment and systems used to diagnose new cases remain relatively underdeveloped. Not only that, but many workers in the construction industry actually own small, one-person businesses, i.e., they are sole proprietors, and for that reason are typically not eligible for (or are less readily able to receive) workers accident compensation insurance even if they are given a diagnosis of an asbestos-linked illness. As with the stock asbestos problem, the problem of asbestos-related diseases among construction workers will be a huge future challenge.

11.3 The Challenge for Measures Against Stock Asbestos in Buildings

Measures have been developed to deal with the issues of stock asbestos and sprayed asbestos coatings in particular. However, due to the difficulties associated with actually removing such asbestos materials and protecting the safety of workers in the process, these measures typically impose enormous costs. Furthermore, building owners, and tenants in general, cannot be described as having developed an accurate understanding of asbestos. The problems associated with stock asbestos in terms of maintenance and methods of removal are described below.

11.3.1 Problems Associated with Maintenance

11.3.1.1 Weak Precautions and Awareness Regarding Asbestos Building Materials

According to a survey by the Ministry of Land, Infrastructure, Transport, and Tourism (MLIT), 7.1% of all privately owned buildings in Japan with a total floor space of 1 000 m^2 or more had exposed surfaces with sprayed asbestos coatings or rock wool containing asbestos.[1] Working through the local governments, the MLIT has relied on administrative guidance to make the owners of large-scale buildings with exposed sprayed asbestos surfaces implement steps for their removal or encapsulation, and comply with the applicable provisions of the Industrial Safety and Health Law and related laws and statutes. However, this policy only applies to large buildings, and not to the private houses and other small-scale buildings that account for the vast majority of buildings. Furthermore, it only addresses sprayed asbestos applications, and does not cover molded asbestos panels or other construction materials that contain asbestos. To date, no surveys or administrative guidance policies have been implemented on a national scale for the overwhelming majority of buildings in Japan that are outside the scope of administrative guidance. To be fair, it would be difficult, if not entirely impossible, to survey or provide guidance covering each and every one of those buildings. Nonetheless, at a minimum, it seems imperative that measures should be taken to alert building owners and tenants, and to foster an accurate awareness of asbestos construction materials. Revisions to the Building Standards Law placed a ban on the use of sprayed asbestos coatings, asbestos-containing rock wool, and other materials that pose a risk of releasing asbestos dust, and mandated that such materials must be removed when existing buildings are expanded or renovated. However, no provisions for routine countermeasures have been established. The government has demanded that relevant industrial groups draw up their own action plans for the implementation of special training programs on asbestos. Among measures for asbestos-related industries, as of March 31, 2006, 147 instructors for training roles related to the demolition of buildings that contain asbestos had been licensed, and approximately 99 000 workers were given special training in the demolition or scrapping of buildings and structures containing asbestos materials.[2] Following the Kubota Shock, reports of asbestos-related diseases and deaths among employees of asbestos product factories and residents in surrounding neighborhoods became major news. Despite that, not enough has been done in terms of broad-based education or publicity campaigns to alert ordinary building owners and tenants, or to disseminate accurate knowledge of asbestos hazards. What is needed is a framework that facilitates ongoing efforts by industry to foster an accurate understanding of asbestos, as well as the exhaustive implementation of appropriate measures to educate ordinary building owners and tenants, and surveys on the current status of asbestos use and management practices.

11.3.1.2 A Lack of Established Strategies for the Management of Stock Asbestos

Essentially four types of strategy exist for the management of stock asbestos: removal, encapsulation, enclosure, and the wait-and-see option when stock asbestos is in an acceptable condition. Appropriate strategies based on an accurate understanding of asbestos building materials are needed.

Removal is a fundamental and permanent strategy. However, an enormous amount of labor and expense is involved in ensuring the safety of workers engaged in the removal of sprayed asbestos coatings and the handling of spent materials. Although huge costs are associated with the removal of large surface areas of sprayed asbestos, even the removal of relatively small-scale surfaces can be out of the question for small building owners with limited finances. Technical issues to do with the removal strategies are discussed later.

The encapsulation strategy involves spraying a hardening agent over the surface of the sprayed asbestos to form a film coating, allowing the hardening agent to penetrate into the sprayed asbestos layer and reinforce the bonding strength of the asbestos fibers, thus preventing their escape as dust particles. Although this approach is less expensive than the removal strategy, some hardening agents are reported to lack durability, and the technology itself is considered to be still in the development stage. Nonetheless, with quality hardening agents, this strategy should be a viable option in situations where removal is not a necessity. The future development of this technology therefore holds promise.

The enclosure strategy comprises covering ceilings, walls, and other sprayed asbestos surfaces with a nonasbestos material that prevents the release of asbestos dust into the air in a room. Although this is generally an inexpensive remedy, it is a temporary fix, not a fundamental, permanent solution.

The "wait and see" strategy would conceivably be limited simply to monitoring on a continual basis whether or not the asbestos material shows signs of fluffing, or is otherwise in a good condition. This strategy would also apply to molded asbestos panels provided that they are in relatively good condition.

Currently, none of these strategies is considered to be viable in practice. Granted current conditions, society must hold out hope for technological advances that improve the viability of these strategies in terms of both the cost and the performance.

11.3.2 Problems Posed by Stock Asbestos When Buildings are Demolished

The management of stock asbestos entails the problems described in the preceding section. However, with proper management practices, the risk of asbestos exposure will be kept low. The risk of airborne asbestos dust dispersal is at its highest when a building is being demolished. The following laws govern the handling of asbestos during building demolition work. The Industrial Health and Safety Law stipulates the performance of a preliminary asbestos survey, the formulation of an operating plan, the submission of a demolition project notification, the utilization of dust-control equipment and measures, and the implementation of special worker training. The Air Pollution Control Law incorporates provisions mandating the submission of a notification concerning working operations that involve emissions of specific types of dust, and sets property boundary benchmarks for compliance. The Waste Disposal and Public Cleansing Law governs the handling of scrap asbestos materials, and the Construction Materials Recycling Law stipulates the preliminary removal of sprayed asbestos and other deposits from specified construction materials. Full compliance with all of these laws is required.

These laws do not contain any provisions which are specific to the handling of molded asbestos panels. However, some local governments have established their own handling guidelines.

Although asbestos paneling is not covered by the provisions of the Air Pollution Control Law, projects for the demolition of buildings in Osaka Prefectural Government premises containing asbestos panels (albeit with a floor space of 1 000 m^2 or more) are subject to the notification requirements mandated by a local ordinance for the preservation of the residential environment. In addition, Tokyo Metropolitan Government has prepared a manual of guidelines on the handling of molded asbestos panels (Tokyo Metropolitan Government, Bureau of the Environment, March 2007), and is striving to inform the public about proper methods of panel removal. This manual lists five key handling guidelines. (1) Moisten the asbestos panels. (2) Use removal methods that avoid breaking or crushing. (3) Use masking methods suited to the work location. (4) Make workers wear protective gear. (5) If a water spray is used during the work, dispose of the asbestos-contaminated waste water correctly.

Special care should be exercised when handling asbestos panels, partly because reports have noted that asbestos fibers tend to accumulate in attics and other closed-off spaces behind asbestos paneling.

When preparing to demolish buildings that contain asbestos paneling, it may well be necessary to develop frameworks for full compliance with all applicable laws, and to apply exhaustive panel handling measures.

11.3.3 Problems Posed by Removal Methods for Sprayed Asbestos Coatings

The MLIT has prepared several measures to assist with the removal of asbestos from ordinary buildings, including developing a quality buildings subsidy (according to the type of renovation of buildings containing asbestos), a local housing subsidy program, and a financing framework for the removal of sprayed asbestos coatings.[3] The framework for the renovation of quality buildings containing asbestos applies to buildings with exposed surfaces comprising asbestos materials which have an unspecified large number of visitors. It provides subsidies to cover one-third of the cost associated with surveys of asbestos content and the cost of sprayed asbestos removal, encapsulation, or enclosure by property owners (including the cost of surveys and design, ancillary office processing fees, and compensation for the cost of tenant relocations required by the asbestos removal process). However, as indicated, this subsidy framework is limited in scope as it only applies to buildings or sections thereof receiving multiple visitors. The local housing subsidy program is a core enterprise for the development of public housing. However, it also facilitates the provision of subsidy funding for the formulation of community housing plans that incorporate assistance for the removal of asbestos materials. The financing framework for the removal of sprayed asbestos coatings comprises funding from the Development Bank of Japan's living environment development promotion program (a program of asbestos-related measures) and the Japan Finance Corporation's environmental fund (related to air pollution). All these programs have limited budgets and scope, and do not incorporate fundamental measures for large-scale projects involving the removal of stock asbestos. Given these circumstances, efforts to remove accumulations of stock asbestos from buildings will probably face several problems. Namely, owners of building properties will bear most of the burden of responsibility for implementation, face enormous costs related to implementation, and have little if any guarantee that the removal measures will be implemented properly.

As noted earlier, the removal of sprayed asbestos coatings imposes a heavy burden of labor and expense. Estimates by the National General Contractors Association of Japan put the cost of removing sprayed asbestos in the range of 20 000–80 000 yen per m^2 for projects involving surfaces of less than 300 m^2, 15 000–55 000 yen per m^2 for projects involving surfaces ranging from 300 to 1 000 m^2, and 10 000–25 000 yen per m^2 for projects involving surfaces exceeding 1 000 m^2 in scale.[4] To give an example, the cost of a project to remove sprayed asbestos coatings from a surface area of around 100 m^2 will total at least 2 million yen, while that for a project dealing with a 1 000 m^2 surface area will come to at least 10 million yen. The aforementioned frameworks alone simply will not be enough to cover the huge expense associated with removal projects in private buildings where significant amounts of asbestos have been used. As such, they lack viability as appropriate strategies.

At present, efforts are under way within construction-related sectors to develop better asbestos removal technologies.

The Taisei Corporation is developing an unmanned asbestos removal and recovery system that utilizes remote-controlled robots. It has already been commercially perfected for warehouse demolition operations involving the removal of dry forms of sprayed asbestos.[5] As designed, the system efficiently peels off and removes asbestos with a jet spray mixture from a nozzle attached to the robot, thus facilitating rapid, automated asbestos recovery and bagging, as well as high levels of safety, with minimal worker entry into areas presenting a health risk. The system removes asbestos at high speed and allows for the compression of waste materials. On average, only about 5 m^2 of sprayed asbestos coating can be removed by hand in 1 h. With this robot-based system, up to 20 m^2 can be removed per hour. As such, it can be expected to reduce the costs associated with enhancements to worker safety and processing speed.

Another new method under development uses infrared irradiation to render asbestos harmless. This technology works by heating installed asbestos construction materials with infrared light, thus melting the asbestos fibers together and rendering them harmless. Furthermore, it is anticipated that this approach will allow processing at a fraction of the cost of conventional methods. However, it is estimated that another 3 years will be needed to improve the processing speed of this approach, develop it for applications to large surfaces, and make it commercially viable.[6]

In the years ahead, it is likely that many of the buildings that were constructed during the era of heavy asbestos consumption (from the mid-1960s forward) will be demolished. Developing asbestos removal technologies of guaranteed effectiveness has thus become all the more urgent.

11.4 Challenges for Measures Against Asbestos-Related Diseases in Construction Industry Workers

Far-reaching studies aimed at identifying the nationwide incidence of asbestos-related diseases among workers in the construction industry have yet to be implemented. The 134 819 businesses that responded to an August and December 2005 MLIT-sponsored questionnaire survey of industries under its jurisdiction together reported 196 workers with asbestos-related diseases, and of those, 131 had already died. However, that total seems unusually small compared with the number of workers reported by the Tokyo Construction Labor Union (hereafter "Tokyo Doken") as having been approved for workers accident compensation for asbestos-related diseases. Drawing from the findings of a Tokyo Doken construction union survey of asbestos-related diseases among construction workers, the following section discusses the challenges for countermeasures against these diseases.

11.4.1 The Incidence of Asbestos-Related Diseases Among Tokyo Doken Union Members

Since 2002, Tokyo Doken has been pursuing full-scale asbestos countermeasures centering around the submission of workers accident compensation applications for members with pneumoconiosis and asbestos-related diseases. Tokyo Doken has been working to identify asbestos patients who are otherwise difficult to detect with conventional diagnostic tests. These efforts are in the process of shedding light on the actual incidence of asbestos-related diseases among workers in the construction industry.

11.4.1.1 Disease-Specific Conditions and Trends

Since 2002, Tokyo Doken has been seriously engaged in efforts to have workers approved for workers accident compensation.

Table 11.1 lists the number of Tokyo Doken-affiliated workers approved for workers accident compensation from FY 2002 to FY 2009, by asbestos-related diseases.[7] Patients with lung cancer accounted for the largest share of the total at 53.5%, followed by 31.8% with pulmonary asbestosis, and 12.1% with mesothelioma. In terms of year-on-year trends, there was a sharp up-surge in the number of cases identified in 2006. This climb in the incidence of asbestos-related

Table 11.1 Number of Tokyo Doken-affiliated workers approved for workers accident compensation from FY 2002 to FY 2007, by asbestos-related diseases

Fiscal year	Mesothelioma	Lung cancer	Asbestosis	Diffuse pleural thickening	Total
2002	1	–	2	–	3
2003	2	5	5	–	12
2004	2	8	3	–	13
2005	3	14	12	2	31
2006	18	57	20	2	97
2007	8	39	20	1	68
2008	7	36	30	4	77
2009	1	26	18	–	45
Total	42	185	110	9	346
Percentage	12.1	53.5	31.8	2.6	100

Source: Tokyo Doken

diseases appears to have two distinct dimensions. One has to do with the development of diagnostic testing frameworks. Asbestos-related diseases are difficult to diagnose without the assistance of a medical specialist, and in many cases, only a specialist can interpret the X-ray findings of lesions of pleural thickening that typically appear after the inhalation of asbestos particles. Some cases of asbestos-induced lung cancer are misdiagnosed as conventional cases of lung cancer. Furthermore, without an understanding of a worker's occupational background, physicians may find it difficult to determine a history of asbestos exposure based on interviews. With the heightened awareness of these issues, in recent years an active emphasis has been placed on having diagnostic tests performed by specialists. This appears to have resulted in the recognition of asbestos-related diseases that were previously not considered to be attributable to asbestos. The nature of symptoms which become manifest after a long period of dormancy appears to be the other dimension behind the increase in cases of asbestos-related diseases. It is believed that the symptoms of asbestos-related diseases begin to appear only after a period of 10–40 years following exposure (Morinaga 2006:4). With 40 years or more having elapsed since the era during which asbestos was in heavy use at many construction sites, it is conceivable that in many patients symptoms are only now beginning to manifest themselves after that long period of dormancy.

However, the frameworks for specialist-led diagnostic testing remain deficient. On top of that, an increasing number of patients can be expected to pass through the period of dormancy and begin to show symptoms. These factors point to a strong upward trend in the incidence of asbestos-related diseases among construction workers in the years ahead, suggesting that the issue of asbestos-related diseases among such workers is likely to become much more severe.

11.4.1.2 Trends in the Incidence of Asbestos-Related Diseases, by Occupation

Table 11.2 and Fig. 11.1 show the number of Tokyo Doken-affiliated workers with asbestos-related diseases who have been approved for workers accident compensation per 1 000 affiliated workers in various construction-related trades.[8] The highest ratio was observed among concrete-block workers, but the data may not be an accurate reflection owing to the small scale of that occupational denominator. The next-highest ratio of incidence was noted among workers in air-conditioning and insulation engineering, who frequently have to crawl into attic spaces and other enclosed areas where asbestos dust has accumulated. High incidence rates were also observed among plasterers, who work in asbestos-contaminated settings, in chipping and demolition occupations, which involve a higher risk of exposure to airborne asbestos dust, in steel-frame workers, who typically have numerous opportunities to be involved in the application of sprayed asbestos coatings, in painters, who usually handle asbestos-contaminated

Table 11.2 Incidence of asbestos-related diseases among Tokyo Doken members approved for workers accident compensation, by occupation

Occupation	Members	Ratio (%)	Average age	Approved workers	Per 1 000 members
Concrete block work	722	0.57	51.0	3	4.16
Air conditioning engineering	3 578	2.82	45.6	13	3.63
Plastering	3 765	2.97	55.0	13	3.45
Chipping, demolition	2 580	2.04	44.6	8	3.10
Steel-frame work	3 205	2.53	52.6	9	2.81
Painting	7 553	5.96	49.5	20	2.65
Carpentry	27 227	21.48	52.2	64	2.35
Roofing	436	0.34	52.6	1	2.29
Sheet metal work	2 158	1.70	54.3	4	1.85
Equipment installation	568	0.45	45.6	1	1.76
Form/mold making	1 418	1.12	45.6	2	1.41
Electrical	10 381	8.19	48.7	13	1.25
Flooring, interior work	9 830	7.76	44.1	12	1.22
Water supply, sanitation, gas	9 315	7.35	48.3	11	1.18
Stone masonry	932	0.74	49.7	1	1.07
Brick and tile laying	2 118	1.67	51.6	2	0.94
Glass work	1 182	0.93	51.0	1	0.85
Doors	3 113	2.46	52.1	2	0.64
Earthwork, paving	6 641	5.24	48.4	3	0.45
Scaffolding	4 361	3.44	45.9	1	0.23
Tatami	695	0.55	58.3	–	0.00
Garden landscaping	1 748	1.38	49.0	–	0.00
Steel-reinforced beam work	1 650	1.30	45.3	–	0.00
Waterproofing	3 032	2.39	43.0	–	0.00
Metal fixtures	1 868	1.47	51.2	–	0.00
Wall hanging	902	0.71	54.7	–	0.00
Communications	1 016	0.80	40.0	–	0.00
Construction machinery operation	4 788	3.78	47.8	–	0.00
Building maintenance	1 438	1.13	45.4	–	0.00
Building services	1 553	1.23	49.3	–	0.00
Design, blueprinting	4 648	3.67	44.3	–	0.00
Civil engineering and building services	972	0.77	48.6	–	0.00
Canopy and billboard installation	900	0.71	47.6	–	0.00
Other occupations	57	0.04	53.9	4	70.18
Clerical	385	0.30	42.7	–	0.00
Total and average age	126 735	100	49.0	188	1.48

As of October 22, 2007
Source: Tokyo Doken

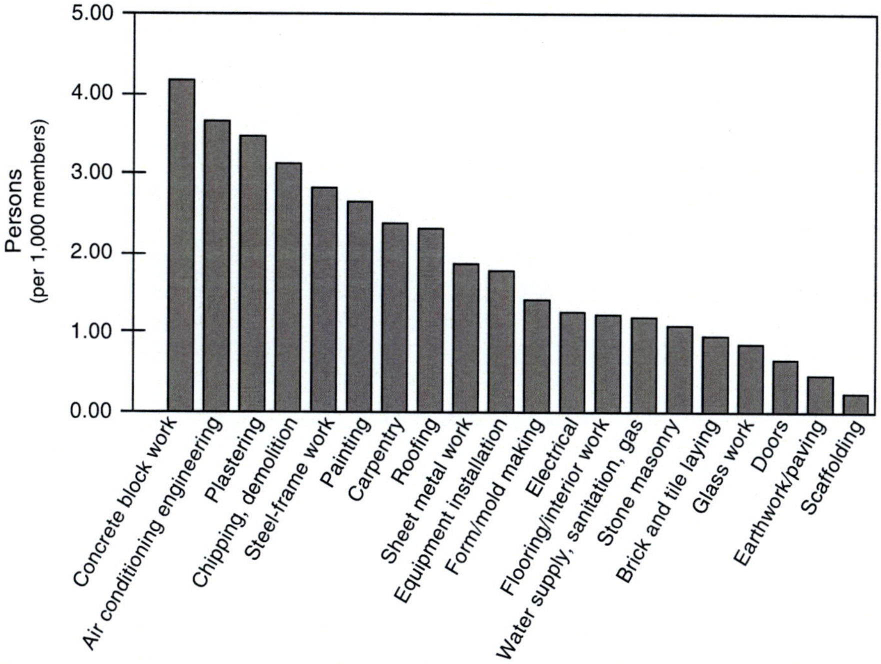

Fig. 11.1 Incidence of asbestos-related diseases among Tokyo Doken members approved for workers accident compensation per 1 000 workers, by occupation. *Source*: Tokyo Doken

construction materials, in carpenters, and in roofers, who often handle asbestos-contaminated roofing materials.

In addition, routine chest X-ray examinations performed by Dr. Isamu Ebihara for Tokyo Doken in 2005 and 2006 detected lesions of pleural thickening in 6 268 workers examined, which is an astonishingly high detection rate (Table 11.3).[9] Lesions of pleural thickening occur only in asbestos-related diseases, and were observed in 6.75% of all workers examined. They were found in 13.64% of plasterers, 11.72% of workers involved in woodworking and doors, over 7% of all workers involved in air conditioning and insulation engineering, painting, tile laying, scaffolding, and demolition trades, and 7.87% of workers in the carpentry trade. Further, Dr. Ebihara noted an almost parallel relationship between the detection rates for lesions of pleural thickening and pulmonary asbestosis (Fig. 11.2). It has been shown that the incidence of pleural hardening, which occurs even with exposure to relatively small amounts of asbestos, and also of asbestosis, which is caused by exposure to asbestos in high concentrations, were detected in construction workers in high ratios and across a broad range of construction-related occupations.

Table 11.3 Detection rate for pleural thickening in construction workers, by occupation (Type-1 or more progressive forms of asbestosis)

Occupation	Worker sample	Workers with detected lesions	Detection rate (%)
Plastering	220	30	13.64
Woodworking/doors	145	17	11.72
Air conditioning/insulation engineering	200	18	9.00
Sheet metal work	130	11	8.46
Tile laying	158	13	8.23
Carpentry/construction	1 194	94	7.87
Electrical	793	60	7.57
Painting	355	26	7.32
Scaffolding/demolition	309	22	7.12
Pipe-fitting/plumbing	573	34	5.93
Steel working/welding	122	7	5.74
Interior work	496	27	5.44
Shingle/roof tile installation, light-gauge steel sheet installation	186	7	3.76
Stone masonry	45	1	2.22
Design, clerical	374	7	1.87
Unknown	89	6	6.74
Other	879	43	4.89
Total	6268	423	6.75

Source: Tokyo Doken

11.4.2 Problems with Workers Accident Compensation Eligibility and Compensation Amounts for Patients with Asbestos-Related Diseases

Workers with mesothelioma are approved for workers accident compensation as patients with asbestos-related diseases. Although lung cancer patients with lesions of pleural thickening, or Type-1, or more progressive forms of asbestosis will also be approved, those without signs of a given concentration of asbestos particles or fibers in their tissues will not be approved. It is difficult to diagnose mesothelioma without a certain level of specialized expertise. Furthermore, in many cases, lesions of pleural thickening can only be identified by a medical specialist in the field.

As these examples suggest, a solid framework for the identification of asbestos-related diseases has not yet been put into place. Until recently, the framework was even less adequate, and conceivably allowed many patients with obvious symptoms of asbestos-related diseases to go undetected and die. Tokyo Doken has been working to identify workers with asbestos-related diseases by examining medical records held by the National Health Insurance Union, re-evaluating worker's X-rays after diagnoses based on union branch health examinations, reviewing applications for Tokyo Doken mutual aid insurance claims, and conducting pneumoconiosis

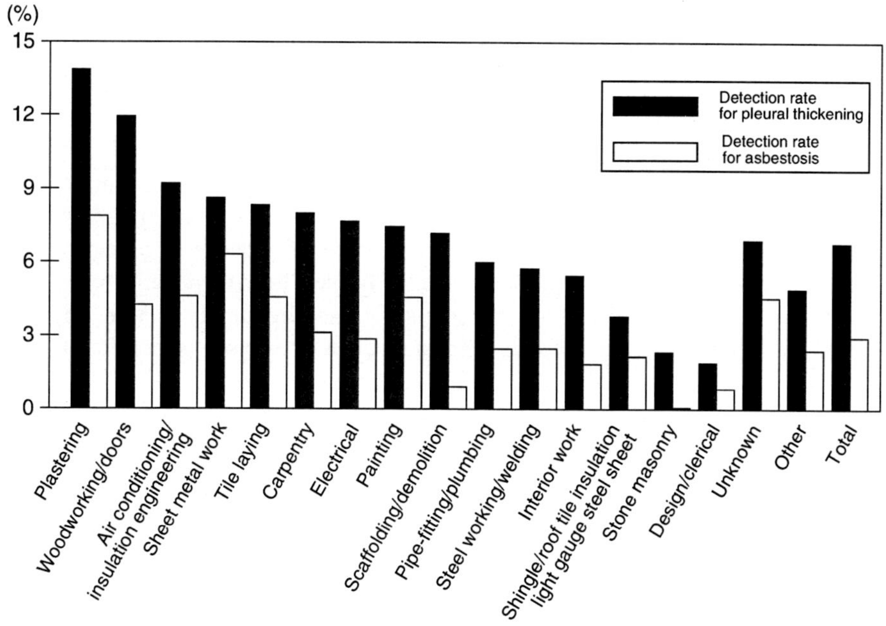

Fig. 11.2 Detection rates for lesions of pleural thickening and asbestosis, by occupation.[10] *Source*: Tokyo Doken

examinations at the union's main and branch offices. Workers with findings indicative of an asbestos-related disease are referred to specialists for further examination, and also have their X-rays re-evaluated by medical specialists. These efforts in detection can be largely credited for the previously cited success in winning workers accident compensation approval for many patients with asbestos-related diseases.

One characteristic shared by the construction trades is that many workers are sole proprietors running their own one-person operations. It is common for such businesses to consist of only a boss with zero employees, and therefore to have no eligibility for workers accident compensation insurance.[11] However, in most cases, it is rare for these sole proprietors to be engaged in work on their own; more often than not, they are likely to be employed from the outset in some arrangement with another business owner. Under the workers accident compensation insurance framework, individuals are treated as workers during the periods they are employed. After they become sole proprietors (or owners of small or medium-scale businesses), they have the option of voluntarily enrolling in the framework provided they pay insurance premiums, and if they do they will be treated as special enrollees. However, unless they are enrolled under this special arrangement, they will not be eligible for workers accident compensation insurance benefits. As a rule, pneumoconiosis or asbestosis patients who apply for workers accident compensation must have been employed as workers, or enrolled as special enrollees, for at least half the length of the period during which they were exposed to asbestos. For lung cancer patients, eligibility is preconditioned on having been employed as a worker for a period of 10 years or more. Mesothelioma patients are granted eligibility if they have been engaged as workers for at least 1 year, and almost all applicants with mesothelioma are approved. Many patients with pneumoconiosis, asbestosis, or asbestos-induced lung cancer are not approved if they are sole proprietors without special enrollment status. Should they be denied coverage under the workers compensation insurance program, they may seek relief under the provisions of the Act on Asbestos Health Damage Relief. Under the New Law, no benefits for medical leave are offered. Further, other relief benefits are frequently smaller then the benefits available through the workers accident insurance compensation framework.

11.4.3 Construction-Related Asbestos Litigation in the Tokyo Metropolitan Area

Against this backdrop, in May 2008 a group of 178 plaintiffs (172 patients), comprising construction workers and bereaved family members from the Tokyo metropolitan area, filed a lawsuit claiming reparation from the national government in the Tokyo District Court (hereinafter, the "Tokyo Metropolitan Asbestos Lawsuit") seeking damages and consolation payments of 35 million yen per plaintiff for a total of 6 billion yen.[12,13] Most of the plaintiffs had worked from the 1960s onward at construction sites mainly in the Tokyo metropolitan area, had inhaled asbestos dust from various construction materials, and were suffering from diseases that ranged from mesothelioma and lung cancer to asbestosis. Their occupations closely resembled those listed in the previously cited Tokyo Doken study of cases of asbestos-related diseases per 1 000 members (Table 11.4).[14] Table 11.5 is a plaintiff breakdown by asbestos-related disease, and Table 11.6 is a breakdown by cause of death.[15] Of the 172 plaintiffs, 82 (47.7%) had already died when the lawsuit was initiated. Further, the average age of the 90 surviving plaintiffs was 70 years.

The national government and 46 companies that were engaged in the manufacture of construction materials containing asbestos have been named as defendants in the Tokyo Metropolitan Asbestos Lawsuit.

According to the formal complaint, the lawsuit seeks to have the following demands fulfilled.[16]

First, both within and outside the court setting, light should be shed on the realities of the asbestos-related diseases afflicting workers in the construction trades, the judges should recognize the diseases and hazards caused by asbestos, and a drastic improvement in the public awareness of those diseases and hazards should be achieved.

Second, light should be shed on the legal liability of the national government and construction material manufacturers, and for the plaintiffs, a formal apology should be sought in good faith, with payments of damage compensation which are commensurate with the harm suffered by the plaintiffs.

Table 11.4 Occupations of plaintiffs in the Tokyo Metropolitan asbestos lawsuit

Occupation	Number	Share (%)
Carpentry	52	30.2
Plastering	19	11.0
Pipe-fitting/plumbing	18	10.5
Electrical	17	9.9
Painting	12	7.0
Demolition	10	5.8
Tile laying	6	3.5
Interior work	6	3.5
Insulation	6	3.5
Sheet metal	5	2.9
Steel-frame work	3	1.7
Waterproofing	3	1.7
Duct installation	2	1.2
Equipment installation	2	1.2
Steel reinforcement work	2	1.2
Scaffolding	2	1.2
Air conditioning	1	0.6
Concrete block work	1	0.6
Site supervision	1	0.6
Other	4	2.3
Total	172	100

Source: Data from the Tokyo Asbestos Lawsuit Legal Team

Table 11.5 Breakdown of plaintiffs in the Tokyo Metropolitan asbestos lawsuit, by disease

Asbestos-related disease	Number	Share (%)
Lung cancer	82	47.7
Asbestosis	62	36.0
Malignant mesothelioma	23	13.4
Diffuse pleural thickening	4	2.3
Benign asbestos pleural effusion	1	0.6
Total	172	100

Source: Data from the Tokyo Asbestos Lawsuit Legal Team

Table 11.6 Causes of death among plaintiffs in the Tokyo Metropolitan asbestos lawsuit

Cause of death	Number	Share (%)
Lung cancer	53	64.6
Asbestosis	9	11.0
Malignant mesothelioma	20	24.4
Total	82	100

Source: Data from the Tokyo Asbestos Lawsuit Legal Team

Third, the national government and construction material manufacturers must create a framework (victim relief fund) for the relief of all construction workers harmed by exposure to asbestos.

Fourth, the national government and construction material manufacturers must implement sweeping restrictions on asbestos, as well as establish and implement far-reaching initiatives to prevent harm from asbestos exposure as a result of building renovation and demolition operations.

Fifth, the national government must establish appropriate measures to protect the rights of asbestos victims and put an end to asbestos-related harm.

As demonstrated in this chapter, many questions surround the issues of workers accident compensation eligibility and compensation amounts for construction workers with asbestos-related diseases. Further, with future increases in the number of asbestos victims predicted, these problems can be expected to become even more serious, and thus demand an appropriate response.

11.5 In Lieu of a Conclusion

Japan has benefited enormously from the intensive use of asbestos, an inexpensive "magic mineral" touted for its excellent properties as a construction material. Conversely, its reliance on this material has also had a darker side, marked by the harm that asbestos exposure has caused many victims to date, and the risk of exposure posed by accumulations of stock asbestos in existing buildings. Measures to deal with asbestos hazards and stock asbestos pose extremely difficult challenges in terms of the related labor, costs, and time required. The negatives of asbestos may ultimately outweigh its benefits, but that is a problem we must deal with correctly in the years ahead. Acting on that understanding, we have no choice but to fully implement measures aimed at cultivating an accurate awareness of the problems, develop the necessary technologies and frameworks, ensure total compliance with the law, and secure the necessary funding to offset the costs. If there is a lack of adequate knowledge and countermeasures, we risk a spread in asbestos-related diseases among building tenants, as well as a secondary wave of asbestos hazards and harm with environmental asbestos exposure from the demolition of buildings and the disposal of scrap asbestos.

Notes

1. MLIT "Survey of sprayed asbestos coatings in privately owned buildings," released on December 18, 2009. http://www.mlit.go.jp/report/press/house05_hh_000138.html. Accessed September 17, 2010.The scope of this survey was limited to private-sector buildings with a total floor space of 1 000 m^2 or more, and built between 1956 and 1989. It was based on local municipal government requests for reports from property owners. Of 274 260 buildings covered

by the survey, reports from property owners were received on 229 779. Of that subtotal, 16 212 had sprayed asbestos coatings on exposed surfaces.
2. MLIT (2006). "Promotion of Measures against Buildings Containing Asbestos." by MLIT (in Japanese).
3. MLIT (2006). *Ibid.*
4. MLIT (2006). *Ibid.*
5. Taisei Corporation website. http://www.taisei.co.jp/giken/topics/1191574716210.html. Accessed September 17, 2010.
6. Kyoto Shimbun Newspaper, January 24, 2008.
7. Tokyo Doken (Tokyo construction labor union) (2007). "Manual for the Application of Workers Accident Compensation for Asbestos Diseases 2007."
8. Tokyo Doken labor union documentation.
9. Tokyo Doken (2007).*Op.cit.*
10. Tokyo Doken (2007). *Ibid.*
11. On June 28, 2007, a decision in the Supreme Court ruled that sole proprietors do not correspond to workers under provisions of either the Labor Standards Law or the Workers Accident Compensation Insurance Law, and thus are not eligible to receive workers accident compensation insurance benefits.
12. Asahi Shimbun Newspaper, January 23, 2008.
13. Interview with Toshihiko Tsukuda, attorney for the Tokyo asbestos lawsuit legal team.
14. *Ibid.*
15. *Ibid.*
16. The headquarters, the plaintiff group, and the team of defense lawyers in the Construction-Related Asbestos Litigation in the Tokyo Metropolitan Area (2008). "The Petition of the Construction-Related Asbestos Litigation in the Tokyo Metropolitan Area."

Reference

Morinaga K (2006) Why an asbestos problem now? In: Morinaga K (ed) Asbestos pollution and its health effect (in Japanese), 2nd edn. Nihon-Hyoron-sha, Tokyo

Chapter 12
Local Government Measures Against Asbestos: Tokyo Metropolitan and Osaka Prefectural Governments as Case Studies*

Kazuhisa Hiraoka

12.1 Introduction

Japan has been slower than most European countries to take action against asbestos. This has clearly aggravated the extent of asbestos exposure and harm, and ensures that Japan will have many more asbestos victims with overt symptoms in the years ahead. Following the Kubota Shock, it was confirmed that a series of incidents involving environmental exposure-induced outbreaks of asbestos-linked illnesses and deaths among residents living near asbestos product factories had taken place. Although this has elevated public concern, the complete picture of the extent of asbestos exposure and harm, including the outlook for future victims, is still not clear. Another problem is the extremely inadequate scope and scale of relief money provided under the Act on Asbestos Health Damage Relief. Furthermore, despite the fact that an all-out ban has been placed on the utilization of asbestos, huge amounts or "stock" of this material have accumulated, primarily in buildings and other structures, and these pose an environmental risk that will only become more serious in the years ahead.

At the national government level, measures against asbestos were held back in favor of the vested interests of industry on the assumption that asbestos was an essential element of buildings and the postwar industrial and social infrastructure. This, combined with bureaucratic sectionalism and a reliance on ad hoc measures, had the effect of delaying national action on asbestos issues (Miyamoto 2007:284–286). Although the nature of asbestos-induced harm as a complex of health-related problems and stock-induced environmental pollution disasters would seem to call

*Translated and revised from the Japanese original, Hiraoka K. (2008). Local Government Measures against Asbestos: Tokyo Metropolitan and Osaka Prefectural Governments as Case Studies. Policy Science (in Japanese), Supplementary Volume, 127–143, with permission of the Policy Science Association of Ritsumeikan University.

K. Hiraoka (✉)
College of Policy Science, Ritsumeikan University, 56-1 Tojiin Kitamachi, Kita-ku, Kyoto 603-8577, Japan
e-mail: hiraoka@sps.ritsumei.ac.jp

K. Miyamoto et al. (eds.), *Asbestos Disaster: Lessons from Japan's Experience*,
DOI 10.1007/978-4-431-53915-5_12, © Springer 2011

for comprehensive countermeasures at the administrative level, the government is still not equipped to pursue adequate policy measures because it has neither developed a full picture of the extent of asbestos-related harm and risk through epidemiological studies or other means, nor clearly identified who should shoulder the blame based on these studies.

Local governments have a critical role to play in handling environmental pollution and industrial disasters. First and foremost, that role comprises protecting the safety and livelihood of local residents, and forming the first line of defense against such contingencies. In a setting where action by the national government has been slow or inadequate, independent measures spearheaded by local governments – as the administrative entities that citizens find most readily accessible – are crucial and deserving of attention. However, that is not to imply that local government measures against asbestos on a nationwide scale have been any more clearly demonstrated than steps taken by the national government.

This chapter focuses chiefly on two local governments, Tokyo Metropolitan Government and Osaka Prefectural Government, as case studies of asbestos measures at the local government level. The reason is that both prefectures have taken the lead in efforts against environmental pollution to date, and have also implemented measures of their own against asbestos. In this chapter, I explore local government responses to and countermeasures against asbestos to date, and on that basis, strive to highlight the importance assigned to the asbestos problem within the context of local government policies, and to a certain extent categorize the key asbestos policy challenges that local governments now face.

12.2 The Local Government Role in the Asbestos Policy Arena and the Status of Current Efforts

12.2.1 Environmental and Disaster Policy Doctrine and Local Government Measures Regarding Asbestos

Kenichi Miyamoto describes environmental policy as a composite of objectives in the following five areas: (1) identifying the actual extent of damage, determining causes, and assigning blame; (2) providing victims with relief, economic compensation, and the restoration of their health and livelihoods; (3) regulations to prevent environmental pollution coupled with environmental protection measures based on social infrastructure and land-use planning; (4) environmentally focused community revival; (5) measures in prevention (including environmental impact assessments, cost–benefit analyses, community and national land-use plans, and international cooperation) (Miyamoto 2007:177).

The core philosophy behind these objectives may also be applied to asbestos pollution and other complex stock disasters that are outside the conventional boundaries of environmental policy. Local government measures against environmental and industrial disasters must be explored in terms of these five ideals. Particularly granted the nature of asbestos exposure and its related hazards as a complex stock pollution disaster affecting segments of the general population, the negatives of asbestos are also of a complex nature, ranging from harm to worker health and safety, pollution of the workplace environment, and harm from exposure during the product distribution and consumption stages, to pollution from discarded or scrap asbestos materials and the risk of dust exposure from stocks of asbestos accumulated in buildings and other structures. Hence, in terms of gauging levels of exposure and harm, identifying sources of contamination, formulating health measures, implementing regulations to prevent pollution and industrial disasters, and pursuing policies shaped by a preventive approach, local governments – as entities with responsibility for all facets of citizen's lives – have a role that can be described as nothing less than momentous.

In addition, judging from the scientific information gathered to date on asbestos pollution and disasters, an all-out ban is warranted, and steps must also be taken to foster the transition to a nonasbestos society.

Understanding the extent of harm from asbestos and the causal relationships will demand efforts to identify victims and implement epidemiological studies. This calls for a collaborative effort by local governments and researchers. Given their access to community health centers and hospitals that are equipped for the provision of specialized care, prefectures and other local governments will have a major role in the implementation of protective health measures. In the context of instituting preventive regulations against environmental pollution and industrial disasters, local governments will be expected to regulate factories that handle asbestos, impose restrictions on building demolition and renovation operations, conduct on-site inspections, and implement measures for the handling and reduction of scrap materials. In the arena of preventive action, local governments will be expected to implement a total ban on asbestos use, maintain registries of health risks, set up registries of buildings and other structures that contain asbestos, and implement measures for asbestos removal. In any case, these efforts must be pursued by enlisting the comprehensive knowledge and expertise of local public servants in each department with an understanding of local conditions within their respective communities. However, concurrent efforts at the national government level to develop related legal frameworks and financial guarantees will be indispensable in facilitating local government efforts of this nature.

12.2.2 An Overview of Local Government Measures Against Asbestos

It is therefore worth asking about the asbestos measures that local governments have been pursuing in practice. To date, those measures have included the following: (1) health counseling, health examinations, and health-care facility development; (2) field surveys of the extent of asbestos-induced illnesses, and surveys of health risks; (3) regulations controlling the demolition of buildings; (4) regulation and on-site inspections of factories that handle asbestos; (5) financial assistance; (6) field surveys of asbestos use in public and private structures, and registry-based management of buildings that contain asbestos materials; (7) measures for the handling of scrap materials that contain asbestos; (8) environmental impact assessments; (9) acceptance of applications for measures for relief; (10) the establishment of counseling centers and the provision of asbestos-related information.

In 2007, the office of the Environmental Dispute Coordination Commission surveyed the asbestos measures taken by prefectures and designated cities (basically cities with populations of 500 000 or more). Using data publicly available on prefectural and municipal government websites, the survey grouped the measures into several categories, including the "establishment of counseling centers," "implementation of field surveys on the incidence of asbestos-induced illnesses," "the provision of financial assistance," "health examinations for local residents," "the establishment of guidelines and ordinances," and "other" actions.[1] Of 61 local governments surveyed, 100% had opened counseling centers, 61% had provided financial assistance, 51% had established guidelines and ordinances, 11% had launched field surveys concerning the incidence of asbestos-induced harm, and 20% had begun conducting health examinations for local residents. Furthermore, most local governments had established inter-agency liaison committees on asbestos countermeasures.

Some local governments have intervened with their own regulations on matters not yet covered by national regulations or laws (acting in advance of the March 2006 revisions to the Air Pollution Control Law), while yet others have pursued their own measures in parallel with revisions to the Air Pollution Control Law. Moreover, some local governments (including Hokkaido, Kagawa, and Fukui Prefectures) have begun preparing asbestos registries. Hokkaido Prefecture, for example, has established its own "Hokkaido guidelines for asbestos registry management," and is moving forward with the preparation of an asbestos registry based on the findings of assorted surveys.

Although not many local governments have implemented field surveys on the extent or incidence of asbestos-induced illnesses, in FY 2007 and FY 2008, Hyogo, Osaka, and Saga Prefectures, as well as the City of Osaka, conducted such surveys in collaboration with relevant municipalities as part of the Ministry of the Environment (MOE) program. Certain other local governments have been conducting their own independent studies into the extent of asbestos-related harm. For example, Ibaraki Prefecture conducted its own 3-year study of the health impact of mesothelioma, relying on death certificates and health examination findings from the years 2003 through 2005. In addition, Oita Prefecture performed a health survey of residents living in the vicinity of a decommissioned facility for the manufacture of asbestos products.

The surveys conducted by the Environmental Dispute Coordination Commission have been limited in their scope chiefly to prefectures and designated cities, and do not necessarily provide consistent information on undertakings implemented prior to the Kubota Shock. Furthermore, they do not shed light on the personnel frameworks or budgets established for the implementation of asbestos countermeasures. These topics will therefore require further study.

The Ministry of Land, Infrastructure, Transport, and Tourism (MLIT) has performed a survey of the progress made by local governments nationwide to establish financial support frameworks for asbestos removal. According to its findings, as of September 2007, 18 prefectures (38% of the total) had already put subsidy frameworks into place, five prefectures (11%) were devoting study to their creation, five more (11%) had no plans to introduce them, and 19 prefectures (40%) were handling asbestos removal with other lending frameworks. Among the designated cities that were surveyed, 13 (77%) had already completed their subsidy frameworks, three (18%) were still studying the idea, and one (6%) was relying on other lending frameworks. Of the municipalities surveyed, 122 (7%) had already put subsidy frameworks for asbestos removal into place, 83 (5%) were still studying the idea, 1 596 (88%) had no plans, and 12 (1%) were reliant on other lending frameworks (Table 12.1) (Administrative Assessment Bureau, MIC (2007):47–48).

The foregoing is an overview of local government-led measures against asbestos. A better understanding of the finer details, as well as the total picture, will demand additional study. In particular, more attention needs to be focused on efforts that have been made in the aftermath of the Kubota Shock. Despite these limitations, the survey of local governments highlights several points. First, after the Kubota Shock, many local governments began pursuing comprehensive asbestos countermeasures, prompted in part by measures under way at the national government level, but not many had taken steps to expand independent efforts of their own prior to that event. Second, after the Kubota Shock, local governments took action to strengthen measures involving counseling centers, the provision of information, and programs for public facilities, but they have not yet done enough in terms of identifying new victims, launching field surveys, or offering health examination services. Third, assistance frameworks for the removal of asbestos from privately owned structures have been set up by only a few local governments, and cannot be described as adequate. This is thought to reflect a deteriorating trend in public finances.

Table 12.1 Local government assistance frameworks for asbestos removal (as of December 2007)

–	Subsidy framework established	Planned to be established	Handled with another lending framework	Not planned to be established
Prefectures	18 (38.3%)	5 (10.6%)	19 (40.5%)	5 (10.6%)
Designated cities	13 (76.5%)	3 (17.6%)	1 (5.9%)	0 (0%)
Municipalities	122 (6.7)	83 (4.6%)	12 (0.7%)	1 596 (88.0%)
Total	153 (8.2%)	91 (4.8%)	32 (1.7%)	1 601 (85.3%)

Notes:

1. The five prefectures without plans to build support frameworks were Iwate, Miyagi, Tokyo, Oita, and Kagoshima
2. Subsidy frameworks include the Quality Building Development Program (for asbestos recovery), Regional Housing Grants, and subsidies exclusively funded by local governments
3. Local governments that have established subsidy frameworks and lending frameworks were categorized as "subsidy framework established"

Source: Ministry of Land, Infrastructure, Transport, and Tourism

The following section discusses measures against asbestos taken by the Tokyo Metropolis (with a focus on Nerima City) and Osaka Prefecture, two local governments that have led with independent undertakings in this area.

12.3 Tokyo Metropolis Measures Against Asbestos

12.3.1 Tokyo Metropolitan Government Measures[2]

12.3.1.1 Tokyo Metropolis Master Plan for Asbestos Countermeasures

In 1987, at a time when spray-on asbestos coatings in schools and other public structures had become a major social issue, the Tokyo Metropolitan Government set up an asbestos liaison committee composed of members from nine related bureaus, launched a field survey of sprayed asbestos use in structures managed by metropolitan Tokyo, and began pursuing strategies for asbestos removal. In May 1989 it established its Tokyo Metropolis Master Plan for Asbestos Countermeasures – a first for a local government – and in December that year inaugurated the Tokyo Metropolitan Conference for the Promotion of Asbestos Countermeasures, a supra-agency body.

Among the core views articulated by the asbestos master plan is the position that because asbestos has a strong tendency to accumulate in the environment, and safe concentrations of the substance are not currently known, more aggressive measures against asbestos are needed, particularly in terms of preventing future incidents of environmental contamination. In effect, this viewpoint portrays asbestos as a source of stock pollution, and stresses the prevention-oriented perspective. Another viewpoint is that Tokyo should take the initiative in transitioning to viable nonasbestos alternatives, thus fulfilling a leadership role, and that the plan's guidelines should also be adapted for use by the private sector.

Among the fundamental principles in the Tokyo plan, curbing the use of asbestos is first and foremost. To that end, the plan calls for measures aimed at curbing asbestos use, promoting alternatives, and requiring that alternative materials should also be used in privately owned structures. Second is the prevention of asbestos dust dispersal. Relevant measures cover building and construction materials, discarded materials containing asbestos, and facilities engaged in the manufacture of asbestos products. The third principle is general environmental monitoring, comprising measures to monitor environmental concentrations of asbestos, as well as levels of dispersal from sources of asbestos dust. The fourth principle has to do with measures to protect human health, and covers surveys and studies on the health effects of asbestos, improved guidance in health matters, and measures to address asbestos in drinking water and baby powder. The fifth principle centers around frameworks for assistance, and calls for a study of programs which finance measures to prevent asbestos dust. The sixth principle consists of lobbying the national government and private industry, and the seventh consists of promotional campaigns aimed at Tokyo's citizens.

12.3.1.2 Measures to Prevent Asbestos Dust Dispersal During Building Demolition or Renovation

In 1989, the Tokyo Metropolitan Government established its Guidelines for the Prevention of Asbestos Dust Dispersal from Construction Projects, and specifically from projects involving building demolition or renovation work, and implemented anti-dust measures in collaboration with municipality governments in the Tokyo Metropolis. The measures applied to structures that utilized spray-on asbestos coatings, asbestos insulation, and molded panels containing asbestos. In 1990, the prefecture next established a set of Administrative Guidelines for the Prevention of Asbestos Dust Dispersal from Construction Projects. In 1994, it revised its Tokyo Metropolitan Pollution Prevention Ordinance (enacted January 1995), adding new provisions for measures against asbestos dust, and thus reinforcing its ordinance-based regulatory framework. In December 2000, the Tokyo Metropolitan Government completely revised this ordinance, and established the Ordinance on Environmental Preservation to Secure the Health and Safety of Tokyoites (the "Environmental Preservation Ordinance"). This was enacted in April 2001, thus superceding the provisions of the Pollution Prevention Ordinance. The Environmental Preservation Ordinance covers fabricated structures in addition to conventional buildings. Notifications are required if the building or structure uses at least 15 m^2 of sprayed asbestos coating, or has total floor space of at least 500 m^2 and uses sprayed asbestos coatings or insulation materials. Although structures with molded asbestos paneling do not need to be reported, compliance with operating standards is mandatory. In addition, builders are obligated to measure environmental concentrations of asbestos prior to, during, and after construction projects.

Later, in 2006, the provisions of the Air Pollution Control Law and its related enforcement ordinances and regulations were revised. However, provisions enacted that March no longer contained scale criteria (thus assuming a broader scope than the Tokyo ordinance), expanded the criteria for categories of special construction materials to include insulation and fireproofing coatings with an asbestos content of 1% or more, and added a work process display requirement to the operating standards. Further, under provisions enacted the following October, the scope of the regulations was widened to cover fabricated structures in addition to buildings, while the asbestos content rate was adjusted to 0.1% or more. Acting on these revisions to the Air Pollution Control Law, in March 2006, the Tokyo Metropolitan Government revised the provisions of its Environmental Preservation Ordinance, and sought to streamline the notification process by consolidating into one office the acceptance of notifications submitted under provisions of either the Law or the Ordinance. With these revisions, the provisions of the Tokyo ordinance, other than those relating to molded paneling and environmental measurements, were ultimately assimilated into the Air Pollution Control Law (Table 12.2). Conversely, this meant that from 1995 until the Air Pollution Control Law was revised in 2006, independent regulations by the Tokyo Metropolitan Government covered areas outside the scope of national law. As such, they may be seen as measures that harnessed the prefecture's historical experience and expertise with pollution countermeasures,

Table 12.2 A comparison of provisions in the Tokyo Environmental Preservation Ordinance and the Air Pollution Control Act for measures against asbestos dust

Scope of regulation				Environmental preservation ordinance			Air Pollution Control Law		
		Scale criteria							
Facility	Construction material	Total floor space (for fabricated structures, horizontal projected area)	Spray-on asbestos surface area	Compliance with work standards	Notification	Environmental measurement	Compliance with work standards	Notification	Measurement of environmental concentrations
Buildings	Spray-on coating	500 m^2 or more	(N/A)	P	P	P	P	P	–
		Less than 500 m^2	15 m^2 or more	P	P	P Visual check	P	P	–
			Less than 15 m^2	P	---		P	P	
	Insulation	500 m^2 or more	(N/A)	P	P	P	P	P	–
		Less than 500 m^2	(N/A)	P	---	Visual check	P	P	
	Molded panel	(N/A)	(N/A)	P	---	Visual check	---	---	---
Fabricated structures other than buildings	Spray-on coating	500 m^2 or more	(N/A)	P	P	P	NP	NP	---
		Less than 500 m^2	15 m^2 or more	P	P	P	NP	NP	---
			Less than 15 m^2	P	---	Visual check	NP	NP	
	Insulation	500 m^2 or more	(N/A)	P	P	P	NP	NP	---
		Less than 500 m^2	(N/A)	P	---	Visual check	NP	NP	
	Molded panel	(N/A)	(N/A)	P	---	Visual check	---	---	---

Notes:
1. Buildings are structures as defined in Article 2 Clause 1 of the Building Standards Law
2. P: Subject to applicable provisions. (NP: denotes items covered in the October 1, 2006, enactment of the revised law.), ---: No applicable provisions

Source: Haga (2007:80)

and enabled it to fulfill its local government responsibilities. Nevertheless, they have been cited for deficiencies in terms of ascertaining the status of asbestos countermeasures for buildings, providing for asbestos concentration measurements and management, and regulating asbestos removal projects. Therefore, these regulations present significant challenges, along with the development of legal frameworks at the national level (Natori and Toyama 2009:61–65).

The Tokyo Metropolitan Government is currently giving priority to measures for the handling of molded asbestos paneling. Judging from current trends in the total floor space of newly constructed buildings in Japan, the volume of asbestos imports, and the number of residential buildings in the Tokyo Metropolis, it is projected that there will be an increase in the number of demolition projects for buildings incorporating asbestos materials that were built during the nation's economic boom years. Measures to address that scenario have become urgent. According to the documents of the Japan Asbestos Association, molded asbestos paneling was used in over 90% of all asbestos-containing buildings constructed in 1972, the year that the use of sprayed asbestos coatings reached its peak. Tokyo has set up a study group on the hitherto inadequate measures to deal with molded asbestos paneling, and completed a Manual of Countermeasures for Molded Asbestos Paneling in March 2007.

12.3.1.3 Asbestos Countermeasures by the Tokyo Metropolitan Government Since 2005

Acting in response to the Kubota Shock, in July 2005 the Tokyo Metropolitan Government convened a Conference for the Promotion of Asbestos Countermeasures. The conference explored strategies against asbestos, and compiled and released a document entitled "The Tokyo Metropolitan Government Policy on Asbestos." Table 12.3 lists those countermeasures along with follow-up actions taken at a later date.

As the table shows, the Tokyo Metropolitan Government has implemented numerous measures since the Kubota Shock. Measures for Tokyoites range from improved access to counseling services (health, environmental, and work-related counseling), expanded publications and information disclosure, and the opening of a specialist asbestos outpatient center at the Tokyo Metropolitan Hiroo Hospital. In addition to the aforementioned expansion of measures against asbestos dust, the measures for buildings include the acceptance of dust-prone scrap asbestos building materials for processing at Tokyo metropolitan landfill sites, supra-agency follow-up surveys and reports on asbestos use in buildings owned by the Metropolitan Government, the implementation of construction projects for the prevention of asbestos dust, field surveys to identify the extent of sprayed asbestos use in privately owned buildings, and the utilization of subsidy frameworks to offset the cost of school construction projects.

To foster closer cooperation with municipalities in the Metropolis, the Tokyo Metropolitan Government has set up an asbestos measures task force, assisted with the provision of counseling services at the municipality level and technical support

Table 12.3 Tokyo Metropolitan Government asbestos measures: frameworks and implementation status (as of August 1, 2007)

		Details	Office with jurisdiction	Implementation status
1. Measures for Tokyo's citizens	Improved access to counseling services (health-, environment-, work-related)	Preparation of counseling center manuals (FAQs)	Environment Bureau	Started August 2005
		Streamlining of government office window services (to minimize red tape)	Environment Bureau; Welfare Insurance Bureau	Ongoing
		Revision of Asbestos Primer and Guidance Manual	Welfare Insurance Bureau	Started May 2006
	Intensive promotional publication activity and information disclosure	Posted on "Asbestos FAQ" website	Environment Bureau	Started August 2005
		Preparation of pamphlet for Tokyo asks, "Do You Know about Asbestos?"	Bureau of Citizens, Culture and Sports; Environment Bureau	Started September 2005 (revised April 2007)
		Launch of Tokyo Asbestos Information Website	Environment Bureau	Started September 2005
	Development of health-care frameworks	Launch of Tokyo Metropolitan Hiroo Hospital Asbestos Outpatient Center	Hospital business headquarters	Started November 2005
2. Building-related measures	Strengthened asbestos dust prevention measures for building demolition, renovation, and scrap material handling operations	Spot inspections of demolition project sites in compliance with the Air Pollution Control Act and other legal provisions	Environment Bureau	Ongoing
		Environmental assessments of areas around demolition sites	Environment Bureau	Ongoing
		Complete guidance during legal filing procedures	Environment Bureau; Bureau of Urban Development	Ongoing
		Revision of Asbestos Dust Prevention Manual (covering sprayed asbestos, insulation and other construction materials containing fluffy or dust-prone asbestos)	Environment Bureau	April 2006

	Preparation of Asbestos Dust Prevention Manual (covering molded asbestos paneling and other construction materials containing non-dust-prone asbestos)	Environment Bureau	March 2007
	Total compliance with proper disposal measures for scrap asbestos materials	Environment Bureau; Bureau of Urban Development	Ongoing
	Acceptance of dust-prone asbestos scrap materials by the Metropolitan Government-run landfill sites	Environment Bureau	Started February 2006
Coverage of privately owned buildings	Field surveys concerning the use of sprayed asbestos applications	Bureau of Urban Development	Implemented July through November 2005
	Counseling, guidance, and recommendations pertaining to specific types of buildings	Welfare Insurance Bureau; Environment Bureau	Ongoing
	Preparation and promotion of Building Asbestos Inspection Handbook	Environment Bureau	Started September 2005
	Provision of financing for asbestos countermeasures taken by small and medium-scale companies	Bureau of Industrial and Labor Affairs	Started November 2005
Coverage of city-owned facilities	Implementation and publication of findings of supra-agency follow-up survey on asbestos use	All offices; Environment Bureau (Secretariat)	September 30, 2005 (announced in October) March 31, 2006 (announced in April) March 31, 2007 (announced in May)
	Implementation of asbestos dust prevention projects reflecting findings of follow-up survey	All offices	Started October 2005

(continued)

Table 12.3 (continued)

	Details	Office with jurisdiction	Implementation status
Coverage of schools and other public facilities	Establishment and operation of subsidy framework to help offset project costs	Bureau of Citizens, Culture, and Sports (private schools); Welfare Insurance Bureau (day-care centers, etc.)	Started FY 2005 Started FY 2006
3. Collaboration with Municipalities	Field surveys of public facility asbestos use; technical assistance with asbestos countermeasures upon building demolition	Environment Bureau	Ongoing
	Counseling support for citizens and businesses	Environment Bureau	Ongoing
	Creation of Asbestos Officers Liaison Conference	Environment Bureau	Started April 2006
4. Environmental surveys	Measurements of airborne asbestos concentration (in Koto City, Shinjuku City, and Tama City)	Environment Bureau	Started October 2005

Source: Tokyo Metropolitan Government

for the implementation of field surveys on asbestos use in public facilities, and the formulation of demolition plans.

However, there are problems with the Tokyo Metropolitan Government's countermeasures, as well as hurdles that must be addressed. First, the measures for identifying new victims, implementing field surveys (epidemiological studies), keeping records on health risks, and providing continued care are either too weak or are nonexistent.[3] Second, the building-related asbestos countermeasures have been marked by various deficiencies, chiefly in terms of identifying conditions in the field, measuring and managing asbestos concentrations, and regulating asbestos removal projects. Furthermore, concerns have been raised about noncompliant building demolition projects, and there is a risk that increasing demand for such projects in the years ahead will only aggravate the problems with asbestos.[4] Third, the Tokyo Metropolitan Government also handles MLIT surveys of privately owned buildings, but their scope is limited to buildings of 1 000 m^2 or larger built between 1956 and 1989, in line with MLIT standards, and as such, they are not adequate.

12.3.2 Nerima City Measures

This section is a case study of Nerima, a municipality in the Tokyo Metropolis that has put special emphasis on asbestos-related countermeasures.[5]

Independent surveys of city-administered facilities are one of the interesting features of the Nerima approach. In 1987, Nerima conducted field surveys of sprayed asbestos use at school facilities (built before 1975) within its boundaries, and by 1988 had launched projects to remove asbestos from local structures. However, in 2002 it was determined that some city-administered facilities had been overlooked by previous sprayed asbestos removal work. This prompted a new survey of facilities built prior to 1981, which identified several in which sprayed asbestos had been used. Further inspections revealed that sprayed asbestos coatings had also been used in certain structures built in 1984. Hence, in September 2003 Nerima established a Nerima City Asbestos Countermeasures Commission headed by the then deputy mayor, and the following month issued a set of Guidelines for the Removal of Materials Containing Asbestos, and launched a follow-up survey of sprayed asbestos coatings and other spray-on materials containing asbestos in all city facilities built before 1996. That survey found 72 city facilities in which asbestos, including materials with less than 1% asbestos content, had been used. Asbestos removal projects were initiated and completed at all these facilities in the period from FY 2003 through FY 2005.

Also worth citing is the fact that Nerima has formulated and implemented its own master plan and ordinances. In May 2004, the city put together its Nerima City Master Plan for Asbestos Countermeasures, the key objectives of which are as follows. First are curbs on asbestos use, with construction materials containing asbestos not to be used at all in principle. Second are preventive steps against the dispersal of asbestos dust, and to that end, the plan includes policies for the prevention of indoor asbestos exposure, and the removal of asbestos materials at city facilities where the use of sprayed asbestos materials has been verified, as well as provisions for the handling of damaged asbestos materials, new construction projects at city facilities, and the proper disposal of scrap materials. One particularly noteworthy feature of the Nerima approach in this context is that it has basically opted for the complete removal of spray-on asbestos materials rather than relying on the modalities of containment or enclosure. The third key objective of the master plan consists of information disclosure and sharing, with efforts to familiarize facility administrators and users with information on asbestos. Fourth, the plan calls for citizens to curb their use of asbestos materials, and for construction contractors to cooperate with the requirements for the notification and proper disposal of asbestos materials. As its fifth objective, the plan stipulates that city agencies should share in the role of administering asbestos countermeasures. (Role-sharing details are listed in Table 12.4.)

To further strengthen its regulatory control over privately owned buildings, in January 2006 the City of Nerima enacted an ordinance against the release of asbestos dust. The key features of this ordinance are as follows. First, it contains regulations that apply to structures in use, and mandates that owners of specified buildings (entertainment facilities, department stores, retail stores, offices, schools, hospitals, and other structures with a floor space of 500 m^2 or more) must examine those structures for spray-on coatings that may present an exposure risk to large numbers of people, and, if asbestos is determined to have been used, implement removal, containment, or enclosure procedures accordingly. Second, it incorporates regulations that apply at the time of building demolition or renovation, but with a broader coverage than comparable Tokyo Metropolitan ordinances. Notification is mandated whenever spray-on materials or insulation materials containing asbestos are used in a building or fabricated structure, regardless of surface area, and whenever molded asbestos paneling is used in buildings or structures with total a floor space of 80 m^2 or more (horizontal projected area for fabricated structures) (Table 12.5). In addition to these notification requirements, the ordinance also mandates that project managers should set up project sign boards, hold project briefing sessions for local residents (if a project is for a structure with an affected floor space of 500 m^2 or more), and, if warranted, measure concentrations of asbestos dust in the air. Third is a provision for the city government to prepare and administer a registry of buildings etc. that utilize sprayed asbestos. This was included on the assumption that such information will be useful in formulating measures to be taken in the event of a disaster. Fourth is the provision to disclose the identity of city ordinance violators publicly when they fail to follow city advisories, are negligent in submitting documentation required by the ordinance, or refuse on-site inspections by city officials.

In April 2007, the City of Nerima implemented a new financial assistance framework for asbestos countermeasures. This was a time-limited, 3-year program lasting from April 2007 to March 2010, which was designed to subsidize a share of the cost associated with projects to remove sprayed and other types of asbestos from privately owned buildings. Subsidies under this program were available to private

Table 12.4 Asbestos countermeasures and agency role-sharing arrangements in Nerima City

Objective	Countermeasure specifics	Methods	Responsible city offices
Curbs on asbestos use	Curbs on use	(1) Add provisions to spec sheets calling for the use of construction materials that do not contain asbestos	Business Dept., Project Management Dept.
		(2) Due consideration at the design stage	Business Dept.
Asbestos dust prevention	Preventive measures against asbestos dust exposure within city facilities	(1) Education of facility administrators (2) Education of facility users (3) Surveys of dust levels (4) Public announcement of asbestos removal projects (5) Periodic inspection of asbestos material degradation until removal	Schools (kindergarten, elementary, and junior high schools): Facility Dept., affected schools, and Business Dept. City facilities for citizens: city offices with jurisdiction, affected facilities, and Business Dept.
	Handling of damaged asbestos materials	(1) Initial measures to be taken by facility administrators (2) Follow-up processing	Affected city facilities, elementary and middle schools, kindergartens
	Measures for city facility projects	Identify locations of asbestos use in advance and conduct sample analyses	Business Dept., Project Management Dept.
	Prevention of air pollution	Familiarize city offices with statutes on asbestos removal projects	Environmental Protection Dept., Business Dept., Project Management Dept.
	Proper disposal of scrap materials	Include comments in spec sheets explaining proper disposal methods for scrap materials that do not contain asbestos	Business Dept., Project Management Dept.
Information sharing	Information disclosure and sharing	(1) Education of facility administrators (2) Education of facility users	Facility Management Dept., affected city facilities, elementary and junior high schools, kindergartens, Business Dept.

(continued)

Table 12.4 (continued)

Objective	Countermeasure specifics	Methods	Responsible city offices
Education of citizens	Curbs on asbestos use by citizens	• Asbestos use plans to be determined upon application for expansion or renovation of existing structures • Requests for asbestos removal and utilization of alternative materials	Construction Dept., Environmental Protection Dept.
	Requests for contract cooperation with notification requirements and proper disposal methods	Educate relevant organizations on details of applicable regulations	Environmental Protection Dept.

Source: Nerima City, *Nerima City Master Plan for Asbestos Countermeasures* (May 2005), 6

citizens and small and medium-scale businesses, as well as condominium and apartment management companies. Although a loan mediation program and an amortization interest subsidy framework to cover the cost of inspections and asbestos removal already existed, the City of Nerima had decided that a new financial assistance framework was needed in order to mitigate the heavy expense associated with the removal of asbestos materials from buildings, a strategy it preferred to the alternatives of containment or enclosure. For single-family homes and condominiums, the new framework subsidized one-half of the cost (up to 3 million yen) of asbestos removal projects implemented in compliance with the provisions of the Building Standards Law, and two-thirds of the cost (up to 4 million yen) of projects implemented voluntarily. For rental apartments and office space, it subsidized one-half of the cost (up to 10 million yen) of removal projects implemented in compliance with the provisions of the Building Standards Law, and two-thirds of the cost (up to 15 million yen) of projects implemented voluntarily. While this rates as an exceptionally proactive approach to implementing a framework that boosts the incentives for voluntary asbestos removal projects in privately owned buildings, the program has yet to build a strong record of accomplishment. Although attention will probably be focused on future outcomes, the record thus far seems emblematic of the difficulties surrounding measures for privately owned buildings.

When it comes to asbestos countermeasures, the City of Nerima has been one of the more enthusiastic local governments, even in terms of budgeting. It has put the emphasis on voluntary programs and asbestos removal, and its subsidy framework comprises actions that deserve attention. However, even Nerima faces obvious challenges and problems. In 2008, tests of public facilities in metropolitan Tokyo detected three "new types" of asbestos, including tremolite, that were previously thought never to have been used in Japan. That prompted the City of Nerima to conduct an investigation of its own public facilities, resulting in the detection not

Table 12.5 Correlations between Nerima City's asbestos dust prevention ordinance and other laws and statutes

Scope of regulations				As of October 1, 2006		
				City ordinance	Air Pollution Control Law	Environmental preservation ordinance
				Established	In force	In force
		Scale requirement				
Facility	Material	Total floor space (Note 1)	Surface area of materials containing asbestos	Notification on project plans		
Buildings (covered by Article 2 Clause 1 of the Building Standards Act)	Spray-on materials	500 m^2 or more	50 m^2 or more	---	✓	✓
			Under 50 m^2	---	✓	✓
		Under 500 m^2	15 m^2 or more	---	✓	✓
			Under 15 m^2	---	✓	---
	Insulation materials	500 m^2 or more	N/A	---	✓	✓
		Under 500 m^2	N/A	✓	✓	---
	Molded asbestos paneling	80 m^2 or more	N/A	---	---	---
		Under 80 m^2	N/A	---	---	---
Fabricated structures other than the above	Spray-on materials	500 m^2 or more	N/A	---	✓	✓
		Under 500 m^2	15 m^2 or more	---	✓	✓
			Under 15 m^2	---	✓	---
	Insulation materials	500 m^2 or more	N/A	---	✓	✓
		Under 500 m^2	N/A	---	✓	---
	Molded asbestos paneling	80 m^2 or more	N/A	✓	---	---
		Under 80 m^2	N/A	---	---	---

Notes:
1. For fabricated structures, horizontal projected area
2. ---: Notification is not required

Source: Nerima City Office materials

only of tremolite, but also of the three "conventional types," including crocidolite, thus calling into question the credibility of past analytical surveys. This problem has arisen not only in Nerima City, but in other local governments as well (Ibe 2008:72–75).

12.4 Osaka Prefecture Measures Against Asbestos

Osaka Prefecture's asbestos problems can be traced back prior to World War II. Its Sennan district had long been a center of textile industry operations, but after the introduction of asbestos spinning and weaving methods by Seiki Sakaeya, the district became a key hub of the asbestos textile industry. From that time to World War II it served as a supplier to the defense sector, but in the postwar era its operations shifted their focus to satisfying the demands of the civilian sector. At its peak in the 1970s, the Sennan district comprised around 200 textile manufacturers, including small-scale operations, and engaged a worker population of around 2 000. Sennan was traditionally a farming district and also underwent residential development. As such, it has been described as a mixed industrial, agricultural, and residential zone, all areas of which faced the risk of exposure to asbestos hazards. Most of the businesses forming the local asbestos industry were small-scale operations that had not installed dust-collection equipment or taken other protective measures. As a consequence, not only workers, but also their families and other local residents faced exposure to the health risks associated with asbestos (Yuoka 2006:21–24).

Also, many new structures were built in Osaka Prefecture around the time it hosted the World Expo, which coincided with a period of heavy sprayed asbestos use. As a consequence, today many structures in the prefecture are contaminated with asbestos materials and will be demolished in the years ahead (Nakajima 2006:8). Granted these factors, Osaka will probably face even more pressure than other local governments to implement asbestos countermeasures.

12.4.1 Asbestos Pollution Countermeasures in Osaka Prefecture to the End of the 1970s

The Osaka Prefectural Government was quicker than other local governments to implement asbestos countermeasures of its own. In 1966, it established an ordinance for the prevention of workplace pollution, added asbestos processing facilities to the scope of its notification mandate, and sought to regulate and provide administrative guidance to asbestos factories by applying a site boundary standard dust concentration of 0.5 mg/m^3 (Pollution Disputes Coordination Commission 2007:148).

In 1971, the prefecture established an ordinance for pollution control that incorporated asbestos countermeasures. This new ordinance sought to regulate smoke and dust emissions by applying site boundary standards, and asbestos dust was included within the scope of the regulations. The standard in effect at that time was based on weight: at 0.1 mg/m^3. However, in itself, the fact that the regulations applied site boundary standards demonstrated that the Osaka Prefectural Government was already aware of the risk of asbestos pollution by that time. Indeed, workplace disasters involving asbestos-dust-induced asbestosis, especially among workers in asbestos factories in the Sennan district, had been reported even before World War II. It thus seems reasonable to assume that Osaka officials were to a certain extent aware of the health risks associated with environmental exposure to asbestos.[6]

Regulations for asbestos factories consisted primarily of guidance aimed at making operating facilities install (bag-filter type) dust-collection equipment. However, the effectiveness of Osaka Prefecture guidance in this area remained questionable. With only a few exceptions, in the 1970s the vast majority of asbestos-related factories in the Sennan district were extremely small business operations, and most lacked the funding, knowledge, and technology to implement measures for workplace health, safety, or pollution control. According to the Sera report, surveys by the Osaka Labor Standards Bureau found that the share of surveyed asbestos manufacturing operations that had localized emissions control equipment had improved from 42.4% in 1967 to 85.8% by 1971. However, they also reportedly found that many asbestos textile workplaces involved in the bale opening and carding processes had measured environmental air asbestos concentrations of 2 mg/m^3. Furthermore, not all localized exhaust ventilators installed in these facilities were equipped with dust-collection systems. In a 1981 survey, the Osaka Labor Standards Bureau reported that half of the surveyed workplaces involved in the manufacture and/or handling of asbestos products installed localized exhaust ventilation systems in only one location. The installation rate for such systems among facilities involved in the cutting or processing of asbestos products was low, and the dust-collection equipment installation rate was close to 50% (Sera 1983:91–93). Also, in an account given by a public official who administered the Sennan district in the second half of the 1970s, Osaka Prefecture's regulation of, and administrative guidance for, asbestos product manufacturing facilities became full-scale in the late 1970s and thereafter, and such facilities were advised to begin measuring dust concentrations in line with site boundary standards, and install bag-filter dust-collection systems. However, pollution regulations were limited to fines for violators and thus lacked force, with some factories even defying administrative guidance, and others removing the hoods from their dust-collection systems, thus rendering the asbestos countermeasures pointless (Yoshida 2006).

Although its policies were marked by the aforementioned deficiencies, the Osaka Prefectural Government nonetheless deserves distinction for pursuing its own independent efforts, and on that basis encouraging more members of the industry to utilize special lending frameworks for the installation of dust-collection systems.

12.4.2 *Asbestos Pollution Countermeasures in Osaka Prefecture from the 1980s*

12.4.2.1 Measures to Prevent Asbestos Dust Dispersal in Factories and Other Structures During Building Demolition Operations[7]

In 1987, the Osaka Prefectural Government set up an asbestos countermeasures study commission in response to growing public concerns over the issue of sprayed asbestos in school structures, and from 1987 to 1988 conducted field surveys to determine the extent of asbestos use in prefecture facilities. In 1989, it established the Osaka Prefecture Master Policy on Asbestos Countermeasures, and decided to pursue a comprehensive range of fundamental initiatives, including controls on manufacturing facility emissions, measures against sprayed asbestos, measures to deal with construction materials, scrap materials, and any other asbestos-containing product, environmental assessments and other steps to identify the health impact of asbestos, field surveys of asbestos use, health counseling services for citizens, financing and subsidy frameworks for asbestos removal projects, steps to promote the use of alternative materials, and educational measures to ensure that construction workers and the general public were better informed about asbestos. Acting in line with a set of interim policies on measures to deal with sprayed asbestos, as well as a set of provisional guidelines for the handling of construction materials that contained asbestos, the prefecture also sought to provide guidance and education for the owners and tenants of affected buildings.

Osaka decided to strengthen its regulations for emissions from asbestos manufacturing facilities after 1989 revisions to the Air Pollution Control Law made notifications and site boundary dust measurements mandatory for industrial sources of specified types of dust (asbestos). Asbestos dust emissions were controlled at levels below the site boundary standard of 10 particles per liter, and from 1989 to 1993, the regulatory framework comprised a dual-layered structure of laws and ordinances. The Osaka Prefectural Government also decided to regulate small-scale pieces of equipment that were at that time outside the scope of applicable law, and utilized a Manual of Guidelines for the Reduction of Asbestos Product Factory Asbestos Dust Emissions, which it had prepared in 1989, to provide factories with guidance and education.

In 1994, the Osaka Residential Environment Protection Ordinance was established, superceding the prefecture's pollution control ordinance. This effectively streamlined procedures by providing a better definition of the scope of applicable laws and ordinances, and brought prefectural standards into line with the asbestos concentration standard (10 particles per liter) applied by the Air Pollution Control Law.

In 1995, the devastation from the Great Hanshin-Awaji Earthquake prompted consideration for the issue of asbestos countermeasures to be taken during structural demolition operations. Under the revised Air Pollution Control Law, in 1996 demolition projects were brought within the scope of asbestos-related regulations. Specifically, provisions were established to regulate "specified dust-emitting operations" engaged in the demolition, renovation, or repair of buildings that utilized sprayed asbestos. However, they were limited in their scope. The buildings subject to regulation were limited to fireproofed and semi-fireproofed structures, and the specified materials for regulation were limited to sprayed asbestos coatings (including rock wool with an asbestos content of over 1%). Scale criteria also limited the scope of the notification requirement to structures with a total floor space of 500 m^2 or more, and sprayed asbestos surfaces of 50 m^2 or more.

12.4.2.2 Asbestos Pollution Countermeasures in Osaka Prefecture Since 2005

In February 2005, Osaka Prefectural Government reorganized its asbestos countermeasures study commission into the asbestos countermeasures liaison committee, and the following April revised its Master Policy on Asbestos Countermeasures. In July 2005, news of the Kubota Shock prompted the prefecture to create an Osaka Prefecture Asbestos Countermeasures Task Force headed by the deputy mayor and including individual departmental managers as members. Three task force subcommittees were then set up and assigned the missions of exploring environmental measures, health measures, and prefecture facility management measures.

The asbestos measures pursued by the Osaka Prefectural Government can be roughly broken down into the following categories: (1) broad-based measures, information dissemination, and education (including the creation of asbestos hotlines and the distribution of leaflets to citizens); (2) environmental measures (including dust-prevention measures for building demolition operations, measurements of asbestos dust concentrations in the air, manufacturing facility spot inspections and guidance, measures for privately owned structures, and measures for scrap materials); (3) health measures (health counseling services, emergency examinations for lung cancer, field surveys on the health impact of asbestos, health risk assessments, and centers for the acceptance of relief applications); (4) measures in prefectural facility management (surveys of asbestos use in prefectural facilities, guidelines for the management of prefectural facilities, implementation of an inspection labeling framework).

Following a review to determine whether the provisions of the Air Pollution Control Law were adequate to the task of controlling the release of asbestos dust during building demolition operations, in October 2005 Osaka Prefecture decided to revise its ordinances and significantly expand their scope of regulation. Regulatory standards were strengthened and site boundary standards were applied to demolition projects. Osaka Prefecture was the only local government then pursuing ordinance-based regulations in this area and subjecting violators to fines.

Although revisions to the Air Pollution Control Law in 2006 incorporated provisions for the regulation of materials other than molded paneling, the prefecture's local ordinance-based regulations for molded paneling remained in force along with local regulations stipulating environmental measurements. The provisions of the Osaka Residential Environment Protection Ordinance and the Air Pollution Control Law are summarized in Table 12.6.

Prior to the revision of these laws and ordinances, the Osaka Prefectural Government received of the order of 50–60 notifications per year. However, that number rose sharply thereafter, reaching 630 notifications (including 156 ordinance-stipulated notifications) in FY 2005, and over 1 000 notifications in FY 2006 (Table 12.7).

The Osaka Prefectural Government has also taken various steps in addition to its independent efforts at regulation. It has launched a conference, with industry participation, on the promotion of measures against asbestos dust, sponsored seminars on related themes, conducted "demolition site patrols" in collaboration with designated cities, and set up a special lending framework for asbestos countermeasures. It has adopted a policy of publishing the names of violators, and has disclosed such names in four cases thus far.

In FY 2005, the Osaka Prefectural Government conducted a field survey of asbestos use in facilities owned and managed by the prefecture, and has been pursuing countermeasures one-by-one at each of the 146 facilities that were determined to have used materials containing sprayed asbestos.

Nevertheless, Osaka Prefecture's measures for buildings face certain challenges. A renewed survey of prefecture-owned facilities for three "new types" of asbestos detected tremolite, as well as conventional chrysotile, in several structures, thus placing the credibility of previous field surveys into question. What is more, as of September 2009, survey reports had yet to be received on 14% of the privately owned buildings that are also subject to asbestos-related field surveys concerning the use of sprayed asbestos. Furthermore, no action had been taken on 46% of the buildings found to have exposed surfaces of sprayed asbestos.[8]

12.4.2.3 Asbestos-Related Health Surveys and Health Risk Surveys in Osaka Prefecture[9]

To mitigate the asbestos-related health concerns of prefecture citizens following the Kubota Shock, the Osaka Prefecture, as well as designated cities and other cities establishing health centers (specified by a cabinet order based on the provisions of the Community Health Law), examined 2 020 citizens during a combined total of 122 clinics held to conduct emergency examinations for lung cancer at prefecture-managed health centers (jointly operated with local municipalities), as well as health centers in the cities of Osaka, Sakai, Higashiosaka, and Takatsuki, from October 2005 through February 2006. Examinations comprised patient interviews, chest X-rays, sputum analysis, and health counseling at prefectural health centers and facilities in Takatsuki, and patient interviews, chest X-rays, and health counseling at health centers in the cities of Osaka, Sakai, and Higashiosaka. Further evaluation was deemed necessary for approximately 20% of all patients examined, which is a rather high share, and of those that underwent more advanced testing, 24% received a diagnosis of asbestos-induced illness (including suspected cases).[10]

Table 12.6 Osaka Prefecture regulations governing the demolition and renovation of buildings and fabricated structures

Scope of regulation			Osaka Residential Environment Protection Ordinance			Air Pollution Control Law		
Facility	Material	Scale criteria (applied surface area)	Compliance with work standards	Notification	Environmental measurement	Compliance with work standards	Notification	Environmental measurement
Buildings	Spray-on materials	50 m² or more	✓	✓	✓	✓	✓	---
		Under 50 m²	✓	✓	✓	✓	✓	---
	Insulation materials	50 m² or more	✓	✓	✓	✓	✓	---
		Under 50 m²	✓	✓	---	✓	✓	---
	Molded asbestos paneling	1 000 m² or more	✓	✓	---	---	---	---
		Under 1 000 m²	✓	---	---	---	---	---
Fabricated structures	Spray-on materials	50 m² or more	✓	✓	✓	✓	✓	---
		Under 50 m²	✓	✓	---	✓	✓	---
	Insulation materials	50 m² or more	✓	✓	✓	✓	✓	---
		Under 50 m²	✓	✓	---	✓	✓	---
	Molded asbestos paneling	1 000 m² or more	✓	✓	---	---	---	---
		Under 1 000 m²	✓	---	---	---	---	---

Note: ✓: Subject to applicable provisions, ---: No applicable provisions
Sources: Osaka Prefecture documentation and interviews with the Osaka Prefecture Environmental Management Office

Table 12.7 Notifications pertaining to building demolition projects and other operations in Osaka Prefecture that generate specified types of dust (asbestos)

Fiscal year	1997	1998	1999	2000	2001	2002	2003	2004	2005	2006
Notifications pursuant to Air Pollution Control Law	45	58	57	57	64	53	52	69	474	980
Notifications pursuant to prefectural ordinance	–	–	–	–	–	–	–	–	156	157
Total	45	58	57	57	64	53	52	69	630	1 137

Note: Prefectural totals including designated cities. FY 2006 totals are for the period from April 1, 2006, to January 31, 2007
Source: Osaka Prefectural Government

Prefecture health centers compiled data on the attributes of patients who underwent emergency examinations for lung cancer. For 1 138 patients (67% men, 33% women) who underwent examinations during one of the 52 clinics held from October through December 2005, those attributes were as follows. A breakdown by age group showed that senior citizens accounted for a large share of the patient total, with 27% aged 70 or older, 68% aged 60 or older, and 86% aged 50 or older. By district of residence, Kitakawachi accounted for the largest share of patients at 37%, followed by Senshu at 30% and Minamikawachi at 15% (these districts exclude the City of Osaka and three other cities which established health centers). Many patients who underwent examinations had subjective symptoms, with cough (47%) and sputum (43%) being extremely common. In response to queries regarding their possible contact with asbestos, 60% stated that they worked in occupations involving the use or handling of asbestos, 10% had family members who engaged in such occupations, 16% lived near an asbestos factory, and 16% had other opportunities for contact with asbestos, thus highlighting the fact that virtually all patients were at risk of exposure to asbestos. By residential history, 20% had lived near asbestos factories or mining operations, 3% near shipbuilding yards, 2% near brake repair facilities, 6% near stockpiles of construction materials or scrap materials from demolished structures, and 66% had no history of residence near asbestos-related facilities or material stockpiles. By occupation (multiple responses counted), 22% worked in asbestos product manufacturing facilities, and the highest share, 24%, worked in the construction trades, followed by those in occupations ranging widely from demolition, warehousing, boiler making, and electrical equipment manufacturing, to automotive repair, shipbuilding, dockyard work, and scrap material processing and recovery.[11]

Widespread anxiety over occupational diseases underlines the importance of cooperation between local governments and labor standards inspection offices. However, because Japan no longer has any asbestos factories in operation, local governments will be expected to provide all health management services for their citizens, including those with health concerns based on occupational histories.

Expectations are that Osaka Prefecture and its municipalities will share information on the findings from emergency examinations for lung cancer, and that follow-up measures at the municipal level will be implemented in the years ahead. Patients with asbestos-related diseases are currently being referred to three hospitals that are equipped to treat their conditions.

Given that the symptoms of asbestos-induced illnesses first appear several decades after exposure, it is critically important that residents with a health risk are registered and have access to health-management services on a continuing basis. However, registration frameworks for citizens at risk of developing an asbestos-related disease could be problematic in terms of protecting personal information, and presumably will be difficult or impossible to implement unless the national government develops the requisite legal structures. Although health risk surveys are currently performed by Osaka Prefecture and other local governments for the MOE, these are on a 5-year schedule and are not part of a permanent framework.

The Osaka Prefectural Government began performing health risk surveys starting in FY 2006. That year, it registered 309 citizens aged 40 and over with histories of residence in the Sennan district prior to 1989, and at risk of asbestos exposure through ordinary environmental pathways. They also issued a handbook on the management of adverse health effects from asbestos exposure, and continued with the follow-up monitoring of registered patients. In FY 2007, more citizens were added to the registry.[12] However, one problem with health risk surveys is that they are designed to obtain medical data, and do not guarantee continued access to health management services. Although the final term for these surveys will be decided through future discussions by the Osaka Prefectural Government and the MOE, the outlook for measures in health risk management seems uncertain. A general framework for the management of adverse health effects from environmental asbestos exposure should be implemented.

12.4.2.4 Osaka Prefecture Measures Against Asbestos: Appraisal and Future Issues

Osaka Prefecture's measures against asbestos should be credited first of all for regulating and providing administrative guidance to asbestos product manufacturing facilities through master policies and ordinances that were implemented prior to actions by the national government, and for subjecting building demolition and renovation projects to the most stringent set of asbestos regulations in effect at the time. Second, they can be credited for the creation of a hotline that placed emphasis on the communication of risks at the time of the Kubota Shock, and for the fast-track implementation of emergency examinations for lung cancer.

However, these approaches still faced limits, among other problems. As indicated earlier, previous efforts to regulate and administer guidance to asbestos product manufacturing facilities were inadequate, partly due to the detrimental effects of a "sectionalist" or insular mentality, and thus have not been completely successful in preventing the adverse health effects of asbestos. Furthermore, in connection with efforts to identify the extent of harm from asbestos exposure and establish effective health measures, urgently needed epidemiological studies and frameworks for the management of health hazards cannot yet be implemented, due in part to underdeveloped frameworks at the national level. Also, in terms of measures for buildings that contain asbestos, Osaka and Tokyo share various challenges, including deficiencies in the understanding of the actual conditions, deficiencies in the measurement and management of asbestos concentrations, deficiencies in their measures for asbestos removal projects, questions surrounding the legal compliance of contractors in building demolition operations, and issues relating to measures for privately owned buildings. Moreover, in Osaka Prefecture, the framework for financing asbestos countermeasures has virtually no record of accomplishment thus far.

12.5 Closing Remarks

This chapter has primarily explored asbestos countermeasures at the local government level, using the examples of the Tokyo Metropolis (including Nerima City) and Osaka Prefecture as case studies. The principal features of the undertakings pursued by these local governments may be summarized as follows.

First, both prefectural governments acted earlier than the national government in devising asbestos countermeasures for asbestos product manufacturing facilities and building demolition and renovation operations on the basis of ordinances and master plans for administrative guidance. Particularly noteworthy is the fact that the Osaka Prefectural Government pursued its own independent efforts at regulation and guidance as a district with a high concentration of asbestos product manufacturing facilities. Second, Tokyo, Osaka, and other local governments developed organizational frameworks for the implementation of countermeasures on a supra-agency level, and pursued comprehensive undertakings on that basis. Granted the complex nature of asbestos disasters, comprehensive frameworks and efforts of this kind deserve to be described as essential. Third, to address the issue of public anxiety stemming from the Kubota Shock, most local governments pressed forward with emergency countermeasures that placed priority on the communication of risks.

Nonetheless, it became increasingly evident that these actions at the local government level had limits. First, not enough has been done to identify the extent of the harm caused by asbestos to human health. The national and local governments have implemented virtually no epidemiological studies of their own, but without a clear view of the total picture and the underlying causal factors, it is doubtful that they will be prepared to implement adequately effective countermeasures.

Second, mechanisms to identify citizens burdened by health risks and provide them with ongoing health-management services have not been put into place. The situation demands that frameworks for registration and the implementation of periodic health examinations should be set up for citizens at risk of environmental exposure.

Third, not enough has been done to address the issue of spray-on asbestos coatings. Building inspections in 2008 detected tremolite and other forms of asbestos that previously were assumed never to have been used in Japan. That finding raises questions about our understanding of the amount of asbestos actually used in buildings, and places in doubt our ability to develop a viable understanding even with regard to public facilities. Within the context of measures for privately owned buildings, assessments of asbestos use have been limited solely to large-scale structures, and thus do not paint a complete picture of the situation. Furthermore, even with local government-backed subsidy frameworks, such measures have shown little progress because building owners and demolition businesses are in principle expected to bear the related costs. As these examples illustrate, current measures to deal with asbestos in buildings are marked by an array of deficiencies, including a lack of insight into the actual extent of asbestos use, inadequate frameworks for measurement and management, and weak measures for asbestos removal projects.

These challenges demand that local governments take ownership of the problems at hand and pursue independent solutions. However, it is imperative that the national government also take action, lay the legal groundwork, and in other ways set the stage for progress. The sharp increase in demolition project notifications submitted after revisions to the Air Pollution Control Law, which included the withdrawal of building scale criteria and an expansion in the scope of coverage, stands as clear testimony to the importance of efforts at the national level to prepare the legal base. As another example, an epidemiological study in Amagasaki was headed by the MOE with cooperation from Amagasaki itself. It must be stressed that the national government has been hugely responsible for facilitating epidemiological studies concerned with the impact of asbestos on factories and businesses that handled asbestos as well as their surrounding neighborhoods. In the context of measures for privately owned buildings as well, suffice it to say national government-led regulations and countermeasures will be equally essential in identifying the scope of the asbestos problem.

Notes

1. Although not evident at the time of the survey, these are believed to have been put into effect at 2006.
2. Information on the Tokyo Metropolitan Government's asbestos countermeasures is based on Haga, M. (2007), Higuchi, Y. (2007), and interviews with the Tokyo Metropolitan Government's Environmental Improvement Division; as well as submitted materials.
3. In December 2007, Ota City announced that eight citizens living in the vicinity of a decommissioned asbestos product factory had become victims of asbestos-induced illnesses. Prior failure to disclose these cases of harm from asbestos highlights the need for efforts by the Tokyo Metropolitan Government to identify more victims and implement field surveys.

4. Information on Nerima City asbestos countermeasures obtained from Nitta, K. (2006) and materials from the Nerima City Environmental Protection Dept. (October 29, 2007).
5. Information on Nerima City asbestos countermeasures obtained from Nitta, K. (2006) and materials from the Nerima City Environmental Protection Dept. (October 29, 2007).
6. In 1968, the former Ministry of Labor designated sprayed asbestos operations as worksites requiring the regular use of localized ventilation systems to curb dust emissions in accordance with provisions of the Pneumoconiosis Act. Again, in 1971 it made prefectural labor standards bureaus provide administrative guidance aimed at encouraging asbestos workplaces to install localized exhaust ventilation systems. An Ordinance for Prevention of Adverse Health Effects from Specified Chemical Substances has also been put into effect.
7. Information on asbestos policies in Osaka Prefecture derived from Nakajima S. (2006) (2007), S Shirotani, S., and Nakajima, S. (2006), and interviews with the Osaka Prefecture Environmental Management Office (December 7, 2007), as well as submitted materials.
8. Based on press materials released by the Osaka Prefectural Government.
9. Information on Osaka Prefecture asbestos health hazard countermeasures based on interviews with the Osaka Prefecture Office for Community Health and Welfare (December 7, 2007), as well as submitted materials.
10. Osaka District Office for Health and Welfare documentation.
11. Osaka District Office for Health and Welfare documentation.
12. For details on FY 2006 health risk assessments, see Osaka Prefecture's Report of the Findings of Asbestos-Related Health Risk Assessments in Osaka Prefecture, May 2005.

References

Administrative Assessment Bureau, Ministry of Internal Affairs and Communications (2007) Present situation and issues of local government measures against Asbestos (in Japanese). Government of Japan

Haga M (2007) Tokyo metropolis measures against asbestos. J Resour Environ (in Japanese) 43(3):78–85

Higuchi Y (2007) Tokyo metropolis measures against asbestos after the Air Pollution Control Act was revised. Ind environ (in Japanese) 36(5):85–88

Ibe M (2008) Problems caused by the lack of proper measures against asbestos. Nikkei Ecol (in Japanese) 111:72–75

Miyamoto K (2007) Environmental economics new edition (in Japanese). Iwanami Shoten, Tokyo

Nakajima S (2006) Undertakings by Osaka Prefecture to resolve asbestos problems. Mon JREI Rep (in Japanese) 342(343):2–12

Nakajima S (2007) Osaka Prefecture measures against asbestos. Ind Environ (in Japanese) 36(12):25–27

Natori Y, Toyama N (2009) Spray-on asbestos coating: surveys, analyses, measurements, management, and removal, plus remaining hurdles. Res Environ Disruption (in Japanese) 38(4):61–65

Nitta K (2006) Asbestos countermeasures and problems in Nerima city. Local Authority Staff Train (in Japanese) 40(10):64–66

Pollution Disputes Coordination Commission (2007) A basic investigation for resolution of asbestos pollution disputes (in Japanese). Research Center for Pollution Countermeasures

Sera Y (1983) Asbestosis in Osaka. In: Research Group on the History of Occupational Health in Osaka (ed) Report on the history of occupational health in Osaka (in Japanese). Research Group on the History of Occupational Health in Osaka, Osaka, pp 88–96

Shirotani S, Nakajima S (2006) Institutionalization of countermeasures to prevent asbestos dust dispersal during building demolition. Environ Conserv Eng (in Japanese) 35(2):82–88
Yoshida M (2006) Asbestos problems and its implication for prevention of global warming. In: Proceedings of sixth annual meeting, Society of Environmental Conservation Engineering (in Japanese)
Yuoka K (2006) History of asbestos industry and latent damage in Sennnan. Roudou Horitsu Jyunpou (in Japanese) 1617

Chapter 13
Asbestos Litigation in Japan: Recent Trends and Related Issues

Katsumi Matsumoto

13.1 Introduction

The benefits of asbestos, along with the human health hazards associated with inhalation of its microscopic fibers, have been known for upwards of 100 years.[1] However, it was not until after World War II that society became aware of the increasing severity of asbestos-induced health problems. In the 1950s, a paper published by Richard Doll in the British Journal of Industrial Medicine (1955) cited the high risk of asbestos-induced lung cancer in pulmonary asbestosis patients, thus earning distinction as the first document in the medical literature to establish a definitive link between asbestos and lung cancer. In addition, the findings of a large-scale epidemiological study by Irving J. Selikoff of Mount Sinai Hospital in New York, namely, that mesothelioma was caused by asbestos exposure, were delivered in 1964 in a presentation at an international conference on the "Biological Effects of Asbestos" sponsored by the New York Academy of Sciences, and received significant attention. Against this background, litigation involving asbestos-related health problems (hereinafter "asbestos litigation") would become a frequent occurrence in the US from the late 1960s onward.[2]

In Japan, the initial momentum for asbestos litigation proved to be slow, but starting in the 1970s, cases began to appear on a sporadic basis, mainly involving damage compensation claims by victims of diseases stemming from occupational exposure to asbestos, and holding their employers liable for a failure to fulfill their obligation to implement safeguards or for tortios act (see the list of court decisions in Table 13.1).

This chapter strives to provide a comparative overview of the background and features of the wave of asbestos litigation in the US, together with the features of asbestos litigation in Japan, and to highlight several theoretical issues for further study.

K. Matsumoto (✉)
School of Law, Ritsumeikan University, 1 Nishinokyo-Suzaku-cho, Nakagyo-ku,
Kyoto 604-8520, Japan
e-mail: kmt07750@law.ritsumei.ac.jp

K. Miyamoto et al. (eds.), *Asbestos Disaster: Lessons from Japan's Experience*,
DOI 10.1007/978-4-431-53915-5_13, © Springer 2011

Table 13.1 Asbestos lawsuits and court decisions

	Case	Filing date	Court	Date of ruling	Publication of notice	Approved/ dismissed	Amount of compensation awarded per victim (in yen)
1	Nagano asbestos dust pneumoconiosis lawsuit	1977	Nagano District Court	1986.6.27	Hanrei Times, Issue 616, p. 34	*A**	16.50–24.20 million
2	US military Yokosuka base asbestos dust pneumoconiosis lawsuit	1999	Yokohama District Court, Yokosuka Branch	2002.10.7	Hanrei Times, Issue 1821, p. 65	*A*	15.40–27.40 million
3	US military Yokosuka base asbestos dust pneumoconiosis lawsuit	2002	Tokyo High Court	2003.5.27	Hanrei Times, Issue 1111, p. 206	*A**	15.40–27.40 million
4	Misawa Resort case	2000	Tokyo District Court	2004.3.25	Hanrei Times, Issue 1210, p. 150	*D*	
5	Kansai Hoon Kogyo case	1999	Tokyo District Court	2004.9.16	Hanrei Jiho, Issue 1882, p. 70	*A*	56.70 million
6	Misawa Resort case	2004	Tokyo High Court	2005.1.20	Hanrei Times, Issue 1210, p. 145	*D*	
7	Kansai Hoon Kogyo case	2004	Tokyo High Court	2005.4.27	Rodo Hanrei, Issue 897, p. 19	*A*	46.77 million
8	Sumitomo Heavy Industries case	1988	Yokohama District Court, Yokosuka Branch	2006.10.30	Unlisted	*A*	
9	Sapporo International Travel case	2004	Sapporo District Court	2007.3.2	Hanrei Jiho, Issue 1978, p. 41	*D*	
10	Sapporo International Travel case	2007	Sapporo High Court	2008.8.29	Hanrei Jiho, Issue 2029, p. 27	*A*	32.35 million
11	Electrical installation project supervisor case	2007	Tokyo District Court	2009.2.16	Hanrei Jiho, Issue 2051, p. 150	*D*	

12	Building demolition project case	2007	Saitama District Court	2009.3.13	Hanrei Jiho, Issue 2044, p. 123	*M*	100 000
13	US military Yokosuka base lawsuit	2007	Yokohama District Court, Yokosuka Branch	2009.7.6	Hanrei Jiho, Issue 2063, p. 75	*A*	76.84 million
14	Chubu Electric case	2007	Nagoya District Court	2009.7.7	Rodo Keizai Sokuho, Issue 2051, p. 27	*A*	30.00 million
15	Kintetsu case involving sprayed asbestos in buildings under elevated tracks	2005	Osaka District Court	2009.8.31	Hanrei Jiho, Issue 2068, p. 100	*A*	49.44 million
16	Nippon Etarnit Pipe case	2007	Takamatsu District Court	2009.9.14	Unlisted	*M*	10–25 million
17	Mitsui Soko case	2007	Kobe District Court	2009.11.20	Rodo Hanrei, Issue 997, p. 27	*A*	33.66 million
18	Kintetsu case involving sprayed asbestos in buildings under elevated tracks	2009	Osaka High Court	2010.3.5	Unlisted	*A*	59.95 million
19	Watanabe Kogyo case	2007	Osaka District Court	2010.4.21	LEX/DB 25442162	*A*	22.90 million (plus 1.10 million for nursing family members)
20	Osaka Sennan asbestos lawsuit	2005	Osaka District Court	2010.5.19	Hanrei Jiho, Issue 2093, P.3	*M*	8.25–27.50 million

Notes:

1. Compensation amounts awarded are rounded down to the nearest 10 000 yen

2. In the approved/dismissed column: *A*, approval; *D*, dismissal; *M*, mixed approval with some plaintiff claims approved and others denied. An asterisk indicates that a plaintiff's claim was denied due to invocation of the extinctive prescription

13.2 A Comparative Review of Asbestos Litigation in Japan and the US

13.2.1 US Trends

One feature of asbestos litigation in the US is that victims of asbestos-related diseases caused by occupational exposure to asbestos dust have typically filed lawsuits not against their employers, but against asbestos companies that manufactured and sold asbestos products. In the US, this stems from a mixture of factors. One is that the victims of occupational diseases are limited in their ability to hold employers liable for damage compensation separate from, and in addition to, benefits the victims may receive through workers accident compensation insurance frameworks. Another factor is that the Second Restatement of Torts in the 1960s established provisions (Section 402A)[3] that recognize the product liability of a seller of manufactured goods for tangible damages suffered by users or consumers of those goods to be a strict liability (i.e., a liability that arises even if the seller has followed all possible precautions in the preparation and sale of the said product). Furthermore, in the 1980s the public was alerted to the danger that exposure to asbestos dust released through the degradation or demolition of structures incorporating asbestos-containing construction materials could cause asbestos-related diseases in building tenants and residents living nearby. This sparked a flurry of lawsuits against asbestos product manufacturers seeking damage compensation to offset the cost of asbestos removal operations (Yamashita 1988:15–18). Hence, by the 1980s, an estimated 30000 or more cases of litigation had been filed against asbestos companies in the US. Many asbestos firms facing heavy burdens of liability for damages went out of business one after the other, leaving the insurance companies that had insured them against such liability in jeopardy.[4] By some accounts, as of 2002, a combined total of some 730000 individuals had participated as plaintiffs in cases of litigation, whereas a total of 8400 companies had been named as defendants (Yoshida 2006:599).

13.2.2 Features of Asbestos Litigation in Japan

Japan has been described as one of the world's largest consumers of asbestos, having used in the range of 10 million metric tons of the material (Kohyama 2006:20). Needless to say, it seems certain that this gave rise to various asbestos-induced diseases.[5] (In Japan, five types of occupational diseases which are considered eligible for workers accident compensation insurance coverage – pulmonary asbestosis, asbestos-induced lung cancer, mesothelioma, benign asbestos pleural effusion, and diffuse pleural thickening – are treated as diseases in which asbestos has a clear causal relationship.[6] Hereafter, these are referred to collectively as asbestos-related diseases.) The following may be cited as the principal features distinguishing asbestos litigation trends in Japan from comparable trends in the US.[7]

13.2.2.1 A Scarcity of Asbestos Lawsuits

First, there have been far fewer asbestos lawsuits in Japan compared with the US, and on top of that, they have been of a sporadic nature.[8] Furthermore, no more than 20 cases are known to have reached a settlement (see the list of court decisions in Table 13.1). The first known case of asbestos litigation in Japan was the Nagano pneumoconiosis lawsuit launched in 1977. In that case, a group of plaintiffs comprising former workers for asbestos product manufacturing companies and the surviving families of deceased workers sought to have liability for damage compensation assigned to their former employers, to the parent companies thereof, and to the national government for its negligence in implementing and enforcing appropriate regulations and controls. The Nagano District Court's ruling of 1986 (court decision 1 in Table 13.1) was a victory for the plaintiffs. Two years later, in 1988, a group of former shipyard workers with pulmonary asbestosis filed a lawsuit seeking damage compensation from their employer, Sumitomo Heavy Industries, Ltd. Seven years later, in 1995, Sumitomo Heavy Industries faced similar charges. After that no other significant lawsuits were filed, and litigation activity did not gain serious momentum until the end of the 1990s. It is believed that litigation became more pronounced around this time because symptoms of asbestos-related diseases were then beginning to manifest themselves after a long period of dormancy in workers who had been exposed to asbestos dust in the 1960s, the period when asbestos came into heavy use.

13.2.2.2 Efforts to Hold Employers Liable for Occupational Exposure

Second, the majority of lawsuits filed in Japan have been for the purpose of seeking damage compensation from employers for harm sustained as a result of occupational exposure to asbestos. This is attributable to the fact that in Japan, unlike the US, citizens are not legally restricted from pursuing cases against employers for claims of civil liability for occupational diseases covered by workers compensation insurance. In a growing number of cases having to do with workers compensation for occupational exposure, employers have been deemed legally liable, and plaintiffs have won favorable settlements. However, to date, not a single case has yet resulted in a decision which recognized the liability of an asbestos company for environmental exposure or the exposure of family members to asbestos. Further, one reason for the lack of progress in pursuing product liability claims against manufacturers of asbestos products in Japan is that legislation providing specific provisions for product liability was not enacted in Japan until 1994, when the Product Liability Act was established, whereas in the US, those provisions were put into effect with the aforementioned Second Restatement of Torts in the 1960s (Japan's product liability law was brought into effect on July 1, 1995). In fact, collective action lawsuits filed in the Tokyo and Yokohama District Courts by metropolitan Tokyo construction workers (Tokyo Metropolitan Asbestos Lawsuit) in 2008 mark the first instances of litigation aimed at pursuing product liability claims against asbestos product manufacturers in Japan. These lawsuits sought damage compensation from the national government and from 44 leading domestic corporations that were engaged in the manufacture and sale of asbestos construction materials.

13.2.2.3 Pneumoconiosis Lawsuits Pursuing Claims of Employer Default of Safety Obligations

As a third feature, early cases of asbestos litigation in Japan took the form of pneumoconiosis lawsuits concerned with asbestosis health problems stemming from exposure to asbestos dust, and sought to hold employers liable for a failure to fulfill their obligation to implement measures for worker safety. This occurred against the following background. In 1960, Japan enacted its Pneumoconiosis Act, which was a special law designed to regulate work environments in workplaces characterized by defined levels of dust, with steps for the prevention of dust-related illnesses as well as to seamlessly link the management of worker health in such workplaces with measures to facilitate applications for workers accident compensation insurance. During the economic boom years from the mid-1950s onward, as workplace accidents and occupational diseases became increasingly common amid industry cuts in spending on worker safety, labor movements put a high priority on disputes over workplace accidents and occupational diseases, and embraced such slogans as "no work without safety" and "no safety without labor resistance." In addition, to secure relief for victims above and beyond the low-level benefits afforded through workers accident compensation insurance, labor unions sought to organize lawsuits seeking civil liability claims against employers involved in workplace accidents and occupational diseases. In this process, employers were considered to have an obligation not only to pay workers decent wages, but also to ensure the safety of worker lives and health, and the debate over employer safety obligations evolved, with workplace accidents and occupational diseases being perceived as an issue stemming from employers failures to meet obligations for worker safety.[9] Several background factors help to explain why lawsuits that can be utilized to seek liability claims for tortious acts took the form of lawsuits pursuing claims of liability for default. The right to seek damage compensation for acts of tort expires 3 years from the date the victim becomes aware of the damage, or the identification of the offender (Civil Code Article 724, first sentence). By contrast, the right to seek damage compensation for default expires 10 years after the date that the right can be first exercised, which is the period of time normally applied to creditor claims (Civil Code Articles 166 and 167). In other words, the benefit in terms of the prescription may be described as the motivating factor.[10]

13.2.2.4 The Kubota Shock as a Major Impetus Behind the Increase in the Number of Lawsuits

Fourth, the so-called Kubota Shock of June 30, 2005 (the announcement made at a press briefing by Kubota executives that workers for, and residents living near, its former Kanzaki factory in Amagasaki city had contracted asbestos-related diseases[11]), prompted a growing public awareness of the grave risks posed by asbestos to human health, and led to an upsurge in the number of asbestos-related lawsuits immediately thereafter. This was also the year that saw the formation of the Kanto and Kansai legal teams for asbestos lawsuits. These groups were formed with the cooperation of the Asbestos Center (Tokyo) and the Kansai Occupational Safety and Health Center (Osaka), and chiefly consisted of lawyers who had been involved in legal counseling activities and lawsuits for the relief of asbestos victims.

13.2.2.5 Building-Related Asbestos Lawsuits as a Future Issue

Fifth, Japan has not yet witnessed any lawsuits resembling the many cases in the US that have been filed for damage compensation claims against asbestos product manufacturers and distributors to offset the cost of removing asbestos materials from buildings. Nonetheless, it is estimated that at least 70% of all asbestos used in Japan has been utilized in construction materials.[12] Moreover, many buildings that incorporate asbestos construction materials are still standing,[13] and there is concern that they will be a source of future asbestos exposure and related health problems.[14] In that context, attention has focused on the court decisions in the case of a shop manager (director of an independent sales business) who had worked for more than 30 years in a leased office building used as a stockroom facility by a stationery store. The shop manager had died of malignant mesothelioma, and the cause was perceived to be spray-on asbestos materials used in the leased structure. Surviving relatives of this man filed a damage lawsuit against Kintetsu and other companies that were lessors and owners of the building in question. This was the Kintetsu case, in which the defendants were declared liable as building owners under Article 717 of the Civil Code (which contains provisions stipulating that building owners ultimately bear liability without fault for buildings and other structures) in both the first and second court rulings (court decisions 15 and 18[15]).

13.2.2.6 Lawsuits Pursuing Claims of National Government Liability

A sixth feature of litigation trends in Japan is that lawsuits aimed at holding the national government accountable for inaction, namely, in terms of neglecting the dangers associated with asbestos, have been pursued as cases seeking reparations from the state (under provisions of the State Redress Act, Article 1), as demonstrated by the Sennan lawsuit (court decision 20). Consider some of the background factors of importance here. Most of the companies directly engaged in the manufacture of asbestos products were small operations without the financial power to pay damage compensation, and many had already gone out of business. In addition, some of the proprietors of those small operations were victims of asbestos-related diseases themselves, and as sole proprietors in legal terms they could not be held accountable as employers. Furthermore, as was the case with the workers who were plaintiffs in the metropolitan Tokyo construction worker-led lawsuit, workers associated with these lawsuits had each been engaged at multiple construction sites, thus from a verification standpoint underscoring the extreme difficulty of pursuing lawsuits against specific employers on claims of default of ensuring worker's safety from the hazards of asbestos.

Drawing on sample cases of litigation in Japan to date that demonstrate these features, the following section will discuss several topics deemed worthy of future study as key theoretical issues of importance.

13.3 Theoretical Issues for Asbestos Litigation in Japan

13.3.1 Causal Relationships

To pursue claims of corporate or national government liability for damages stemming from asbestos-linked health problems, it is considered necessary that causal relationships should be demonstrated between the incidence of those health problems and tortious acts or default. On the subject of the extent to which a causal relationship should be established for civil litigation purposes, Japan's Supreme Court has ruled that evidence need not be scientifically verifiable beyond a single iota of doubt, but that it should at least demonstrate a probability strong enough to endorse a relationship between specific facts and the specific outcomes of a comprehensive, empirical assessment of all evidence, and further, that conclusions based on that assessment *need to be capable and sufficient to the task of convincing the average layperson of the truth beyond a shadow of a doubt* (Supreme Court 1975/10/24, Minshu (Supreme Court Reports, Civil Cases), Vol. 29, No. 9, p. 1417. Emphasis in italics added by the author here and thereafter).

In terms of demonstrating causal relationships, since the 1960s, efforts have been made to add to the theoretical depth of lawsuits involving cases of environmental pollution. The issues for causal relationships in such cases have revolved around whether the pathogenesis of the health problems stemmed from acts of tort (collective causal relationships) and, even assuming that to be the case, whether the health problems of individual plaintiffs were caused by acts of tort by the defendants (individual causal relationships). Judging from legal precedents and theoretical perspectives derived from air pollution-related lawsuits in recent years, collective causal relationships have been accepted by demonstrating pathogenic relationships between acts of tort and the incidence of observed health problems. Further, taking those pathogenic relationships and the personal circumstances of individual plaintiffs into general consideration, the prevailing view has been that even the individual causal relationships can be deduced as fact provided that there are no extenuating circumstances suggesting that the plaintiff's health problems were attributable to other valid causes (Yoshimura 2002:243–244).

Unlike certain nonspecific forms of disease that are caused by air pollution, one feature of asbestos-related diseases, and especially mesothelioma, is that they can readily be shown to be caused by exposure to asbestos dust.[16] However, as will be discussed below, the causal relationships presented in cases of asbestos litigation varied depending on the modality of the asbestos exposure itself.

13.3.1.1 Occupational Exposure

Consider the case in which a victim of a disease was clearly exposed to asbestos dust in one workplace location, and that the victim's disease is recognized to be one of the previously cited asbestos-related diseases. In this scenario, the cause of the asbestos-induced health problems can be readily identified, and the causal relationship does not present much of a problem (court decisions 1, 2, 3, 7, 8, 17, 18, and 19).

However, if the victim has worked at multiple asbestos-related workplaces, it may no longer be clear whether he contracted his asbestos-related disease while employed by the company named as a defendant in the lawsuit or while employed by some other company. One example that perfectly illustrates this dilemma is the case involving an electrical installation project supervisor (designated "Worker A" here) who had died of malignant mesothelioma. In that case, Worker A's surviving family (wife and child) filed a lawsuit seeking damage compensation from Company Y, Worker A's employer. The court decision in this case (court decision 11 in Table 13.1) notes that prior to employment with the named defendant company, Worker A was engaged from 1975 to 1986 in electrical installation duties for an electrical equipment trading company, Company B, and is known during that time to have been involved in projects for the removal of sprayed asbestos used as an electrical insulating material on steel ceiling support frames. Accordingly, the court concluded "*that it is conceivable that Worker A was exposed to significant quantities of asbestos dust during electrical construction work performed prior to his employment with Company Y*, that even if Worker A was not exposed to asbestos during his employment by the defendant, that is not inconsistent with the fact that Worker A contracted mesothelioma … cannot accept that Worker A's death from malignant pleural mesothelioma was undoubtedly an outcome of exposure to asbestos during electrical construction work performed at worksites while employed by Company Y," and thus rejected the claim of a causal relationship between the health problems of Worker A and default or a tortious act by Company Y.

In the Sennan lawsuit representing former asbestos factory workers and surviving kin of deceased workers who had been employed by asbestos factories in the Osaka community of Sennan,[17] which had the highest concentration of asbestos operations in Japan, the plaintiffs filed damage compensation claims against the national government. Of those plaintiffs, 24 had contracted asbestos-related diseases as a result of occupational exposure, and of that subgroup, approximately 70% had been engaged in work at multiple asbestos facilities. Eleven members of this subgroup had histories of employment with three or more asbestos product manufacturers, and of those, several had been engaged in operations at numerous asbestos facilities; two had worked at five different facilities, one at seven, and one at 17. In the Sennan lawsuit, the plaintiffs sought damage compensation not from employers, but from the national government as the named defendant. One reason that may be cited for this was that because many of the plaintiffs had worked at numerous factories, it would have been difficult, if not impossible, to establish the causal relationships in a lawsuit that pursued damage compensation claims against specific defendant companies. In fact, on the subject of the national government's involvement, the court decision (court decision 20) in the Sennan case observed no specific difficulty in establishing a causal relationship between the plaintiffs and deceased workers who were recognized to have contracted asbestos-linked diseases through occupational exposure and the failure of the national government to exercise its regulatory authority after regulations had been put into effect.[18]

Because many of the workers had career histories spanning multiple corporate employers, it would be difficult to determine exactly which employer they were working for when they contracted their illness. This was a situation that had also presented a challenge for earlier pneumoconiosis-related lawsuits against coal mining facilities. Legal precedents in this matter have established that claims of joint acts of tort may be made against multiple corporate employers in chronological order.[19] It may be useful to explore whether this legal framework also applies to asbestos lawsuits.

13.3.1.2 Secondary Exposure in the Home Setting

If a worker at an asbestos factory returned home wearing work clothing contaminated with asbestos dust, family members in the same household were at risk of being exposed indirectly to asbestos dust despite the fact that they themselves did not work at an asbestos factory. In the US, victims of asbestos-related diseases caused by such secondary exposure have filed and won cases involving product liability claims against asbestos product manufacturers (Osaka 2010:175–180).

In Japan, there have been no known cases of product liability lawsuits filed against asbestos product manufacturers on claims of secondary exposure. However, there have been several cases of lawsuits filed by workers against their employers on claims of liability for damages. One such legal battle was the Misawa Resort case, in which the plaintiffs complained that their relative, A, had died of malignant mesothelioma as a result of inhaling asbestos dust contaminating the work clothes and dust masks brought home by A's father, B, who worked for the defendant, an asbestos product manufacturer. Their lawsuit claimed that the defendant was liable for a tortious act because it had been negligent in its obligations to thoroughly educate employees about the dangers of asbestos, or adequately caution employees to avoid taking home contaminated work clothes and dust masks and causing consequent health problems for family members. However, in both the first and second rulings (court decisions 4 and 6, respectively), the courts concluded that malignant mesothelioma could not be confirmed as the cause of A's death, and that even if it was the cause, and there was sufficient reason to assume that A had been exposed to asbestos dust at home, at that time the defendant was not capable of foreseeing the risk of harm to the health of family members of employees as a result of secondary exposure in their homes, and thus rejected the plaintiffs' claim of liability.

Similarly, in the Nippon Etarnit Pipe case, the plaintiffs pursued a claim of legal liability for violations of the defendant's obligation to ensure the safety of former employees and their family members who had been afflicted with malignant mesothelioma. The ruling in that case (court decision 16) approved the part of the plaintiffs' claim relating to the former employees, but rejected the part relating to the family members on the grounds that it was not possible to foresee that family members would experience health problems through secondary exposure in their homes.[20]

If a person is determined to be a victim of an asbestos-related disease, and all potential sources of exposure have been ruled out except for secondary exposure in the home, asbestos dust contaminating work clothes will probably be considered to be the cause of the victim's disease. Accordingly, the focus of contention can then be expected to shift to whether the asbestos company was capable of foreseeing that health problems would be caused by secondary exposure to dust in employees' homes.

13.3.1.3 Environmental Exposure

With the Kubota Shock, the risk that airborne asbestos dust from asbestos product factories could create health problems for residents in surrounding neighborhoods became an issue of public concern. Lawsuits seeking to hold asbestos companies liable for such forms of environmental pollution have only just begun. In the Sennan lawsuit against the national government, one of the plaintiffs was an individual engaged in farming operations near an asbestos factory. He asserted that he had been exposed to airborne asbestos dust from that facility and had contracted an asbestos-linked disease as a result. However, the court ruling (court decision 20) rejected the assertion of a causal link on the grounds it was not clear whether asbestos was actually the cause of the plaintiff's symptoms. In another case, local residents sought damage compensation from a building demolition contractor on the grounds that intolerable levels of noise and vibration, as well as airborne asbestos dust from demolition work in their neighborhood, had caused them symptoms ranging from hypertension to insomnia. In the ruling in that case (court decision 12), the Saitama District Court agreed that the defendant was liable for acts of tort relating to noise pollution that affected plaintiffs residing within a given radius of the demolition worksite, but dismissed claims of defendant liability for harm from asbestos dust owing to uncertainties as to whether dust had been released from the worksite to surrounding areas, the quantities released, and the damage that the dust may have caused.

Judging from the insights on causal relationships that previously cited pollution litigation in Japan has provided to date, if a plaintiff is suffering from symptoms of an asbestos-related disease and has otherwise not been at risk of exposure to asbestos dust, it should be assumed that their symptoms are attributable to asbestos dust from nearby asbestos factory operations.[21]

In May 2007, the surviving children of a deceased man filed a damage compensation lawsuit against Kubota and the national government on the claim their father had died of mesothelioma due to environmental exposure to asbestos from the Kubota factory in Amagasaki city. That lawsuit is still under way.

13.3.2 Predictability

Of all the asbestos-related litigation filed to date in Japan, the following rulings dismissed claims of defendant liability on the argument that it was not possible to foresee that harm would be caused: the first ruling (court decision 9) in the Sapporo International Travel occupational exposure case, the first and second rulings (court decisions 4 and 6, respectively) in the previously cited Misawa Resort home exposure case, and the ruling (court decision 16) in the Nippon Etarnit Pipe home exposure case.

The Sapporo International Travel case involved a damage compensation lawsuit against an employer by the surviving family of a worker who had died of malignant mesothelioma attributed to exposure to asbestos dust from sprayed asbestos coatings in the hotel engine room and boiler room where the worker was assigned. In the first trial, the Sapporo District Court observed that in view of the construction contract for the building in question, it would normally be difficult to assume that the building purchaser or owner would have sufficient knowledge of the safety of the construction materials or the installation methods that were utilized.

The Court also noted that primary responsibility for the safety of the construction materials used in the building without any doubt rested with the material manufacturers, construction contractors, and the national government as the entity with regulatory authority over structural inspections and the safety of construction materials, and on this point concluded that *the building purchaser and building owner should not be expected to bear an overly heavy obligation to caution tenants or users*. Given these considerations, it declared that in the light of the circumstances that prevailed in the period up to 1985, the defendant could not be criticized for the causal relationship with Worker C's death, namely, the outcome of making Worker C perform duties without any precautionary measures in the engine room or the tenth floor attic space where the air was contaminated with asbestos dust, and thus dismissed the claim that the defendant was liable for a failure to fulfill its obligation of implementing measures for worker safety. However, in its appeal ruling (court decision 10), the Sapporo High Court did not accept any exemptions of this kind, and declared that the defendant was liable for violations of its obligation to take measures for worker safety, noting that there was not enough evidence to demonstrate that the defendant had implemented legally mandated safety measures (air ventilation with on-site exhaust ventilation equipment, the utilization of protective breathing gear, the humidification of work environments, the designation of off-limits areas, and the monitoring of worker health status through routine health examinations) for workers engaged in duties that involved the handling of materials containing asbestos. The defendant was not in the simple position of a building purchaser or owner, but rather an employer with an obligation to protect the lives, physical health, and well being of its employees. (As indicated earlier, under the provisions of Civil Code Article 717, building owners are subject to liability without fault for buildings and other structures.) In the first place, the point is that the courts recognize that business owners have an obligation to implement measures aimed at ensuring safety in work environments that are regulated by law precisely because of the risk of contracting an asbestos-related disease after exposure to asbestos dust. Accordingly, it does not seem fair to lighten an employer's obligations to show foresight or the avoidance of undesirable consequences, as the ruling in the first trial (court decision 9) did, simply on the grounds that they are not the manufacturer or the national government. Given that perspective, the ruling in the second trial (court decision 10) seems more appropriate.

In the Kansai Hoon Kogyo case (court decision 5), the court noted that as a precondition of the obligation to protect worker safety, in the light of the grave importance of the legal interests of life and health, *it is sufficient that the employer's accepted obligation to predictability be applied as an abstract apprehensiveness equivalent to suspicions about safety; it need not necessarily require that the employer be specifically aware of the nature, extent, or frequency of health problems presenting a risk to human life or health.* This is a satisfactory criterion for judgment from the perspective of preventing danger.[22] On the subject of exposure in the home setting, it is worth noting the point made by Naoki Ikeda, namely, that materials that are hazardous to workers should also be understood as being hazardous to their families (Ikeda 2006:45).

13.3.3 Prescription

One of the distinctive features of asbestos-related diseases is that they require long periods of dormancy before their symptoms become manifest. That period of dormancy, or incubation, for mesothelioma reportedly may extend anywhere from 30 to 50 years (Japan Industrial Safety and Health Association 2006:93–96). One question in relation to extinctive prescription or exclusion periods for the right to damage compensation is whether the long-term dormancy of these diseases forms a bottleneck to relief for the victims.[23]

13.3.3.1 Damage Compensation Claims of Liability for Default

In the cases that have been taken to court to date, many workers that were exposed to large amounts of asbestos dust, as might be expected when working in dusty settings at asbestos factories, ended up contracting pulmonary asbestosis and suffering from a long-term illness. In the Nagano pneumoconiosis case (court decision 1) and the second trial in the US Yokosuka military base case (court decision 3), for example, the plaintiffs included surviving family members of workers who had already been dead for at least 10 years when the lawsuits were filed. The families sought damage compensation on claims that the defendants were liable for default of ensuring worker safety. The courts in these two instances dismissed the families' damage compensation claims on the grounds that the 10-year period of extinctive prescription on their right to file those claims had already expired. In this author's view, it is objectively unclear to what extent symptoms may have progressed in a patient with a progressive disease like pneumoconiosis, and thus in terms of the extinctive prescription, plaintiffs should become able to exercise their right to a claim seeking compensation for damages at the time that their lawsuit is filed. On that understanding, I have advocated a time-of-death interpretation whereby the prescription does not start until the death of the patient.[24] Nonetheless, even if the time of death of the patient concerned is utilized as the starting point, the period of extinctive prescription on employers' liability for failure to fulfill their obligations to ensure worker safety will still have expired if a period of 10 years or more has already elapsed. In this situation, we should consider the possibility of utilizing some interpretations. For example, the prescription will not start at the time of death even if it is then recognized that the patient died as an outcome of exposure to asbestos dust as long as the plaintiff is unaware that the defendant may be held liable for failure to fulfill obligations to ensure worker safety, which means that there is no reasonable expectation for the plaintiff to exercise the right to file, or the defendant's invocation of extinctive prescription may be limited under certain conditions as an abuse of rights, as the ruling in the first US Yokosuka military base case (court decision 2) concluded.[25]

13.3.3.2 Damage Compensation Claims of Liability for Tortious Acts

In the event that 3 years have elapsed following the diagnosis of a patient's condition as the symptoms of an asbestos-linked disease, or following the death of the patient from that disease, and a plaintiff files a lawsuit seeking damage compensation on a claim of defendant liability for a tortious act, the defendant may conceivably invoke short-term extinctive prescription (in Civil Code Article 724, first sentence) stipulating that claims should be filed within 3 years from the date that the damage or identity of the offender becomes known to the plaintiff.[26] However, it is also noted that "granted the spirit of the special provisions concerned with the starting point of the prescription, as discussed in Civil Code Article 724, the text in that Article pertaining to 'the date the identity of the offender is known' *is understood to mean the time that it becomes known, to the extent that that is possible, by the potential plaintiff that they are actually in a position to file a damage compensation claim against the offender*" (Supreme Court, 1973/11/16. Minshu (Supreme Court Reports, Civil Cases), Volume 27 No. 10, p. 1374). Furthermore, the possibility of being aware of the damage is not a sufficient substitute for that awareness itself, and "the time that the victim learned of the damage *should be understood to mean the time that the victim actually became aware that the damage had occurred*" (Supreme Court, 2002/1/29. Minshu (Supreme Court Reports, Civil Cases),

Volume 56 No. 1, p. 218). Hence, it follows that the starting point of the applicable prescription should be interpreted as the time that the victim or plaintiff is actually aware that an asbestos-related disease caused the patient's health problem or death, knows the identity of the party responsible for causing that asbestos-related illness, and is aware that the legality of the actions or inactions of the said party may be challenged by pursuing a damage compensation claim against that party for tortious act.[27]

In addition, given the scenario where a patient has contracted an asbestos-related disease due to the failure of his employer to fulfill its obligation of implementing workplace safety measures, and the symptoms of that disease have become manifest 20 years or more after the patient left that dust-contaminated workplace, an issue will arise with the prescription for the exercise of one's rights: within 20 years (same Article, second sentence) from the date of the tortious act concerned (Civil Code Article 724, second sentence – under legal precedent, the exclusion period shall not be limited by interruption or suspension, or good-faith or abuse-of-rights restrictions, which is different from prescription – Supreme Court 1989/12/21, Minshu (Supreme Court Reports, Civil Cases), Vol. 43, No. 12, p. 2209[28]). Formerly, determining "the date of the tortious act" was a controversial issue among legal academics and lower court rulings – some claim that "the date of tortious act" coincides with the time when the offense took place, while others claim that it coincides with the time when harm resulted. In the appeal hearing for the Chikuho pneumoconiosis lawsuit, the Supreme Court issued a landmark decision, ruling that in view of the nature of the harm resulting from the tortious acts in question, such as harm due to a substance known to cause human health problems should it accumulate in human bodily tissues, or harm with symptoms that manifest after a long period of dormancy, in situations where harm occurs after a significant period of time following the offensive act that led to that harm, *the time when the symptoms of the said harm have become manifest, in full or in part, should be understood as the starting point* of the exclusion period. The reason, it added, was that allowing the exclusion period to progress in this scenario without waiting for symptoms of harm to appear *would be excessively harsh for the victim*, and furthermore, given the nature of the potential harm caused by actions of the offender, that offender should conceivably be able to foresee that they may face demands for damage compensation from a victim who comes forward at a significantly distant time in the future (Supreme Court 2004/4/27, Minshu (Supreme Court Reports, Civil Cases), Volume 58, No. 4, p. 1032[29]). In the Supreme Court ruling in the Chikuho pneumoconiosis case, the reference to "harm due to a substance known to cause human health problems should it accumulate in human tissues, or harm with symptoms that manifest after a long period of dormancy" applies perfectly to asbestos-related diseases. For that reason, the starting point of the exclusion period can be understood as the time when disease symptoms begin to appear. More specifically, for pneumoconiosis and other progressive disease conditions, this means that the starting point for the exclusion period will be the time when the disease has reached its most advanced stage in the patient, or when the patient dies.

However, if a plaintiff files legal proceedings aimed at seeking damage compensation for an tortious act more than 20 years after the patient has died, the question will be whether the time limit to exercise that right of claim may have expired because the harm itself has already occurred. In this scenario, several points should be given consideration. For example, if it were not realized that an asbestos-related disease had caused the patient's death (a highly likely scenario in the event of the secondary exposure of family members), would it be acceptable to concede that the rights of the claimant had been forfeited purely as a result of the passage of a 20-year time span? Or, in the event that the victim or surviving relatives thereof were not able to recognize the cause of the victim's disease because the defendant took no action to protect the safety of workers or alert them to the risk of such health problems, despite being able at least to foresee that such health problems would arise, should a good-faith restriction be placed on the right of the defendant to declare exemption from a damage compensation claim due to the passage of time as the defendant's negligence could be regarded as an act equivalent to obstruction of filing?

13.3.4 National Government Liability and the Product Liability of Asbestos Companies

In the Osaka Sennan lawsuit, plaintiffs filed liability claims against the national government under circumstances explained earlier. The Osaka District Court ruled (court decision 20) that the national government was liable for damage compensation to workers who had contracted asbestos-related diseases as a result of occupational exposure to asbestos dust. As justifications for that ruling, the court cited at a minimum the illegality of the government's failure in 1960, when the Pneumoconiosis Act was enacted, to mandate factory installation of on-site exhaust ventilation systems, and also its failure in 1972, when the Ordinance on Prevention of Hazards Due to Specified Chemical Substances was enacted, to mandate that factories should report their measured levels of workplace asbestos dust and pursue improvements. National government liability was also sought in the aforementioned metropolitan Tokyo construction workers' lawsuit, but the ruling in the Sennan lawsuit was extremely significant in terms of recognizing that the government was liable for inaction as early as 1960.

From the findings of prewar research, the national government knew, or was in a position to know, about the hazards of asbestos, yet it took no regulatory action against those hazards and failed to share that hazard-related information with the public. Instead, it sided with industry and contributed to the expanded utilization of asbestos. These points have been cited as factors behind the extent of health problems caused by asbestos in Japan to date, and a state of affairs that suggests that those problems will continue to grow in the years ahead.[30] Katsuya Uga asserts that if government administrative agencies knew that companies in a regulated industry had sold dangerous products to the public and thus caused serious health problems for citizens, and yet placed priority on industrial development, withheld information that could have substantiated the dangers of the product, and conversely declared that the product was safe, thus delaying public awareness of those dangers, then those administrative agencies were actively complicit in an illegal offense, and the national government bears liability for a joint tortious act.[31] In the Tokyo Metropolitan Asbestos Lawsuit, plaintiffs have insisted that the government was jointly liable with asbestos construction material manufacturers for acts of tort contributing to the expanded distribution of asbestos construction materials because it failed to exercise its regulatory authority and was actively involved in promoting the use of such materials. The liability of the national government for a joint tortious act will be acceptably established if the circumstances were as described by Uga.

The same lawsuit is also seeking to establish the complicity of 46 construction material manufacturers in joint tortious acts. If the complicity of these defendants is established, plaintiffs will then be able to file claims of total liability against each defendant, and demand damage compensation in full. Accordingly, this will enable the plaintiffs to avoid the risk of being unable to collect compensation should the group of defendants include parties that lack the financial strength to pay. Further, it is understood that even if individual causal relationships linking the actions of individual defendants with the final outcome of harm from asbestos cannot be demonstrated, the complicity of the defendants in joint tortious acts will be established by demonstrating the causal relationships between asbestos-induced health problems and joint conduct that the defendants engaged in through association. Hence, in terms of demonstrating cause and effect as well, the plaintiffs have the upper hand.

Traditionally, many of the lawsuits involving joint tortious acts have had to do with cases of environmental pollution caused by waste-water effluent and smoke emissions from factories, or exhaust gas emissions from automobiles (The Sanno River case, Supreme Court 1968/4/23; Minshu (Supreme Court Reports, Civil Cases), Vol. 22, No. 4, p. 964; the Yokkaichi petrochemical complex case, Yokkaichi Branch of Tsu District Court, 1972/7/24; Hanrei Jiho, Issue 672, p. 30; Nishiyodogawa air pollution case, Osaka District Court 1991/3/29; Hanrei Jiho, Issue 1383, p.22, etc.). In the former pollution cases, emissions of industrial waste containing substances that presented a health hazard were the by-product, not an objective, of corporate operations. However, in the later cases, the problem had to do with asbestos in construction materials. In fact, the asbestos content was one of the prime selling points behind the marketing and distribution of these materials, and in terms of attributes, it differed from water and air pollution in that it was specifically designed to generate more profit (through the act of sales).

Most construction material manufacturers produce and distribute products only for those applications in which they are specialized, not for all building construction applications. Even so, their business is conditioned on the expectation that their products will be integrated with various other materials to create a complete building structure. Furthermore, were it the case that industry groups of construction material manufacturers manufactured and promoted the sale of construction materials containing asbestos, then naturally they were able to foresee that the construction materials produced by other manufacturers would also contain asbestos. Assuming that to be the case, it could have been foreseen that construction workers would later become victims of asbestos-related diseases after exposure to asbestos dust from a variety of asbestos-containing construction materials during the construction, renovation, and demolition of buildings that incorporated those materials. Nonetheless, assuming that the industry sought to supply asbestos construction materials indiscriminately to the marketplace, the result would have been not only the creation of a stronger, integrated whole (a community of weak affiliations), but would also have enabled individual manufacturers in the industry to boost their profits by exploiting their respective activities (the sale and distribution of the asbestos construction materials in which they specialized) in a way that could be expected to contribute to asbestos exposure and consequent harm. On this point, they were in a position to establish an even stronger integrated whole (an industrial community of strong cross-affiliations).[32]

13.4 Conclusions

This chapter has compared some of the distinguishing features of asbestos lawsuits in the US and Japan, and explored several theoretical issues that deserve further study. The previously cited Kubota Shock set the stage for the eventual 2006 establishment of the Act on Asbestos Health Damage Relief, a law designed to provide relief for victims of home exposure or environmental exposure to asbestos, and other victims who were not eligible for compensation from the workers accident compensation insurance system, as well as the surviving kin of deceased workers who had received no workers accident compensation insurance benefits despite being exposed to asbestos dust in their workplaces. This law was brought into force on March 27, 2006. However, it has not escaped criticism. In particular, even if a person is approved for relief compensation under this law, the monthly allowance they receive for medical treatment will be only slightly more than 100000 yen. Also, the surviving kin of a victim who has died of an asbestos-linked disease will receive special condolence money or an adjusted lump sum of 2.8 million yen. These amounts are an order of magnitude smaller than comparable amounts paid through the workers accident compensation insurance framework.[33] As the court decisions listed in Table 13.1 illustrate, of the asbestos lawsuit rulings issued to date, those that acknowledged the liability of defendants for certified cases of asbestos-related diseases awarded plaintiffs compensation amounts of 76 million yen or more (court decision 13). Of course, as this problem has been described as the worst complex stock disaster of the twenty-first century, the total scale of asbestos harm is not something that can be settled through individual lawsuits. Legal and administrative solutions to this problem, including the establishment of a victim relief fund based on contributions from asbestos product manufacturers as well as affiliated companies that utilized those products, must also be explored from a broad-based perspective.[34] Nonetheless, to cultivate the basic perspectives that facilitate progress toward a lasting solution to this huge problem, it will also be critically important to clarify in detail the fundamental legal liabilities of private companies and the national government for the harm suffered by the individual victims of asbestos-related diseases.

This chapter is a modest endeavor to that end.

Notes

1. For detailed information on the historical development of awareness of asbestos hazards, see Furukawa (2009).
2. For an instructive overview of asbestos litigation trends and background factors in the US, Yamashita (1988), 11–20 provides a credible introduction shaped by a comparative perspective on trends in Japan. For an introduction to product liability litigation involving the US asbestos industry, see Koga (1995), 214–240. For an introduction to recent cases of litigation, see Osaka (2010), 161–185.
3. For a Japanese primer on Section 402A, see Yasuda Research Institute (1989), 15–16.
4. For works that introduce Japanese readers to the drastic increase in product liability litigation in the US and its relationship with the insurer crisis, see Yasuda Research Institute (1989), 284–332, and Kobayashi (1993), 15–17.
5. Ministry of Health, Labour, and Welfare (2005). "About Accreditation Criterion Diseases by Asbestos (Ordinance of the Labor Standards Bureau, Nr. 0209001, February 9, 2005)." In addition to this notification, Study Group on the Asbestos Problem (2007) contains asbestos-related laws and statutes. For an overview of asbestos-linked diseases, see Morinaga (2008), 91–228.
6. In 1960, a total of only 11 applications from victims of asbestos-induced lung cancer and mesothelioma were approved for workers accident compensation insurance benefits. Forty years later, in 2000, that total had risen to 52. However, in 2006, the year following the Kubota Shock, the number of approved applicants exploded to 2 195. (In 2009, the combined total came to 1 153. Based on the Ministry of Health, Labour, and Welfare statistics.
7. Yoshida, K. (2006) compares trends in Japan and the US using a somewhat different perspective than this chapter.
8. On the reasons for the scarcity of asbestos lawsuits in Japan, Naoki Ikeda, who provided an overview of asbestos litigation in Japan 4 years ago, cites the following causal factors: the "time barrier" (long-term dormancy of disease symptoms), the "information barrier" (no information was available from the health-care industry or administrative bureaucracy on asbestos-induced health problems), and the "corporate barrier" (corporations did not share information internally on workplace disasters. (Ikeda 2006: 41–47).
9. For a discussion of the evolution of the debates over workplace accident liability and the obligation to ensure worker safety in Japan, see Matsumoto (1986/1987).
10. On the benefit in terms of the prescription in the context of the issue of employer obligations to ensure worker safety, see Matsumoto (2002), 11–182.
11. On the Kubota Shock, see Morinaga (2006), 1–11. In the 4 years from news of the Kubota disaster to early 2009, a combined total of close to 300 former workers and local residents living near Kubota's defunct Kanzaki factory were said to have died (Kato 2009:40).
12. In 1982, construction materials accounted for approximately 70% of all asbestos utilized in Japan. By 1996, that share had reportedly risen to 93% (Kamei et al. 2007:2).
13. According to a survey report released by the Ministry of Land, Infrastructure, and Transport in March 2006, 5.8% of all private structures with floor space of 1 000 m^2 or more had exposed surfaces of spray-on asbestos and asbestos-containing rock wool (Ishihara 2006:2).
14. A comprehensive exploration of the issue of asbestos materials in buildings is pursued in Ozawa (2006). Also, see Ishihara (2009).
15. The significance of, and issues relating to, court ruling 15 were discussed elsewhere (Matsumoto 2010).
16. Kenji Morinaga notes that asbestos can safely be assumed to be the causal factor behind the vast majority of new cases of mesothelioma that have multiplied sharply in Japan in recent years (Morinaga 2006:7).
17. For discussions concerning the history of asbestos industry development in the Sennan district and the scale of harm from asbestos-induced disease there, see Mori (2009) and Minami (2008).
18. Court decision 20 treats 1960, the year that the Pneumoconiosis Act was established, as the time that the national government's failure to exercise its regulatory mandate would have

become illegal. For that reason, the claim by one plaintiff who had been employed in an asbestos dust-contaminated workplace only prior to 1960 was thrown out because no causal relationship was established with illegal inaction by the national government.

19. Matsuo pneumoconiosis lawsuit, Tokyo District Court 1990/3/27 decision. Hanrei Jiho, 1342, 16 contains the following statement: "As the defendants note, the plaintiffs had records of working in other dusty workplaces. Even if there was a risk of inhaling dust at worksites other than the Matsuo rock quarry, it must be concluded by analogous application of the provisions in the second sentence of Civil Code article 719 paragraph 1 that a causal relationship may be legally inferred between the current symptoms of the plaintiffs and the aforementioned breach of obligations by the defendants for failure to fulfill their obligations. Unless the defendants declare and demonstrate that no causal relationship exists either in full or in part between their respective breaches of obligations and the current symptoms of the plaintiffs, it should be stated that said defendants cannot be exempted from their liabilities either in full or in part." See also the Hosokura pneumoconiosis lawsuit, Sendai District Court 1996/3/22 decision, Hanrei Jiho, 1565, 20. Yoshida (2006), 620 also implies this point.
20. For an introductory study of this court decision, see Osaka (2010), 166–168. After the first ruling in this case, it is reported that the defendant (the former Nippon Etarnit Pipe, now Resort Solution Co., Ltd.) on September 28, 2009, formally apologized to all plaintiffs, including family members exposed to asbestos, and settled the case with a conciliation payment of 542 million yen.
21. Yoshida (2006), 629 notes that in cases involving asbestos, "a categorical treatment in keeping with epidemiological and statistical approaches may be called for" Additionally, Tsuji (2007/2009) of the same publication treats the issue of demonstrating a causal relationship between the epidemiological evidence and the symptoms of individual asbestos victims as a topic for research. Substantive progress with that research is eagerly anticipated.
22. As one paper stressing this point, see Miyamoto Kenzo (2006), 27, and Yoshida (2006), 629 also focuses on court decision 5 as something that can be expected to ease the burden of proof for victims.
23. Ozawa (2006) 112 states that it is difficult to predict whether the liability of asbestos construction manufacturers will be recognized for tortious acts relating to asbestos-linked diseases that manifest symptoms 20 years or so following the inhalation of asbestos fibers by victims.
24. Matsumoto (2002) 309–311. Yoshida (2006), 616 and 652 reaches the same conclusion
25. On the subject of placing good faith or abuse-of-right restrictions on the invocation of the extinctive prescription or exclusion periods, see Matsumoto (2002), 143–170.
26. In the Yokosuka asbestosis lawsuit (court decision 3), the defendant's invocation of the short 3-year extinctive prescription in the first sentence of Article 724 of the Civil Code was recognized but left open questions. Also, in the case concerning the Sennan lawsuit (court decision 20), four of the claims were from surviving family members who had filed more than 3 years after the deaths of workers who were victims of asbestos-linked diseases (one of whom had died 17 years prior to the filing by his surviving family), but the national government, in its position as defendant, has not invoked the extinctive prescription. It is conceivable that the plaintiff families did not consider the national government to be liable for criminal inaction at the time their loved one died. (Asbestos did not evolve into a public issue until after the Kubota Shock of June 2005 at the earliest.) Although it is unlikely that the 3-year prescription has expired anyway, the national government should not declare itself exempt from liability due to the passage of time by invoking the prescription in a case of such tragic proportions that is the outcome of its own gross inaction.
27. For a personal opinion on the starting point stipulated in the first sentence of Article 724, refer to Matsumoto (2003).
28. Interpretations of the exclusion period based on legal precedent have been criticized from both personal and theoretical standpoints. For specifics, see Matsumoto (2006a).
29. For more on the implications and issues surrounding this ruling, see Matsumoto (2006b) and Matsumoto (2007).
30. Yoshimura (2010), 54–56. In reference to these points, Kenichi Miyamoto refers to "flaws of government." Miyamoto Kenichi (2009), 6. Additionally, on the Japanese government's negligence

in the regulation of asbestos imports, Yoshida (2006), 614 points out that the government is responsible for willful, or at least gross, negligence. See also Obata (2006).

31. Uga (2008), 269. On this point, the legal team for the plaintiffs in the metropolitan Tokyo construction workers' lawsuit has been paying attention.
32. On that point, see the arguments presented in Matsumoto (2009), 2. The author plans to pursue a more substantive theoretical study of this topic in the future.
33. Tsuji (2007/2009), Paper (3), 152. In addition, for a discussion of problems with the relief law, see Murayama (2008).
34. Yoshida (2006), 623 offers a wealth of suggestions regarding the responsibility for contributions into funds of this kind.

References

Furukawa T (2009) Awareness of health problems caused by asbestos. In: Asbestos Center (ed) Why did the asbestos whirlwind spread? A history of Japan's asbestos industry and the involvement of government (in Japanese). Nihon-Hyoron-sha, Tokyo

Ikeda N (2006) Asbestos litigation in Japan: current status and future challenges (in Japanese). NBL 827:40–47

Ishihara K (2006) Future subjects of countermeasures against asbestos in construction industry. Policy Sci (in Japanese), supplementary volume: 1–11

Ishihara K (2009) Asbestos damage and remedial measures in the construction industry. Res Environ Disruption (in Japanese) 38(4):55–60

Japan Industrial Safety and Health Association (ed) (2006) Why is asbestos dangerous? (in Japanese). Japan Industrial Safety and Health Association, Tokyo

Kamei T et al (eds) (2007) Asbestos and mesothelioma (in Japanese). Shinoharashinsha Inc, Tokyo

Kato M (2009) Asbestos Shock by Kubota Corp. Disaster in Amagasaki: lessons from Asia's worst asbestos damage. Res Environ Disruption (in Japanese) 38(4):40–45

Kobayashi H (1993) The product liability law: enactment and countermeasures (in Japanese). Chuo Keizai-Sha, Tokyo

Koga T (1995) A study of product liability (in Japanese). Koyo Shobo, Kyoto

Kohyama N (2006) What is asbestos? In: Morinaga K (ed) In asbestos pollution and its health effect (in Japanese), 2nd edn. Nihon-Hyoron-sha, Tokyo

Matsumoto K (1986/1987) A theoretical study of the debate over the obligation to ensure worker safety, with a focus on the relationship with the debate over liability for workplace accidents (1) to (3). Ann Waseda Univ Grad Sch Law (in Japanese) 38:95–115; 40:275–298; 43:243–267

Matsumoto K (2002) Prescription and justice: the newly evolving perspectives on the extinctive prescription and exclusion periods (in Japanese). Nihon-Hyouron-sha, Tokyo

Matsumoto K (2003) The starting point of the prescription stipulated in the first sentence of Article 724 of the Civil Code: from a realized starting point to a normative starting point. Ritsumeikan Hogaku (in Japanese) 286:243–287

Matsumoto K (2006a) The beginning of the end for the exclusion period concept in Civil Code Article 724, 2nd sentence: a reassessment of the supreme court 1989 ruling unifying legal precedents in accordance with the exclusion period concept. Ritsumeikan Hogaku (in Japanese) 304:316–339

Matsumoto K (2006b) The 'time of the tortious act' in the second sentence of Article 724 of the Civil Code and the prospect of exercising one's rights: the trajectory of the supreme court's 2004 ruling the Chikuho Pneumoconiosis Lawsuit. Ritsumeikan Hogaku (in Japanese) 307:148–205

Matsumoto K (2007) Torts with delayed outcomes and the starting point of the 20-year prescription stipulated in Civil Code Article 724, 2nd sentence: advocacy of a normative concept of damage and a comparison with the prescription for public prosecution. Ritsumeikan Hogaku (in Japanese) 310:424–454

Matsumoto K (2009) Asbestos victims and Joint Tortious Act. Rodo Hanrei (in Japanese) 975:2

Matsumoto K (2010) Sprayed buildings with asbestos and liability for structure and land of lesson of building: about the judgment of Osaka District Court 31.8.2009 on the case of Kintetsu. Ritsumeikan Hogaku (in Japanese) 327–328:880–927

Minami S (2008) The development of the asbestos industry and the incidence of occupational disasters: a focus on Osaka Prefecture's asbestos industry. Policy Sci (in Japanese), supplementary volume: 145–165

Miyamoto K (2006) Employee death due to asbestos inhalation and corporate liability for failure to take steps to protect worker safety. Shiho Hanrei Remarks (in Japanese) 32:24–27

Miyamoto K (2009) Relief for asbestos damage: responsibility and remedial measures for multiple stock disaster. Res Environ Disruption (in Japanese) 38(4):2–7

Mori H (2009) Problems and responsibility for asbestos disaster in Japan: the Sennan case. Res Environ Disruption (in Japanese) 38(4):34–39

Morinaga K (2006) Why an asbestos problem now? In: Morinaga K (ed) Asbestos pollution and its health effect (in Japanese), 2nd edn. Nihon-Hyoron-sha, Tokyo

Morinaga K (ed) (2008) Asbestos exposure and asbestos-related diseases: fundamental knowledge and compensation/relief expanded edition (in Japanese). Sanshin Tosho, Tokyo

Murayama T (2008) Current and future relief for asbestos victims. Res Environ Disruption (in Japanese) 37(3):8–13

Obata N (2006) Inaction and the slow response to asbestos victims. Policy Sci (in Japanese), supplementary volume: 13–27

Osaka E (2010) "Take home" asbestos exposure cases and tort liability: comprehensive study of United States and Japan. Toyohogaku (in Japanese) 54(1):161–185

Ozawa H (2006) Asbestos in buildings and the law (in Japanese). Hakuyosha, Tokyo

Study Group on the Asbestos Problem (ed) (2007) Asbestos countermeasures handbook (in Japanese). Gyosei, Tokyo

Tsuji H (2007/2009) A study on asbestos litigation: focusing on epidemiologic causality (1) to (3). Okayama Law J (in Japanese) 57(2):65–78; 58(2):23–59; 59 (1):133–156

Uga K (2008) Right of intervention by public administration and liability of risk management. In: Isobe T et al (eds) New concepts of administrative law, vol III (in Japanese). Yuhikaku, Tokyo

Yamashita T (1988) The legal handling of asbestos-induced health problems: the court ruling in the Nagano asbestosis lawsuit as a turning point. Hanrei Times (in Japanese) 660:4–20

Yasuda Research Institute (ed) (1989) Product liability: a challenge for multinationals (in Japanese). Yuhikaku, Tokyo

Yoshida K (2006) Property, housing welfare, and reparation problems in the multi-cultural age (in Japanese). Yuhikaku, Tokyo

Yoshimura R (2002) Environment private law: its development and contemporary problems (in Japanese). Houritsu Bunkasha, Kyoto

Yoshimura R (2010) Damage by asbestos and their liability of the state: problems of Sennan asbestos case (in Japanese). 82(2): 52–56

Chapter 14
The French Indemnification Fund for Asbestos Victims: Features and Formative Historical Factors: Preliminary Observations for a Comparative Analysis of Asbestos Relief Frameworks*

Gakuto Takamura

14.1 Introduction

"FIVA" (Fonds d'indemnisation des victimes de l'amiante), a relief compensation fund for asbestos victims that France established in 2000, has been described as a historic step in labor–management (equating here to victims and offenders) conciliation that compares with the Workers Accident Compensation Law enacted in 1898.

The FIVA framework is designed to compensate all asbestos victims through social insurance premiums[1] and allocations from the national government. FIVA has received attention as a model that deserves to be emulated by asbestos victim relief frameworks in Japan and other countries because it uses a quick application process and pays out higher compensation amounts than the system in Japan.

Several contributing factors may be cited for the formation of this outstanding system. One is the welfare state regime underlying the administration of the social security system that encourages labor–management dialog and cooperation. Another is a judicial system that assumes that employers are at fault ("faute,"[2] negligence or misconduct) in cases involving workplace accidents, and has set a precedent of deciding in favor of workers.

This chapter strives to explain the principal features of the FIVA compensation framework, including related historical factors, and follows up with a discussion of current issues as identified through recent interviews[3] conducted in France.

* Translated and revised from the Japanese original, Takamura, G. (2009). France's Compensation Fund for Asbestos Victims, with a Focus on the Mutually Regulating Relationship Between the Judicial System and the Welfare State Regime. *Research on Environmental Disruption* (in Japanese), 38(4), 14–19, with permission of the Editing Committee of *Research on Environmental Disruption*.

G. Takamura (✉)
College of Policy Science, Ritsumeikan University, 56-1 Tojiin Kitamachi, Kita-ku, Kyoto 603-8577, Japan
e-mail: takamura@sps.ritsumei.ac.jp

K. Miyamoto et al. (eds.), *Asbestos Disaster: Lessons from Japan's Experience*, 303
DOI 10.1007/978-4-431-53915-5_14, © Springer 2011

14.2 Historical Factors

14.2.1 Delayed Restrictions on Asbestos Use

France banned the large-scale use of asbestos in 1997. This was later than comparable steps taken by other European countries.[4] Despite having adequate knowledge of the hazards of asbestos by the mid-1960s, the French government did not act to implement timely and suitable curbs on its use. For this reason, it was held accountable for inaction in a decision by Conseil d'État, the highest administrative tribunal in France.[5]

The asbestos problem became a major focus of news media attention around 1994, prior to implementation of the ban on asbestos use. In 1994, it was disclosed that five professors affiliated to the University of Paris VII (Jussieu) had contracted mesothelioma as a result of environmental exposure to asbestos. An epidemiological study released in 1995 measured the scale of harm from asbestos, and projected that close to 100000 citizens would die in France over the ensuing 30 years (1995–2025) from asbestos-linked illnesses.

These revelations over the extent of the asbestos crisis prompted the creation of a unified national alliance of asbestos victims (ANDEVA) in 1996. ANDEVA launched an active campaign calling for a halt to asbestos use, ultimately leading to the imposition of the legal ban.

14.2.2 Principles of the Workers Accident Compensation Law of 1898

ANDEVA sought not only to bring about improvements in workplace environments through a ban on asbestos use, but also to ensure – with legal assistance – that asbestos victims received adequate damage compensation through the judicial system.

In contrast to the US approach that is based on product liability lawsuits against asbestos product manufacturers, asbestos-related lawsuits in France chiefly seek to hold employers liable for violations of their obligation to ensure worker safety.

The structure of employer legal liability in workplace accidents is rooted in three principles established by the Workers Accident Compensation Law of 1898.

Those three principles are as follows. First is the principle of liability without fault, whereby an employer is still held liable for accidents in the workplace even if negligence or fault cannot be demonstrated. This principle prompted significant changes to the civil code, which was principled on the notion of liability for negligence.

Second, in exchange for these favorable terms for workers, is the principle of having damage compensation amounts computed using a fixed-percentage method which is exempt from applicable provisions of the civil code. The adoption of the fixed-rate approach had the benefit of facilitating better predictions of damage compensation amounts, and encouraging insurance enrollment by employers who calculated their future risk. One downside, however, is that victims were eligible only for partial compensation that covered their loss of wage income, but did not take into account other losses of a psychological or financial nature.

The third principle is that of allowing victims to seek damage compensation in addition to the amount computed with the aforementioned fixed-rate if it is demonstrated that the employer is guilty of an inexcusable fault ("faute inexcusable"). To obtain full rather than partial compensation, it is necessary that plaintiffs demonstrate "inexcusable fault" in a court of law.

It was the Workers Accident Compensation Law that demanded compromises of this kind by employers as well as workers. Several contributing background factors may be cited for concessionary settlements. (1) The increasing sophistication in manufacturing operations, particularly among large-scale factories, eventually ruled out the assignment of blame for workplace accidents to individual intent. (2) It was probably assumed that accidents of this nature should be accepted by society as an occupational risk. (3) The insurance system had become technically advanced enough to function as a framework for the coverage of such risk (Iwamura 1984).

According to Ewald, who is known for his knowledge-based sociological analysis of the history behind the establishment of the Workers Accident Compensation Law, "risk" is not the actual occurrence of a hazardous event itself, but rather a technical strategy for the avoidance of unfortunate events that can affect groups of people (Ewald 1994).

As a type of risk, workplace accidents and compensation decisions were thus handled not in courts of law, but under the provisions of the workers accident compensation framework.

From the outset, fixed-rate compensation was readily approved for workplace victims of asbestos exposure, a new risk that had emerged in the closing years of the twentieth century. However, this allowed for no more than the small amounts of compensation that the insurance system provided as relief to victims of unfortunate events. That is why ANDEVA sought to test employer liability in the courts, demonstrate their inexcusable faults, and ensure that victims received levels of compensation which were consistent with principles of human dignity. Lawsuits were nothing less than acts aimed at clearly assigning blame and recovering a sense of morality, in contrast to the system for relief compensation, which allowed offenders to escape blame and shirk their moral duties under the principle of liability without fault.

The establishment of ANDEVA led to a subsequent increase in litigation aimed at holding employers accountable for "inexcusable faults." Although it was rare in the early days for the courts to recognize cases of "inexcusable faults," the meaning of the French "faute" gradually evolved from that of "subjective negligence," whereby an employer has failed to adequately perform its obligations despite awareness of a risk, to "objective misconduct," whereby an employer has predictably caused harm to its workers. With this development, in more and more cases, the courts increasingly decided in favor of claims charging "inexcusable faults."[6]

This favorable shift toward the recognition of "inexcusable fault" claims in the court setting had a stimulating effect on the trend in litigation. Prior to the creation of FIVA, the "deluge of asbestos lawsuits" became a hot topic even in France.

14.2.3 The FIVA Challenge

The creation of FIVA in 2000 was driven by the objective of providing asbestos victims with fast resolutions and full compensation through a mechanism outside the judicial system, which at that time was already confronted by a flood of litigation. Its creation as a special institution[7] was also something that ANDEVA had urged.

FIVA is an independent administrative institution. Although it is not a judicial body, the "I" in the FIVA acronym is for the French "Indemniser," the legal term for the English "indemnify," meaning to "legally compensate for damages." That is not to imply that there has been any desire to undermine the transparency of administrative decisions on legal liability in asbestos disaster cases. Certain provisions give FIVA the mandate to file lawsuits against liable individuals or corporations on behalf of victims even after damage compensation has been paid.

Aside from accident compensation, the scope of FIVA compensation included compensation for mental stress, the loss of happiness, harm to personal appearance, and other nonfinancial damage of a moral or ethical nature.[8]

Established against the backdrop of a mounting asbestos crisis, increased news coverage, aggressive campaigning by victims' groups, and an upward trend in cases of litigation, FIVA had the mission of alleviating the complications of lawsuits and helping victims restore moral losses that could not be offset with monetary compensation for damages from workplace accidents.

14.3 The FIVA Framework

This section details the way FIVA has been set up.

As noted earlier, FIVA is an independent administrative institution (Établissement public administratif). However, 90% of its budget is financed through the social security system, with the remaining 10% financed through allocations from the national government. Consequently, releases of FIVA-prepared lists on compensation amounts for victims of various illnesses have to be approved in advance by the Minister of Social Welfare and the Minister of Finance.

The social security system in France is independent of the national treasury, and depends for its funding on the social insurance premiums paid by all corporations and employees nation-wide. Consequently, the burden of funding for FIVA is spread thinly and broadly as a cost of dealing with social risk; asbestos-related companies actually do not bear more of a burden than other companies based on a polluter-pays principle.

The FIVA board of directors has 22 members: eight directors from the National Health Insurance Fund for Salaried Workers (CNAMTS: Caisse nationale de l'assurance maladie des travailleurs salariés), five labor representatives and three corporate representatives, four directors from victims' organizations, five directors serving as representatives of the national government, and four expert directors including lawyers, physicians, and so on. The chairperson's position is filled by the chief justice of the Court of Cassation.

Reflecting its diverse makeup, the FIVA board of directors serves as a forum for the active exchange of views and opinions on administrative affairs.

The following are qualified to submit applications to FIVA and receive compensation: (1) patients with a certified asbestos-linked disease listed on the occupational disease list; (2) patients with mesothelioma or pleural plaque; (3) patients with other diseases that can be demonstrated to be linked to asbestos exposure; (4) surviving family members or other legal heirs of deceased patients. Furthermore, FIVA-based compensation is available not only to patients who experienced occupational exposure, but also to those who experienced environmental exposure[9] to asbestos.

Individuals who have already received workers accident compensation are still eligible to apply to FIVA. If approved, they will receive the difference between a FIVA-calculated compensation amount and the workers accident compensation they have already received. Normally, FIVA-based compensation packages are more substantial than workers accident compensation packages because FIVA also covers nonfinancial damages.

Plaintiffs who have filed lawsuits against their employers may also apply for FIVA-based compensation. However, they cannot receive FIVA compensation in addition to any damage compensation won through a court judgment. Should FIVA approve compensation packages in such cases, applicants must decide whether or not to accept FIVA compensation and drop their lawsuits. Furthermore, if they accept compensation from FIVA, they will not be able to file a future lawsuit for damage compensation purposes.

FIVA applications are prepared and filed with prescribed forms and documentation. Usually, the screening process is completed only with a review of the submitted materials. Should there be any doubts about causal relationships with asbestos, the applicant will be subject to an interview at FIVA's offices. Also, should an applicant be unhappy with the amount of compensation offered, they may file a motion to have the matter reviewed by a court of law.[10]

In terms of the amount of time required for application processing, FIVA has to remit a compensation payment to an approved applicant no later than 6 months after the date when the application was submitted. In reality, though, only around half of all applications are completely processed within 6 months. FIVA assigns higher application screening priority to applicants with more serious medical conditions. Although it has implemented reforms aimed at speeding up decisions for those who are not expected to live much longer, due to the strong upsurge in applications in recent years and judging from the average screening periods for applications in 2004 and 2007, FIVA has not made significant progress in this area.

14.4 Compensation Specifics

FIVA compensation packages are determined in line with the idea of restoring the victim's life to the condition it would have been in had the victim not suffered any damage.

Damages covered by FIVA compensation include financial damages as well as nonfinancial damages. Financial damages would include damages in the form of lost occupational income owing to functional disability and related expenses for medical treatment. Nonfinancial damages would include damages in the form of mental stress, physical suffering, the loss of happiness, and harm to personal appearance.

To calculate compensation amounts, FIVA has prepared and released a table of compensation amounts for specific medical conditions.[11] The table was prepared using three criteria: disease name, severity, and victim age. Annual benefits vary depending on the extent of functional disability (a percentage-based measure of inability to engage in work). Victims with cancer are considered to be 100% disabled, while a disability rate of 5% applies to compensation for those with pleural plaque.

Table 14.1 lists the average compensation amounts for specific diseases that FIVA has paid out since its foundation.

Pleural plaque is not an accepted condition for compensation eligibility in other countries. FIVA's basis for extending eligibility to pleural plaque victims derives from the recognition of pleural plaque, in a May 2005 ordinance, as an asbestos-induced medical condition. Although this might be described as the outcome of a political decision by a certain government minister at that time, the reasoning was that even if pleural plaque does not confront patients with a life-threatening medical condition at present, it is still very likely to become the cause of a serious medical condition in the future, and as such, causes victims of asbestos exposure mental suffering.

FIVA compensations total 10 000 euro or more for 87.7% of patients with benign illnesses, including pleural plaque. Compensation reaches 40 000 euro or more for 91.7% of patients with malignant illnesses, with 38% receiving amounts of 100 000 euro or more. From its founding date until December 31, 2007, FIVA had received

Table 14.1 FIVA compensation amounts, by medical condition (average since FIVA founding) (currency unit: euro)

Disease	Surviving victims	Victim deaths	Average
Asbestosis	22 662	74 544	35 427
Lung cancer	89 668	134 992	120 121
Pleural thickening	19 068	26 131	19 490
Mesothelioma	97 114	121 333	115 360
Other condition	22 729	104 417	47 714
Pleural plaque	18 741	20 078	18 777
General average	26 035	115 634	45 779

Source: FIVA

a cumulative total of 47 210 applications for compensation, and to date it has paid out more than 1.64 billion Euro in compensation benefits.

To date, no more than about 2% of the total number of applications have involved cases that were not approved for compensation on the grounds that the applicant's illness was not linked to asbestos. In addition, applicants unhappy with the amount of compensation offered by FIVA have filed a total of 3 341 court claims, while 91% of all applicants accepted FIVA-offered compensation packages without complaint.

As these statistics illustrate, compared with the relief benefits paid to victims of asbestos-linked illnesses in Japan, compensation packages from FIVA are larger, and a higher percentage of applicants are approved for compensation.

14.5 Current Issues

Although the FIVA compensation system seems outstanding compared with the workers accident compensation or the relief framework in Japan, that is not to say that it does not have its own problems.

The first of those problems is that FIVA has not been fully effective in meeting its initially expected objective of curbing litigation.

Founded in 2000, FIVA began accepting compensation applications in 2002. Since then, though, the number of cases of court litigation has actually risen, even exceeding 1 000 cases per year (FIVA 2008). In 2007, 10% of all asbestos victims chose to seek damage compensation through the courts rather than gain relief through FIVA.[12]

There were at least a couple of reasons for this. One was the high court decision 16 that facilitated judgments of "inexcusable fault." The other was that damage compensation amounts won in court cases generally tend to be larger than the compensation packages provided by FIVA.

The differential is particularly large in terms of the amounts calculated for nonfinancial damages. FIVA compensation packages for victims with a disability rate of 10% average no more than 67% of the compensation awards that comparable victims may win in court.[13]

Working through the FIVA board of directors and through independent outside campaigns, ANDEVA has called for improvements in the compensation amounts awarded by FIVA, but to date, no changes have been observed in the compensation benchmarks that FIVA applies.

This is why the number of applicants who are unhappy with FIVA-tendered compensation packages has been rising in recent years (FIVA 2008:57).

In 2007, 1 495 applicants filed court claims demanding that FIVA damage compensation amounts should be recalculated. That represented a 2.16-fold increase on the number of such claims filed the preceding year. In 2003, no more than 1% of applicants expressed dissatisfaction with the compensation amounts calculated by FIVA. In 2007, however, fully 14% filed court claims against FIVA, citing objections to the FIVA-calculated amounts.

A second issue for FIVA is that only rarely has it ever actually filed litigation on behalf of an applicant against their employer for "inexcusable fault." In effect, the "polluter pays" principle has hardly ever been put into practice.

The explanation given at the time of FIVA's founding was that the funding burden would be spread thinly and broadly, with contributions coming out of the national treasury and social security premiums. However, it was also expected that FIVA would actively exercise its powers against polluters through proxy litigation, and enforce the "polluter pays" principle after the fact by making offenders pay compensation awards to FIVA.

In reality, though, FIVA has only filed 790 cases of proxy litigation to date, which is equivalent to barely 0.02% of the cumulative total of applications received (FIVA 2008:69). Although recent years have witnessed a mild upward trend in the annual totals of proxy litigation cases filed, those numbers are still far less than even 1% of the corresponding annual totals for applications received.

The presence of corporate employer delegates on the FIVA board of directors has been cited as one factor behind the institution's hesitance to pursue proxy litigation proceedings. However, these tiny percentages invite the criticism that FIVA-based relief merely ignores the question of offender liability, and shoulders taxpayers with the bulk of the compensation burden.

The number of court cases citing "inexcusable fault" has not subsided because FIVA has failed to hold offenders accountable.

A third issue for FIVA has to do with the question of whether the proportion of funding for its operations which is born by the national government is appropriate. In the 2004 decision by Conseil d'État, the national government was deemed to be liable for inaction owing to its failure to impose restrictive regulations on asbestos use.

As the basis for the 10% contribution to be made by the national government to the budget for FIVA upon its founding, it was assumed that the state also would share in the burden of "social risk" associated with the harms of asbestos.

Granted that its liability for damages had already been recognized, from a legal theory perspective, it followed that the state bore liability as an offender, not merely as an actor, in the sharing of risk. Nonetheless, the state's 10% share of funding for FIVA has not been altered, and does not reflect the decision by Conseil d'État.

Placing more emphasis on state liability does have a dimension that runs counter to the previously discussed notion of enforcing the polluter-pays principle and holding corporate employers to account, and as such, it is a point that must be weighed with care.[14] Increasing the burden of funding from the national treasury merely adds support to the argument that the bill for liability is being placed on the shoulders of taxpayers.

One plausible alternative to the strategy of increasing the taxpayer burden in ways which are not transparent to the public would be for FIVA to become more active – through proxy litigation – in lawsuits aimed at holding the national government liable for damages,[15] thus demonstrating government responsibility to the public at large, and boosting its own position of independence in the process.

14.6 Concluding Remarks

The foregoing sections began with a discussion of the historical factors behind the formation of FIVA, and continued with an exploration of key issues confronting that institution. The FIVA framework for asbestos victim relief was defined by the Workers Accident Compensation Law of 1898, a historic framework that combines the notion of "liability without fault" with a framework for the provision of fixed-rate damage compensation.

Lawsuits run counter to this framework as endeavors that seek to hold employers accountable for "inexcusable fault," enforce liability, and restore moral values. FIVA may be described as a new system that complements the workers accident compensation framework, and strives to have aid to asbestos victims underwritten within the welfare state regime as a social risk.

However, litigation is not counter to the goals of FIVA. FIVA's creation is itself a product of litigation. Moreover, it is readily conceivable that the achievements of litigation can be integrated into the damage compensation methodology applied by FIVA. FIVA was designed to have the ability to exercise its powers of proxy litigation, and assimilate the morals of offender liability into its own operations through utilization of the courts.

However, in actual practice, FIVA has not adequately integrated the power of litigation into its compensation strategy or operations, and as a consequence, the potential complementarity from that combination has not been harnessed.

In the years ahead, ensuring that FIVA continues to evolve into an even better institution than it is today will demand that it become more receptive to the benefits of litigation. Furthermore, fostering the advancement of the relief compensation framework itself will demand that fresh gains be pursued relentlessly through litigation independent of FIVA.

The ongoing study[16] of foreign asbestos relief frameworks should be pursued from a perspective that focuses not on the content of advanced relief systems per se, but rather on the questions of how and whether the historical systems of litigation and the welfare state regime that together defined the current relief framework have achieved their dynamic development in a complementary fashion.

Notes

1. In France, the social security system ("sécurité sociale") evolved from a mutual aid association-based group relief framework for beneficiaries in specific occupational fields. Its core feature is that it is based in civil law, and thus cannot be equated with the frameworks of a welfare state (Dupeyroux et al. 2008).
2. In French tort law, the meaning of "faute" is closer to the negligence of an offender in accepting objectively applied responsibilities than the psychological feelings of negligence in the performance of one's duties. As such, it is not translated as "negligence" but expressed as "faute." (Noda 1975).

3. Interviews were conducted from March 5 through March 8, 2007, with FIVA, the Welfare Ministry's Asbestos Countermeasures Office, labor inspectors in Europe, the attorney Jean Paul Teissonnière, and the victims support group ANDEVA. In addition, Teissonnière provided updates on the latest developments in a televised lecture for the International Asbestos Symposium hosted in November 2008 by Ritsumeikan University.
4. For an analysis of the historical record of the asbestos problem and regulation, and why France was slow to implement regulations, see the French Senate's report, Les rapports du Sénat (2005).
5. For additional information on the decision by Conseil d'État recognizing government liability for damages, see Kitamura (2007).
6. According to the Cour de Cassation, Chambre sociale, du 28 février 2002, objective liability for the consequences, and not subjective employer awareness of risk, is the criterion for the demonstration of "inexcusable fault." Based on this interpretation, many asbestos lawsuits asserting employer liability almost automatically resulted in decisions recognizing "inexcusable fault" (Teissonnière and Topaloff 2002).
7. In addition, Fonds de cessation anticipée d'activité des travailleurs de l'amiante (FCAATA) (an early retirement fund for workers exposed to asbestos) was also established in 2000. Employees of companies that were involved in the hazardous utilization of asbestos will receive smaller retirement allowances should they retire under normal circumstances. This fund encourages early retirement, facilitates earlier starts for the payment of retirement benefits, and ensures that retirees receive adequate retirement allowances.
8. FIVA's website. http://www.fiva.fr. Accessed September 3, 2010.
9. However, the number of applications from victims of environmental exposure has been small, accounting for only 6.3% of all applications received in 2007.
10. Reviews are conducted by the court of appeals with jurisdiction in the applicant's district of domicile.
11. The list is available on FIVA, http://www.fiva.fr. Accessed September 3, 2010.
12. However, comparisons of the data for 2006 and 2007 reveal a slight downward trend both in the total number of lawsuits and the percentage of asbestos victims opting to file for litigation.
13. However, compared with records for 2002, the gap is 9% smaller, indicating that some improvement has been made.
14. Attorneys in Japan have applauded the court decision finding the French national government liable for damages. However, attorneys involved in asbestos-related litigation in France have taken a more critical view of the decision as something that could be exploited to exempt companies and employers from charges of "inexcusable fault."
15. FIVA is not prohibited from pursuing damages from the state through proxy litigation. In fact, FIVA launched four such lawsuits for this purpose in 2006.
16. Earlier studies include the following by Tanase (2006).

References

Dupeyroux J, Borgetto M, Lafore R (2008) Droit de la sécurité sociale. Dalloz, Paris

Ewald F (1994) L'Etat providence. Grasset, Paris

FIVA (2008) 7ème rapport d'activité au parlement et au gouvernement année 2007

Iwamura M (1984) Workers accident and damage compensation (in Japanese). University of Tokyo Press, Tokyo

Kitamura K (2007) Harm from asbestos and state liability for damages in France. Ritsumeikan Hougaku (in Japanese) 311:218–235

Les rapports du Sénat (2005) Le Drame de l'amiante en France, Paris

Noda Y (1975) The concept of 'faute' in French civil law. In: Kawashima T (ed) Commemoration of professor wagatsuma's 60th birthday. A study of damage compensation, vol 1 (in Japanese). Yuhikaku, Tokyo

Tanase T (2006) The design of the compensation system for asbestos victims. NBL (in Japanese) 826:20–27

Teissonnière J, Topaloff S (2002) L'affaire de l'amiante. Semaine sociale Lamy, Paris, Supplément 1082

Index

K. Miyamoto et al. (eds.), *Asbestos Disaster: Lessons from Japan's Experience*,
DOI 10.1007/978-4-431-53915-5, © Springer 2011